COMPUTER CIRCUIT CONCEPTS

McGraw-Hill Series in Electrical Engineering

Consulting Editor
Stephen W. Director, Carnegie-Mellon University

Circuits and Systems
Communications and Signal Processing
Control Theory
Electronics and Electronic Circuits
Power and Energy
Electromagnetics
Computer Engineering
Introductory
Radar and Antennas
VLSI

Previous Consulting Editors

Ronald N. Bracewell, Colin Cherry, James F. Gibbons,
Willis H. Harman, Hubert Heffner, Edward W. Herold,
John G. Linvill, Simon Ramo, Ronald A. Rohrer,
Anthony E. Siegman, Charles Susskind, Frederick E. Terman,
John G. Truxal, Ernst Weber, and John R. Whinnery

Computer Engineering

Consulting Editor
Stephen W. Director, Carnegie-Mellon University

Bartee: Digital Computer Fundamentals
Bell and *Newell*: Computer Structures: Readings and Examples
Garland: Introduction to Microprocessor System Design
Gault and *Pimmel*: Introduction to Microcomputer-Based Digital Systems
Givone: Introduction to Switching Circuit Theory
Givone and *Roesser*: Microprocessors/Microcomputers: An Introduction
Hamacher, *Vranesic*, and *Zaky*: Computer Organization
Hayes: Computer Organization and Architecture
Kohavi: Switching and Finite Automata Theory
Levine: Vision in Man and Machine
Peatman: Design of Digital Systems
Peatman: Digital Hardware Design
Peatman: Microcomputer Based Design
Ritterman: Computer Circuit Concepts
Sandige: Digital Concepts Using Standard Integrated Circuits
Sze: VLSI Technology
Taub: Digital Circuits and Microprocessors
Wiatrowski and *House*: Logic Circuits and Microcomputer Systems

COMPUTER CIRCUIT CONCEPTS

Saul Ritterman
Bronx Community College
City University of New York

McGraw-Hill Book Company
New York   St. Louis   San Francisco   Auckland   Bogotá
Hamburg   Johannesburg   London   Madrid   Mexico   Montreal
New Delhi   Panama   Paris   São Paulo   Singapore   Sydney
Tokyo   Toronto

COMPUTER CIRCUIT CONCEPTS

Copyright © 1986 by McGraw-Hill, Inc. All rights reserved. Printed in the United States of America. Except as permitted under the United States Copyright Act of 1976, no part of this publication may be reproduced or distributed in any form or by any means, or stored in a data base or retrieval system, without the prior written permission of the publisher.

1234567890DOCDOC898765

ISBN 0-07-052952-3

This book was set in Times Roman.
The editors were Sanjeev Rao and J. W. Maisel;
the designer was Elliot Epstein;
the production supervisor was Phil Galea.
The drawings were done by J & R Services, Inc.
R. R. Donnelley & Sons Company was printer and binder.

Cover photograph courtesy of Advanced
Micro Devices, Inc., Sunnyvale, California.

Library of Congress Cataloging in Publication Data

Ritterman, Saul A.
    Computer circuit concepts.

    (McGraw-Hill series in electrical engineering.
Computer engineering)
    Includes index.
    1. Computers—Circuits. I. Title. II. Series.
TK7888.4.R58 1986 621.3819′5835 85-6687
ISBN 0-07-052952-3

For Kathy—
the world's greatest wife, who patiently typed and retyped this book until I got it right.

CONTENTS

Preface   xiii

1 LOGIC   1

1.0 Introduction, 1
1.1 Boolean Algebra, 1
1.2 Writing Boolean Equations, 6
1.3 Simplifying Boolean Equations, 11
1.4 Standard Boolean Forms, 15
1.5 Karnaugh Maps, 19
1.6 Three-Variable Maps, 22
1.7 Four-Variable Maps, 26
1.8 Electronic Simplification, 32
Summary, 37
Problems, 38

2 NUMBER SYSTEMS   43

2.0 Introduction, 43
2.1 Binary Numbers, 43
2.2 Octal Numbers, 51
2.3 Hexadecimal Numbers, 57
2.4 Addition, 62
2.5 Subtraction, 67
2.6 Multiplication and Division, 71
2.7 Signed Numbers, 75
2.8 Complementary Arithmetic, 80
Summary, 86
Problems, 87

3 GATES   91

3.0 Introduction, 91
3.1 The Switch, 91

3.2 AND Gates, 95
3.3 OR Gates, 98
3.4 Inverters, 103
3.5 Three-State Gates, 107
3.6 Bipolar Gates, 111
3.7 MOS Gates, 120
3.8 Gate Comparison, 125
Summary, 133
Problems, 134

4 FLIP-FLOPS 139

4.0 Introduction, 139
4.1 Regeneration, 139
4.2 NOR Latch, 143
4.3 NAND Latch, 147
4.4 Flip-flops, 153
4.5 *D*-Type Circuits, 158
4.6 *T*-Type Circuits, 163
4.7 *J-K* Flip-flops, 168
4.8 Waveshaping Circuits, 172
Summary, 178
Problems, 178

5 COUNTERS 183

5.0 Introduction, 183
5.1 Asynchronous Counters, 183
5.2 Down Counters, 187
5.3 Synchronous Counters, 191
5.4 Modulo Counters, 197
5.5 Hybrid Counters, 203
5.6 Serial Input Registers, 210
5.7 Parallel Input Registers, 214
5.8 Counter and Register Displays, 219
Summary, 224
Problems, 225

6 MATHEMATICAL OPERATIONS 229

6.0 Introduction, 229
6.1 Half-Adders, 229
6.2 Full-Adders, 235
6.3 Serial Adders, 240
6.4 Parallel Adders, 244
6.5 Signed Arithmetic, 249
6.6 Multiplication and Division, 256

6.7 Floating-Point Arithmetic, 263
6.8 Binary-Coded Decimal, 268
Summary, 274
Problems, 275

7 MEMORY CIRCUITS 279

7.0 Introduction, 279
7.1 Drum Memory, 281
7.2 Core Memory, 284
7.3 Core RAMs, 288
7.4 Bipolar RAMs, 293
7.5 MOS RAMs, 299
7.6 Dynamic RAMs, 304
7.7 ROM Principles, 310
7.8 ROM Methods, 314
Summary, 321
Problems, 321

8 PERIPHERAL DEVICES 325

8.0 Introduction, 325
8.1 Paper Input/Output, 325
8.2 Processing Paper Data, 328
8.3 Magnetic Input/Output, 335
8.4 Magnetic Recording Techniques, 339
8.5 CRT Terminals, 344
8.6 D/A Conversion, 349
8.7 A/D Conversion, 356
8.8 Data Links, 361
Summary, 365
Problems, 366

9 A COMPLETE COMPUTER 369

9.0 Introduction, 369
9.1 Computer Architecture, 370
9.2 CPU Operation, 374
9.3 Control Unit, 378
9.4 Microprogramming, 383
9.5 Stored Programs, 389
9.6 Codes, 393
9.7 Instruction Sets, 398
9.8 Programming Techniques, 404
Summary, 409
Problems, 410

Index 415

PREFACE

The first industrial revolution began in the 1750s. Human labor, in many cases very skilled, was gradually replaced by machines. Changes caused by the first industrial revolution affected more than manufacturing processes. The entire social and economic fabric of civilization changed permanently. The second industrial revolution began some 200 years later, when computers began taking over operations performed by humans. This revolution began in the 1960s with the introduction of integrated circuits (ICs), which shrink the size and power requirements of electronic circuits. Results have been astounding.

Until quite recently diverse activities such as home entertainment, marketing, business office operations, and manufacturing were conducted without benefit of computers. This is no longer the case. Stereo tuners are digital and TV sets are remote control. Talking cash registers read package codes, present itemized bills, and announce correct change. Typewriters have become word processors, and business decisions are based on computer analysis of data. In manufacturing, computer-controlled robots are becoming the work force. Clearly, computers have and will continue to have profound effects on our lives.

Proliferation of computer texts is a by-product of the computer revolution. On the surface these texts plow similar ground with equal attention to equally important topics. However, equal attention is an illusion. Chapter lengths as well as numbers of sections per chapter vary considerably. As a result, equally important topics cannot receive equal treatment. One text stresses systems while another stresses circuits. Some texts emphasize computer mathematics while others are in effect catalogs for a specific family of chips. Still other texts delve into minute detail about the author's favorite programming language. These topics are important, but not equally important.

Unbalanced treatment results in an unbalanced understanding of computer operation. By selecting a specific text, one becomes proficient in machine language programming but has only a hazy understanding of how a computer performs calculations. Similarly, another text stresses Boolean algebra but fails to demonstrate that the utility of microcomputers lies in ease of interfacing with the outside world. A sense of proportion is needed.

This book presents a balanced approach. There are nine chapters of equal importance and approximately equal length. Moreover, each chapter contains eight sections. Equal treatment of equally important topics is logical and also simplifies studying. Furthermore, topics are arranged in sequential order.

Each chapter builds on the previous:

Chapter 1: Introduces Boolean algebra, the basis of computer operation.

Chapter 2: Extends Boolean algebra to binary numbers because digital computers are really binary computers.

Chapter 3: Describes gates, the basic circuits for implementing Boolean and binary operations.

Chapter 4: Introduces flip-flops, groups of gates connected for storage and transfer functions.

Chapter 5: Introduces counters and registers. These circuits use flip-flops and are the lowest level of complex ICs used in modern computers.

Chapter 6: Describes arithmetic operations. Arithmetic circuits are extensions of counters and registers.

Chapter 7: Describes the memory circuits. Storing arithmetic results and instructions is vital to computer operation.

Chapter 8: Discusses peripheral devices. Previously described techniques are used to connect computers to the outside world.

Chapter 9: Combines material developed in all previous chapters to obtain a complete computer.

Each chapter begins by stating an objective and ends with summary; this approach emphasizes the thrust of each chapter. As a further study aid, answers to odd-numbered problems are presented along with the problems. Typically, answers are in the rear of a book, but there is no educational benefit to flipping pages back and forth between problems and answers.

In conclusion, this book presents a chain of equally important computer topics. The reader comes away with a balanced understanding of what a computer does and how it does it.

ACKNOWLEDGMENTS

In a real sense my students wrote this book. The subject matter and treatment are in response to their needs. I would like to express my thanks for the many useful comments and suggestions provided by colleagues who reviewed this text during the course of its development, especially to R. Dale Anderson, Iowa State University; R. H. Berube, Community College of Rhode Island; David Beyer, Middlesex County College; A. O. Brown; Denton Dailey; S. W. Director, Carnegie-Mellon University; Frank Duda, Grove City College; Antulio Gomez, El Camino Community College; Bernard Harris, Manhattan College; Martin Kaliski, Northeastern University; W. H. McDonald, U.S. Merchant Marine Academy; Michael Miller, DeVry Institute of Technology; John L. Morgan, Neury Institute of Technology; John Pavlat, Iowa State University; Lee Rosenthal, Farleigh Dickinson University; and M. Silevitch, Northeastern University.

Saul Ritterman

One who is afraid to ask is ashamed of learning.
Danish proverb

COMPUTER CIRCUIT CONCEPTS

1
LOGIC

1.0 INTRODUCTION

Electronic devices are useful in many applications. In particular, electronic computers perform all sorts of computations rapidly.

Modern computer circuits are based on a single characteristic of semiconductor devices, the ability to switch between two states in extremely short times. Every computer circuit only has two possible states, on or off. A special set of rules exists for working with two-state devices. These rules, developed by George Boole in the 1840s, are known as *Boolean algebra*. Originally Boolean algebra was developed for analyzing the validity of philosophical arguments, but in the 1930s Claude Shannon adapted Boolean algebra to designing electronic switching circuits. Shannon's work can be considered as the beginning of the modern electronic computer.

Since Boolean algebra is the basis of computer operation, this book begins with a discussion of Boolean algebra. In effect, computer circuits are the electronic version of Boolean equations. Since simpler equations result in simpler circuits, techniques for reducing Boolean equations are investigated; both algebraic and graphical reduction methods are practical. This chapter discusses constructing circuits to perform Boolean operations. The methods presented here are used throughout the entire book.

1.1 BOOLEAN ALGEBRA

Boolean algebra only considers two possibilities. A quantity either exists or it does not exist. Existence is represented by the number 1 and nonexistence by the number 0. For example, if A represents airplanes then

$$A = 1 \tag{1.1a}$$

means there are airplanes, while

$$A = 0 \tag{1.1b}$$

means there are no airplanes.

**TABLE 1.1
Developing a Truth Table**

A	A	$\bar{A}$	A	$\bar{A}$	$\bar{\bar{A}}$
0	0	1	0	1	0
1	1	0	1	0	1

(a) (b) (c)

Boolean algebra relies on precise definitions of the three words NOT, OR, and AND, which are capitalized to emphasize Boolean rather than ordinary meaning.

In Boolean algebra NOT means *opposite*. Boolean equations indicate NOT as a bar over the quantity. If

$$A = 1 \quad \text{then} \quad \bar{A} = 0 \tag{1.2a}$$

Similarly if $A = 0$ then $\bar{A} = 1$ \quad (1.2b)

These Boolean equations apply to airplanes or any other quantity represented by the symbol A. If a quantity exists, then its opposite does NOT exist, and vice versa.

A NOT statement is sometimes called an *inverse*. Double inverts are possible. For example,

$$\bar{\bar{A}} = A \tag{1.3a}$$

This equation says that the inverse of an inverse is the original quantity. Again, if A represents airplanes, then $\bar{A}$ represents no airplanes. Similarly, $\bar{\bar{A}}$ represents the inverse of no airplanes, which is airplanes. Equation (1.3a) can be extended to more than two inverts:

$$\bar{\bar{\bar{A}}} = \bar{A} \tag{1.3b}$$

Boolean equations can be displayed as *truth tables*. A truth table is a tabulated presentation of all possible values of the quantity. A truth table for Eq. (1.3a) begins with A as the first column; as shown in Table 1.1a, the only possible values of A are 0 and 1. Table 1.1b shows the corresponding values of $\bar{A}$, which are given by Eqs. (1.2). Table 1.1c, the completed truth table, shows the corresponding $\bar{\bar{A}}$ values, which are also determined from Eq. (1.2). Since the A and $\bar{\bar{A}}$ columns are identical, the truth table verifies Eq. (1.3a).

OR is the second Boolean term. The meaning of OR is that *at least one member* of the class *exists*. The Boolean symbol for OR is a plus sign. For example

$$A + \bar{A} = 1 \tag{1.4a}$$

means that either A OR its inverse exists.

An OR truth table can be used to verify Eq. (1.4a). The first column of Table 1.2 contains all possible values of A, and the second column contains the corresponding values of $\bar{A}$. In the last column, the results of ORing A with $\bar{A}$ are shown. Since OR means the existence of at least one member, if a single

TABLE 1.2
$A + \bar{A}$ **Truth Table**

A	$\bar{A}$	$A + \bar{A}$
0	1	1
1	0	1

1 exists in the classes being ORed, the result of ORing is 1. Since Table 1.2 shows all 1s in the $A + \bar{A}$ column, Eq. (1.4a) is valid.

Another Boolean relationship is ORing A with 0.

$$A + 0 = A \tag{1.5a}$$

This equation may seem trivial, but it is important. The validity of Eq. (1.5a) can also be demonstrated with a truth table.

Any Boolean quantity can be ORed with itself

$$A + A = A \tag{1.6a}$$

Table 1.3 is the truth table for this equation. Since A is ORed with A, the first and second columns are identical. The third column shows the result of ORing A with itself. The first and third columns are identical, and thus Eq. (1.6a) is valid.

$A + \bar{A}$ can be considered as adding $\bar{A}$ to A. In fact ORing is sometimes called *logical addition*. However, in complicated Boolean expressions the similarity to addition breaks down. This truth table also shows why the OR symbol should be read as OR and not a plus sign: 1 + 1 does not equal 1 in ordinary arithmetic, but 1 OR 1 does equal 1 in Boolean algebra.

AND is the last Boolean word. AND means *all* members of the class. Thus AND has a more restricted meaning than OR. While OR can be used in the sense of one, several, or all, AND requires the inclusion of each and every quantity. The symbol for AND is a dot between the quantities being ANDed

$$A \cdot A = A \tag{1.6b}$$

AND statements can also be verified with truth tables. The result of ANDing can only be 1 if all quantities are 1. Table 1.4 shows the truth table for this equation.

Table 1.4 resembles ordinary multiplication, and ANDing is sometimes referred to as *logical multiplication*. As with ORing, there is only partial agreement. It is safer to read a dot in Boolean algebra as AND instead of a times sign.

TABLE 1.3
$A + A$ **Truth Table**

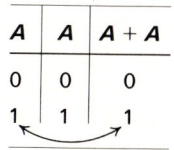

A	A	$A + A$
0	0	0
1	1	1

TABLE 1.4
$A \cdot A$ **Truth Table**

A	A	A·A
0	0	0
1	1	1

It is convenient to group Eqs. (1.6a) and (1.6b) together because in both cases a quantity is combined with itself. Similarly, the AND relationships below can be grouped with Eqs. (1.4a) and (1.5a):

$$A \cdot \bar{A} = 0 \tag{1.4b}$$

$$A \cdot 0 = 0 \tag{1.5b}$$

These results differ from the corresponding OR equations because OR and AND have different meanings. If necessary, Eq. (1.4b) and (1.5b) can be verified with truth tables.

There is another pair of equations which involve a single quantity. One equation ORs the quantity with 1, and the other ANDs it with 1.

$$A + 1 = 1 \tag{1.7a}$$

$$A \cdot 1 = A \tag{1.7b}$$

Since A and $\bar{A}$ are opposite members of the same class, Eqs. (1.1) through (1.7) contain only one variable. When Boolean statements contain more variables, truth tables contain more conditions. If two variables exist, there are four possible combinations. Table 1.5 lists the possibilities for variables called A and B.

Actually Table 1.5 is the beginning of counting in the binary number system, a topic discussed in the next chapter. For now it is sufficient to present a pattern which guarantees all possible combinations of variables. In Table 1.5 the top row column begins with 0 and alternates 0s and 1s. The bottom row column also begins with 0 but alternates pairs of 0s and 1s.

Equations (1.8a) and (1.8b) are the Boolean *commutative* OR and AND equations.

$$A + B = B + A \tag{1.8a}$$

$$A \cdot B = B \cdot A \tag{1.8b}$$

It happens that conventional algebra is also commutative. But this is not a proof that Boolean ORing and ANDing can be performed in any order, since differences between Boolean and conventional algebra have been observed. A truth table verifies a Boolean equation; the table for Eq. (1.8a) is shown In

TABLE 1.5
Two-variable Combinations

A	0	0	1	1
B	0	1	0	1

TABLE 1.6
Commutative OR Table

A	B	A + B	B + A
0	0	0	0
0	1	1	1
1	0	1	1
1	1	1	1

Table 1.6. The first two colums list all possible combinations of two variables, the third column ORs A with B, and the fourth column ORs B with A. If at least one quantity is 1, the result of ORing is 1. Since the third and fourth elements of Table 1.6 are identical, Eq. (1.8a) is valid. Equation (1.8b) can be verified in the same manner.

Having admitted that two variables are possible, we can also consider three variables; Eqs. (1.9a) and (1.9b) are the three-variable *associative* equations.

$$A + (B + C) = (A + B) + C \tag{1.9a}$$
$$A \cdot (B \cdot C) = (A \cdot B) \cdot C \tag{1.9b}$$

Similarly, Eqs. (1.10a) and (1.10b) are the three-variable *distributive* Boolean equations:

$$A \cdot (B + C) = A \cdot B + A \cdot C \tag{1.10a}$$
$$(A + B) \cdot (A + C) = A + B \cdot C \tag{1.10b}$$

When three variables occur, there are, as shown in Table 1.7, eight possible combinations. Listing all combinations of three variables is an extension of the two-variable method.

- C alternates between 0 and 1 each time
- B alternates in pairs
- A alternates in fours

TABLE 1.7
Three-Variable Combinations

A	B	C
0	0	0
0	0	1
0	1	0
0	1	1
1	0	0
1	0	1
1	1	0
1	1	1

6 Computer Circuit Concepts

TABLE 1.8 Three-Term Truth Table

A	B	C	(A + B)	(A + C)	(A + B) · (A + C)	B · C	A + B · C
0	0	0	0	0	0	0	0
0	0	1	0	1	0	0	0
0	1	0	1	0	0	0	0
0	1	1	1	1	1	1	1
1	0	0	1	1	1	0	1
1	0	1	1	1	1	0	1
1	1	0	1	1	1	0	1
1	1	1	1	1	1	1	1

This table begins as any truth table for a three-variable Boolean equation. For example, to test the validity of Eq. (1.10b), begin with all possible combinations of three variables. Then combine variables from each side of the equation. Finally, compare both sides of the equation. Table 1.8 demonstrates this situation. Similarly, combining terms of other three-variable equations such as Eqs. (1.9a), (1.9b), and (1.10a), tests their validity.

The equations in this section are the theorems of Boolean algebra and are shown in Table 1.9. Boolean theorems are used to write Boolean equations. Such equations are the basis of computer circuit design.

1.2 WRITING BOOLEAN EQUATIONS

A systematic procedure exists for designing computer circuits. The first step is problem definition. In other words, the start is a description of circuit requirements: The circuit must . . ., but the circuit must not . . . . After circuit requirements are specified, the specifications are translated into a truth table, and the

TABLE 1.9 Boolean Theorems

	(a)	(b)
(1.1)	$A = 1$	$A = 0$
(1.2)	$A = 1 \rightarrow \overline{A} = 0$	$A = 0 \rightarrow \overline{A} = 1$
(1.3)	$\overline{\overline{A}} = A$	$\overline{\overline{\overline{A}}} = \overline{A}$
(1.4)	$A + \overline{A} = 1$	$A \cdot \overline{A} = 0$
(1.5)	$A + 0 = A$	$A \cdot 0 = 0$
(1.6)	$A + A = A$	$A \cdot A = A$
(1.7)	$A + 1 = 1$	$A \cdot 1 = A$
(1.8)	$A + B = B + A$	$A \cdot B = B \cdot A$
(1.9)	$A + (B + C) = (A + B) + C$	$A \cdot (B \cdot C) = (A \cdot B) \cdot C$
(1.10)	$A \cdot (B + C) = A \cdot B + A \cdot C$	$(A + B) \cdot (A + C) = A + B \cdot C$

**TABLE 1.10
Darkroom Truth Table**

L	A
0	0
1	1

truth table is converted into a Boolean equation. Finally, the circuit which performs the Boolean equation is constructed.

We are not yet in a position to design computer circuits, but we can relate Boolean algebra to everyday occurrences. Writing Boolean equations for ordinary situations simplifies writing Boolean equations for computer circuits.

Consider a photographic darkroom. There is a warning lamp above the darkroom door. When the lamp is on the darkroom is in use, and the door must remain closed. But when the lamp is off, the darkroom is not in use, and it is safe to open the darkroom door. These conditions define the problem. Only two conditions are possible: the lamp is either on or off. Boolean existence and nonexistence represent the lamp conditions. The number 1 means the lamp (L) is on, and the number 0 means the lamp is off. Table 1.10 shows the truth table for the alarm (A) that indicates the lamp condition.

Next the Boolean equations represented by this truth table are written by combining all conditions which result in a 1. In this case there is only one condition

$$L = A$$

Similarly, the inverse Boolean equation combines all conditions which result in a 0. In this case

$$\bar{L} = \bar{A}$$

Boolean equations with one variable are important, but applications are limited. Much more is accomplished when equations contain two or more variables.

Example 1.1 Andy A and Bruce B both like the movies, but they don't like each other. As a result Andy will not go to the movies M if Bruce goes. Feelings are mutual, and Bruce will not go if Andy goes. Show Boolean equations for: (*a*) somebody going to the movies; and (*b*) nobody going to the movies.

Solution The two variables are Andy and Bruce. There are four possible combinations:
- Neither Andy nor Bruce goes
- Andy does not go but Bruce does
- Andy goes but Bruce does not
- Both Andy and Bruce go

**TABLE 1.11
Movies Truth Table**

A	B	M
0	0	0
0	1	1
1	0	1
1	1	0

We represent going to the movies by a 1 and not going by a 0. Of the four possibilities, only two result in a moviegoer. The truth table for this Andy-Bruce-movie situation is shown in Table 1.11.

(a) The possibilities for somebody going to the movies have a 1 in the M column. These are

    NOT Andy AND Bruce

which in Boolean algebra is $\bar{A} \cdot B$

    OR Andy AND NOT Bruce

which is $+ A \cdot \bar{B}$.

The Boolean equation for somebody going to the movies is the ORing of conditions for which someone goes to the movies

$$M = \bar{A} \cdot B + A \cdot \bar{B}$$

(b) The possibilities for nobody going to the movies have a 0 in the M column. These are

    NOT Andy AND NOT Bruce

which is $\bar{A} \cdot \bar{B}$.

    OR Andy AND Bruce

which is $+ A \cdot B$.

The Boolean equation for nobody going is

$$\bar{M} = \bar{A} \cdot \bar{B} + A \cdot B$$

This example is a trivial case of two variables. Whether or not anyone goes to the movies is unimportant. However, a circuit with the same truth table is the basic addition circuit in real computers.

Situations involving Andy and Bruce can be more involved. A three-variable problem delves further into Boolean equations.

Example 1.2   Andy and Bruce both want to take Cindy to the movies. Andy is shy and will only ask Cindy if Bruce is not present. However, Bruce is not shy; he will ask

TABLE 1.12 Dating Cindy

A	B	C	D
0	0	0	0
0	0	1	0
0	1	0	0
0	1	1	1
1	0	0	0
1	0	1	1
1	1	0	0
1	1	1	1

Cindy whether or not Andy is present. Write Boolean equations which give Cindy's chances of (*a*) being invited to the movies; and (*b*) not being invited.

Solution   Table 1.12 lists all eight possibilities involving Andy, Bruce, and Cindy. The 1s represent yes and the 0s represent no.

(*a*) Since Andy is shy, he has only one chance to invite Cindy, $A \cdot \bar{B} \cdot C$. Bruce is bolder and has two chances, $\bar{A} \cdot B \cdot C + A \cdot B \cdot C$. The Boolean equation for inviting Cindy ORs all possibilities which yield a 1 in the date *D* column.

$$D = A \cdot \bar{B} \cdot C + \bar{A} \cdot B \cdot C + A \cdot B \cdot C$$

(*b*) Cindy cannot be invited when she is not present: $\bar{A} \cdot \bar{B} \cdot \bar{C} + \bar{A} \cdot B \cdot \bar{C} + A \cdot \bar{B} \cdot \bar{C} + A \cdot B \cdot \bar{C}$. Also, Cindy cannot be invited when she is alone: $\bar{A} \cdot \bar{B} \cdot C$. Cindy's chances of not being invited OR all possibilities which yield a 0 in column *D*.

$$\bar{D} = \bar{A} \cdot \bar{B} \cdot \bar{C} + \bar{A} \cdot \bar{B} \cdot C + \bar{A} \cdot B \cdot \bar{C} + A \cdot \bar{B} \cdot \bar{C} + A \cdot B \cdot \bar{C}$$

These illustrations of writing Boolean equations result in a single outcome: opening a darkroom door, going to the movies, and dating. Truth tables and resulting Boolean equations can represent more complicated situations.

Example 1.3   Cindy is not interested in Andy or Bruce, but she does like to jog. Rain or shine, Cindy jogs 6 days a week, usually alone. However, she is willing to jog with others. Write: (*a*) a truth table which includes Cindy's dating and jogging prospects; and (*b*) the Boolean equation for jogging or going to the movies.

Solution   Cindy's jogging possibilities are another situation added to the truth table.
(*a*) This is shown in Table 1.13. Cindy can jog whenever there is a 1 in the *J* column.

$$J = \bar{A} \cdot \bar{B} \cdot C + \bar{A} \cdot B \cdot C + A \cdot \bar{B} \cdot C + A \cdot B \cdot C$$

TABLE 1.13
Cindy's Busy Life

A	B	C	D	J
0	0	0	0	0
0	0	1	0	1
0	1	0	0	0
0	1	1	1	1
1	0	0	0	0
1	0	1	1	1
1	1	0	0	0
1	1	1	1	1

(b) Similarly, Cindy can jog OR go to the movies whenever 1s occur in either J or D.

$$D + J = \bar{A} \cdot \bar{B} \cdot C + \bar{A} \cdot B \cdot C + A \cdot \bar{B} \cdot C + A \cdot B \cdot C + \bar{A} \cdot B \cdot C + A \cdot \bar{B} \cdot C + A \cdot B \cdot C$$

The second part of this example combines two independent relationships among the same variables. It is possible to include as many variables and relationships as a situation requires.

Regardless of the number of variables and the number of relationships the procedure is the same:

1. The problem must be defined in two-state terms. In dealing with darkrooms, dating, or computer circuits, problems must be stated so that there is only one of two possible choices. If the problem is not in such a form, it must be reformulated. Answers must be in yes/no, all/nothing, on/off, or 1/0 form.

2. All possible combinations of variables must be included. The orderly scheme of alternating 0s and 1s ensures listing each combination.

3. Each combination is evaluated separately. The result of each combination is either 1 or 0.

4. Write the Boolean equations. There are two equations for each relationship, one of which ORs all combinations which result in a 1 while its inverse ORs all combinations which result in a 0. Both the equation and the inverse are equally valid.

These four steps are the same regardless of the nature or complexity of the problem. Computer circuits make decisions and are designed in the same way. Boolean algebra can also determine if a correct circuit is the most efficient circuit. In other words, Boolean algebra also simplifies complicated circuits.

1.3 SIMPLIFYING BOOLEAN EQUATIONS

After a truth table is completed, the corresponding Boolean equation is written. The next step could be construction of the circuit to implement the Boolean equation. However, simplification of the Boolean equation is often possible. From a mathematical standpoint, a simpler equation is more elegant, and from a circuit standpoint, a simpler Boolean equation has the practical advantage of requiring fewer parts. Equation simplification results in smaller, less expensive, and more reliable circuits.

Boolean equations are simplified with the aid of theorems listed in Table 1.9. Simplification is accomplished by observing standard forms or else by converting terms into standard forms.

Example 1.4 Simplify $L = A + B + \bar{A}$.

Solution Boolean theorem (1.9a) permits rearrangement. In particular, A and $\bar{A}$ can be grouped as

$$L = (A + \bar{A}) + B$$

From theorem (1.4a) the terms inside the parentheses reduce to 1

$$L = 1 + B$$

The only difference between this equation and theorem (1.7a) is the name of the variable. Since names are arbitrary, symbols can be changed. $1 + B$ has the same sense as $1 + A$. Thus the equation reduces to

$$L = 1$$

In this case simplification is based on the associative OR theorem. Other equations can be simplified with the distributive OR theorem.

Example 1.5 Simplify the Boolean equation $E = C + C \cdot D$.

Solution Theorem (1.10a) allows regrouping as

$$E = C(1 + D)$$

and with theorem (1.7a)

$$1 + D = 1$$

Therefore the equation reduces to

$$E = C$$

In the two previous examples simplification is a result of removing a common variable. Removing more than one common variable is also possible. Cindy's problems are a convenient illustration.

Example 1.6 Simplify the equation for Cindy's jogging possibilities.

Solution Cindy's jogging opportunities have been determined as

$$J = \bar{A} \cdot \bar{B} \cdot C + \bar{A} \cdot B \cdot C + A \cdot \bar{B} \cdot C + A \cdot B \cdot C$$

The first two quantities contain $\bar{A} \cdot C$ and the last two quantities contain $A \cdot C$. Removing these common terms yields

$$J = \bar{A} \cdot C(\bar{B} + B) + A \cdot C(\bar{B} + B)$$

Using theorem (1.4a), reduces $(\bar{B} + B)$ to 1.

Therefore $\quad J = \bar{A} \cdot C \cdot 1 + A \cdot C \cdot 1$

and by the theorem of Eq. (1.7b) the ANDing with 1 can be simplified. This leaves

$$J = \bar{A} \cdot C + A \cdot C$$

C is common and can be removed from both terms

$$J = C(\bar{A} + A)$$

and again using theorem (1.4a), the equation reduces to

$$J = C \cdot 1$$

which further simplifies to

$$J = C$$

Cindy is the only required person if Cindy wants to jog. This information was given in setting up Example 1.3. However, the important point is that Boolean simplification is possible. Simplification is important regardless of the source of the equation.

Another method of simplifying Boolean equations is by expanding terms. Expanding terms is the opposite of removing common terms. Both techniques are justified by theorem (1.10a); any theorem is valid in either direction. Expansion results when going from left to right, and removing common terms results when going from right to left.

Example 1.7 Simplify the Boolean equation $L = A(\bar{A} + B)$.

Solution Expanding $L = A(\bar{A} + B)$ results in

$$L = A \cdot \bar{A} + A \cdot B$$

and from Boolean theorem (1.4b)

$$A \cdot \bar{A} = 0$$

Logic 13

The equation reduces to
$$L = 0 + A \cdot B$$
By using the Boolean theorem of Eq. (1.5a) the equation reduces to
$$L = A \cdot B$$

Again, the reduced equation is simpler than the original version.
Expansion of terms has some subtle applications, as illustrated by the next example.

Example 1.8 Simplify the Boolean equation $L = A \cdot B + A \cdot B \cdot \bar{C}$.

Solution Each quantity contains $A \cdot B$. If both quantities contained three terms, common variables could be removed. Theorem (1.7b) permits increasing the number of terms since
$$1 \cdot A = A$$
It is always possible to AND any Boolean quantity with 1. Therefore the original equation can be rewritten
$$L = A \cdot B \cdot 1 + A \cdot B \cdot \bar{C}$$
Removing the common term $A \cdot B$ yields
$$L = A \cdot B(1 + \bar{C})$$
There is no Boolean theorem for $1 + \bar{C}$. But a truth table, as shown in Table 1.14, resolves the situation.
Table 1.14 demonstrates that $1 + \bar{C} = 1$. Therefore
$$L = A \cdot B(1 + \bar{C})$$
can be further reduced to
$$L = A \cdot B \cdot 1$$
and again applying theorem (1.7b) yields
$$L = A \cdot B$$

In this example expansion is achieved by ANDing with 1. In other cases it is

TABLE 1.14
Truth Table for $1 + \bar{C}$

C	$\bar{C}$	1	$1 + \bar{C}$
0	1	1	1
1	0	1	1

helpful to expand an expression by ORing an existing term. This is permissible since $A + A = A$.

Inverted variables are a normal part of Boolean equations. Quantities such as $\bar{A}$, $\bar{A} \cdot \bar{B}$, $\bar{A} \cdot \bar{B} \cdot \bar{C}$, etc., appear regularly. Regardless of how many inverts are involved, such terms contain individual inverts.

Another condition is the *extended* invert, in which two or more variables are included beneath the same invert sign. $\overline{A \cdot B}$ and $\overline{A \cdot B \cdot C}$ are examples of the extended invert. At first glance individual and extended inverts such as $\bar{A} \cdot \bar{B}$ and $\overline{A \cdot B}$ seem identical. However, these terms are different, and the difference can be demonstrated by comparing truth tables, as shown in Table 1.15.

Boolean expressions are only equal if truth tables are identical. Therefore $\bar{A} \cdot \bar{B} \neq \overline{A \cdot B}$. Table 1.15 demonstrates a situation in which a term with individual inverts is different from the extended invert version. However, cases occur in which individual and extended inverts are equal. In particular

$$\bar{A} \cdot \bar{B} = \overline{A + B} \qquad (1.11a)$$

and $\quad \bar{A} + \bar{B} = \overline{A \cdot B} \qquad (1.11b)$

These equations are known as *De Morgan's theorem*; their validity can be determined from truth tables. Equation (1.11a) contains an extended inverted OR, called a NOR, which is a contraction of NOT OR. In words, Eq. (1.11a) says the NOR is the same as ANDing individually inverted variables. Similarly, an extended inverted AND is called a NAND. Equation (1.11b) says that a NAND is the same as ORing individually inverted variables.

These equations contain two variables. De Morgan's theorem can be extended beyond two variables. The three-variable versions are

$$\bar{A} \cdot \bar{B} \cdot \bar{C} = \overline{A + B + C}$$

and $\quad \bar{A} + \bar{B} + \bar{C} = \overline{A \cdot B \cdot C}$

and so forth as required.

De Morgan's theorem is also used to simplify Boolean equations, and its use is not limited to entire equations. Simplifications can be realized by De Morganizing some but not all terms.

TABLE 1.15 $\bar{A} \cdot \bar{B}$ versus $\overline{A \cdot B}$

A	B	$\bar{A}$	$\bar{B}$	$\bar{A} \cdot \bar{B}$	AB	$\overline{A \cdot B}$
0	0	1	1	1	0	1
0	1	1	0	0	0	1
1	0	0	1	0	0	1
1	1	0	0	0	1	0

Example 1.9 Simplify $L = \overline{A \cdot B} + \bar{C} + \bar{D}$.

Solution Equation (1.11b) permits replacing $\overline{A \cdot B}$ with $\bar{A} + \bar{B}$. Thus the original equation becomes

$$L = \bar{A} + \bar{B} + \bar{C} + \bar{D}$$

This is the four-variable version of Eq. (1.11b) and reduces to

$$L = \overline{A \cdot B \cdot C \cdot D}$$

It is even possible to De Morganize an equation which does not contain inverts. Since $\bar{\bar{A}} = A$, any Boolean equation can be doubly inverted.

Example 1.10 Rewrite $L = A + B + C$ in a form which eliminates OR.

Solution The inverted form of

$$L = A + B + C$$

is $\bar{L} = \overline{A + B + C}$

Inverting both sides of this version yields

$$\bar{\bar{L}} = \overline{\overline{A + B + C}}$$

Since $\bar{\bar{L}} = L$ this can be rewritten as

$$L = \overline{\overline{A + B + C}}$$

The lower invert is independent of the upper invert. Now use De Morgan's theorem; a NOR can be replaced with a string of individually inverted ANDs. Thus

$$L = \overline{\bar{A} \cdot \bar{B} \cdot \bar{C}}$$

which satisfies the requirement.

In this case, the final equation is more complicated than the original. However, the opposite is true when simplification of the final version is required. The important feature illustrated in Example 1.10 is elimination of OR terms. Similarly, De Morgan's theorem can completely eliminate AND terms. In many cases eliminating one of the three Boolean terms results in simpler circuits.

1.4 STANDARD BOOLEAN FORMS

Computer circuits implement Boolean equations, and simplifying these equations is usually desirable. However, many circuits of the same type can be less expensive than many unrelated but simplified circuits. Thus, standard Boolean forms should be investigated. Two standard forms exist and both begin with truth tables.

Each row of a truth table ANDs all variables: $\bar{A} \cdot \bar{B} \cdot \bar{C}$, $\bar{A} \cdot B \cdot C$, etc. These forms are sometimes called the *product of variables*. Thus, consider a Boolean equation that ORs all products of variables resulting in a 1. Because OR is similar to addition, ORing is sometimes called taking a *sum*. An equation written directly from the truth table is called the *sum of products* (SOP) form.

For example the Boolean equation

$$M = \bar{A} \cdot B + A \cdot \bar{B}$$

is the SOP equation for Andy or Bruce going to the movies and is obtained from Table 1.11. An alternative form is obtained by working with the inverse Boolean equation. The inverse equation results when terms resulting in 0 are ORed. The inverse equation for Table 1.11 is

$$\bar{M} = \bar{A} \cdot \bar{B} + A \cdot B$$

If both sides of the inverse equation are inverted, the result is

$$\bar{\bar{M}} = \overline{\bar{A} \cdot \bar{B} + A \cdot B}$$

Applying De Morgan's theorem to the right-hand side of this equation yields

$$\bar{\bar{M}} = (\bar{\bar{A}} + \bar{\bar{B}}) \cdot (\bar{A} + \bar{B})$$

Theorem (1.3a) permits reduction to

$$M = (A + B) \cdot (\bar{A} + \bar{B})$$

Again because of the similarity to conventional algebraic operation, this form is called the *product of sums* (POS). Any Boolean equation can be written in either SOP or POS form; these are called the *standard* forms or sometimes the *canonical* forms.

Example 1.11 Find the (*a*) SOP and (*b*) POS form of a 2-variable AND.

Solution Table 1.16 shows the 2-variable AND truth table.
(*a*) The SOP form ORs all ANDs resulting in a 1. In this case, a single 1

TABLE 1.16 AND Truth Table

A	B	L
0	0	0
0	1	0
1	0	0
1	1	1

exists. Therefore the SOP of a 2-variable AND is

$$L = A \cdot B$$

(b) Finding the POS form begins with the inverse equation, which is obtained by ORing all ANDs resulting in a 0

$$\bar{L} = \bar{A} \cdot \bar{B} + \bar{A} \cdot B + A \cdot \bar{B}$$

and then inverted to yield

$$\bar{\bar{L}} = \overline{\bar{A} \cdot \bar{B} + \bar{A} \cdot B + A \cdot \bar{B}}$$

Applying De Morgan's theorem

$$\bar{\bar{L}} = (\bar{\bar{A}} + \bar{\bar{B}}) \cdot (\bar{\bar{A}} + \bar{B}) \cdot (\bar{A} + \bar{\bar{B}})$$

which reduces to the POS form

$$L = (A + B) \cdot (A + \bar{B}) \cdot (\bar{A} + B)$$

In this case the SOP is simpler than the POS, but the opposite case also occurs. The SOP can be simpler, equal in complexity to, or more complicated than the POS. Nevertheless, the SOP is usually called the *minterm* and the POS is called the *maxterm*. The truth table yields the minterm form, and the maxterm is obtained by De Morganizing the inverse equation.

When the starting point is a truth table, it is easy to find the minterm. However, some effort is usually required to find the maxterm. Comparing minterm and maxterm complexity determines which form is simpler. Thus, a less involved method of finding the maxterm is helpful, such as a "cookbook" application of De Morgan's theorem, which yields the POS directly from a truth table. De Morgan equivalents require:

1. Changing ANDs to ORs and vice verse

2. Inverting uninverted terms and vice versa
   This approach is shown in Table 1.17, which gives general two-variable expressions in the SOP and POS.

Thus the minterm is found by ANDing the variables, and maxterms are the De Morgan equivalents of the minterms. Example 1.12 illustrates the cookbook approach to finding maxterms.

TABLE 1.17 Equivalent Forms

A	B	Min	Max
0	0	$\bar{A} \cdot \bar{B}$	$A + B$
0	1	$\bar{A} \cdot B$	$A + \bar{B}$
1	0	$A \cdot \bar{B}$	$\bar{A} + B$
1	1	$A \cdot B$	$\bar{A} + \bar{B}$

TABLE 1.18 3-Input OR

A	B	C	L	Min	Max
0	0	0	0	$\bar{A} \cdot \bar{B} \cdot \bar{C}$	$A + B + C$
0	0	1	1	$\bar{A} \cdot \bar{B} \cdot C$	$A + B + \bar{C}$
0	1	0	1	$\bar{A} \cdot B \cdot \bar{C}$	$A + \bar{B} + C$
0	1	1	1	$\bar{A} \cdot B \cdot C$	$A + \bar{B} + \bar{C}$
1	0	0	1	$A \cdot \bar{B} \cdot \bar{C}$	$\bar{A} + B + C$
1	0	1	1	$A \cdot \bar{B} \cdot C$	$\bar{A} + B + \bar{C}$
1	1	0	1	$A \cdot B \cdot \bar{C}$	$\bar{A} + \bar{B} + C$
1	1	1	1	$A \cdot B \cdot C$	$\bar{A} + \bar{B} + \bar{C}$

Example 1.12 For $L = A + B + C$ find (a) the minterm, (b) the maxterm form.

Solution The truth table for a three-variable OR equation in both minterm and maxterm form is shown in Table 1.18.

(a) The minterm form is found by ORing all minterms resulting in a 1:

$$L = \bar{A} \cdot \bar{B} \cdot C + \bar{A} \cdot B \cdot \bar{C} + \bar{A} \cdot B \cdot C + A \cdot \bar{B} \cdot \bar{C} + A \cdot \bar{B} \cdot C + A \cdot B \cdot \bar{C} + A \cdot B \cdot C$$

(b) The maxterm form is found by ANDing all maxterms resulting in a 0. In this case, there is only one maxterm, and

$$L = A + B + C$$

is an illustration of a maxterm that is simpler than a minterm.

A truth table is not always available, but it may be necessary to convert a Boolean equation into either SOP or POS form. This is accomplished by ANDing the appropriate terms with 1 in the form of a variable ORed with its invert.

Example 1.13 Compare minterm and maxterm forms of

$$L = \bar{A} + B$$

Solution AND both terms with 1 in order to combine the variables:

$$L = \bar{A} \cdot 1 + B \cdot 1$$
$$L = \bar{A}(B + \bar{B}) + B(A + \bar{A})$$

Expansion results in

$$L = \bar{A} \cdot B + \bar{A} \cdot \bar{B} + A \cdot B + \bar{A} \cdot B$$

Since $A + A = A$, reduce $\bar{A} \cdot B + \bar{A} \cdot B$ to $\bar{A} \cdot B$. Thus the minterm form of the

TABLE 1.19
$L = \bar{A} \cdot \bar{B} + \bar{A} \cdot B + A \cdot B$

A	B	L
0	0	1
0	1	1
1	0	0
1	1	1

original equation is

$$L = \bar{A} \cdot \bar{B} + \bar{A} \cdot B + A \cdot B$$

Next set up the truth table for this equation, which is Table 1.19.

The inverse equation ORs all 0 terms:

$$\bar{L} = A \cdot \bar{B}$$

In maxterm form

$$L = \bar{A} + B$$

which happens to be the original equation.

1.5 KARNAUGH MAPS

Reducing Boolean equations requires skill and experience. Nevertheless, it is difficult to determine if the simplest reduction has been obtained—as a matter of fact, it is difficult to determine if any reduction is possible. Algebraic reduction presents a simpler form but does not guarantee the best reduction. Similarly comparing original and reduced truth tables demonstrates equivalence but not ultimate simplicity. Luck and skill are required to simplify Boolean equations.

In 1953 Maurice Karnaugh developed a pictorial method of simplifying Boolean equations. Pictures eliminate a good deal of algebraic manipulation. The pictorial is called a *Karnaugh map*, and Fig. 1.1 shows a two-variable example.

Columns show the possible states of variable A: the left-hand column represents $A = 0$ and the right-hand column represents $A = 1$. Similarly, rows show the possible states of B: the top row represents $B = 0$ and the bottom row represents $B = 1$. Each square ANDs the respective variable

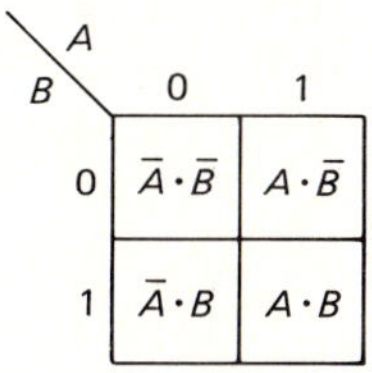

Figure 1.1 Karnaugh map representation.

B \ A	0	1
0	$\bar{A} \cdot \bar{B}$	$A \cdot \bar{B}$
1	$\bar{A} \cdot B$	$A \cdot B$

Figure 1.2
$L = \bar{A} \cdot \bar{B} + \bar{A} \cdot B$. (*a*) Truth table; (*b*) Karnaugh map.

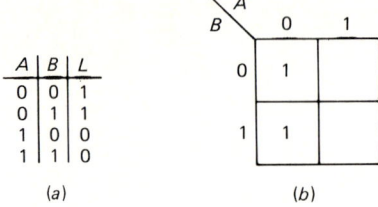

states. For example, the square on the top to the left is the intersection of $A = 0$ with $B = 0$. Therefore this square is $\bar{A} \cdot \bar{B}$. Similarly, the right top square is the intersection of $A = 1$ with $B = 0$; this square is $A \cdot \bar{B}$. Similarly, the bottom left square is $\bar{A} \cdot B$, and the bottom right square is $A \cdot B$.

Karnaugh maps are similar to truth tables. A 1 in the output column of a truth table show a combination of variables which exist. Truth table rows resulting in 1 are ORed together to form a complete Boolean equation. With a Karnaugh map, squares with 1s are ORed together to present a Boolean equation. Similarly, combining 0s of a truth table or Karnaugh map yields the inverse equations.

Combinations which result in 0s are usually left blank. Blank squares have the same meaning as 0s, but make a clearer map. Figure 1.2 compares the truth table and Karnaugh map for the same equation.

Karnaugh map simplification is accomplished by grouping adjacent squares with 1s together, and eliminating any variable in a group which changes state. Eliminating variables which change state is the basis of Karnaugh map reduction. Adjacent squares with 1s are equivalent to ORing. Thus, a variable which changes state is in $A + \bar{A}$ form and equals 1. Karnaugh map reduction is the graphical version of Boolean theorem (1.4*a*). Figure 1.3 compares Karnaugh with Boolean algebraic reduction.

Skill and luck are needed when using algebraic reduction. It is never clear whether removing common terms, ANDing with 1, ORing in an existing form, or some other complex operation yields the simplest form. On the other hand, Karnaugh reduction is almost by inspection.

An existing map square can be adjacent to more than one square. In this case, the best reduction uses multiple groupings with the common square.

Figure 1.3
Equation simplification. (*a*) Karnaugh map; (*b*) Boolean algebra.

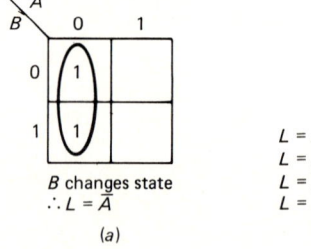

Multiple grouping is the Karnaugh equivalent of ORing in an existing Boolean equation variable. In other words, multiple grouping is the graphical version of Boolean theorem (1.6a).

Example 1.14 Use a Karnaugh map to simplify $L = \bar{A} \cdot B + A \cdot B + A \cdot \bar{B}$.

Solution As shown in Fig. 1.4a, 1s are placed in each existing square. Next, group adjacent squares together. In this case, $A \cdot B$ is common to both groups.

1. In the horizontal group A changes state. Therefore the horizontal adjacency reduces to B.

2. In the vertical group B changes state. Therefore the vertical adjacency reduces to A.

3. The simplified Boolean equation is obtained by ORing reduced terms. Thus

$$L = A + B$$

In some cases, the inverse equation is an acceptable substitute. Recall that a lamp over a darkroom door gives equivalent information whether on or off. When an inverse equation is acceptable, the Karnaugh map shows which form is simpler. In Example 1.14 the inverse equation can be taken directly from the map; there is only one blank square, and therefore the inverse equation is

$$\bar{L} = \bar{A} \cdot \bar{B}$$

Observe that the reduced equation from a Karnaugh map is the minterm form of the equation for which the inverse equation is the maxterm. Duality of minterm and maxterm forms has been mentioned.

Karnaugh maps have another useful feature which is more important than Boolean reduction. As equations become more complicated, it is difficult to determine if any reduction is possible. Regardless of effort, there is no assurance that any reduction method leads to a simpler equation. Boolean equations can be complicated but irreducible. Inspecting a Karnaugh map determines if simplification is possible.

Figure 1.4 Simplifying $L = \bar{A} \cdot B + A \cdot \bar{B} + A \cdot B$. (*a*) Map; (*b*) adjacent groups.

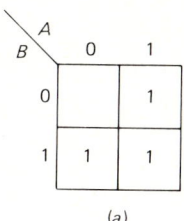
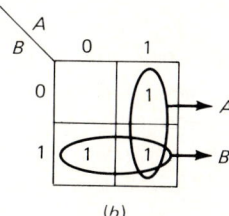

Figure 1.5
$M = A \cdot \bar{B} + \bar{A} \cdot B$.

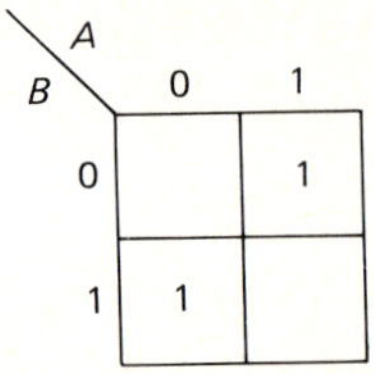

Example 1.15 Use a Karnaugh map to simplify the equation for Andy or Bruce going to the movies.

Solution The equation for this situation as developed in Example 1.1 is

$$M = \bar{A} \cdot B + A \cdot \bar{B}$$

and the map of this equation is shown in Fig. 1.5. The squares with 1 do not have an adjacent side, so no simplification is possible. Similarly, the blank squares do not have an adjacent side, and no simplification of the inverse equation is possible. Neither M nor $\bar{M}$ of the Andy-Bruce-movie situation can be reduced.

In summary, the features of a two-variable Karnaugh map are:

1. Each square ANDs a unique combination of both variables.

2. A 1 represents a combination which exists, and a 0, or blank square, represents a combination which does not exist.

3. The Boolean equation is obtained by ORing all squares with 1s. The inverse equation ORs all blank squares.

4. If two adjacent squares both contain 1s, the variable which changes state is eliminated.

5. A single square can be included in more than one adjacency.

6. Only adjacent squares can be simplified.

1.6 THREE-VARIABLE MAPS

By their very nature three-variable Boolean equations are more complicated than those with two variables. Therefore when three-variable equations must be reduced, a three-variable Karnaugh map eliminates much effort.

A three-variable Karnaugh map must be larger than a two-variable map. The two-variable map contains four squares because there are four possible combinations of two variables. Since there are eight possible combinations of three variables, a three-variable map must contain eight squares, each of which ANDs three variables. Three variables can be shown on a two-dimensional map if two variables are combined. Precisely which variables are com-

Figure 1.6
Three-variable map.

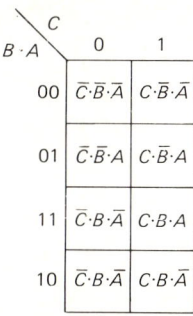

bined is unimportant. Typically, as shown in Fig. 1.6, C is on top while the $B \cdot A$ combination is on the side.

This map shows an important but subtle point. A Karnaugh map is only useful if one variable changes state between adjacent squares. With a two-variable map merely numbering the squares in sequence satisfies this condition, but care is required for three-variable maps. The lone variable C changes from 0 to 1 in the normal manner, but the ordering of 0s and 1s presented in Table 1.5 does not quite work for the $B \cdot A$ combination. The conventional sequence for two variables is 00, 01, 10, and 11; in going from 01 to 10 both variables change state at the same time. Such a condition cannot exist on a Karnaugh map, and therefore the sequence is rearranged to 00, 01, 11, 10. This arrangement preserves the single-variable change between adjacent squares.

Whichever variable changes state between two adjacent squares can be eliminated. For example, if the top two squares of Fig. 1.6 both contain 1s, C is the only variable which changes, and these two squares can be reduced from $\bar{C} \cdot \bar{B} \cdot \bar{A} + C \cdot \bar{B} \cdot \bar{A}$ to $\bar{B} \cdot \bar{A}$. On a three-variable map two adjacent 1s reduce to a two-variable quantity.

Example 1.16 Use a Karnaugh map to simplify

$$L = \bar{A} \cdot B \cdot C + A \cdot B \cdot C + A \cdot B \cdot \bar{C}$$

Solution The map for this equation is shown in Fig. 1.7a, and the combined adjacencies are shown in Fig. 1.7b.

1. In the horizontal adjacency only C changes state. Thus this combination reduces to $A \cdot B$.

2. In the vertical adjacency only A changes state, and this combination reduces to $B \cdot C$.

3. The simplified Boolean equation is obtained by ORing the reduced terms:

$$L = A \cdot B + B \cdot C$$

24 Computer Circuit Concepts

Figure 1.7
$L = \bar{A} \cdot B \cdot C + A \cdot B \cdot C + A \cdot B \cdot \bar{C}$.
(*a*) Basic map; (*b*) combined adjacencies.

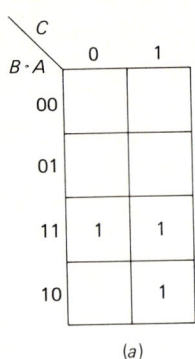
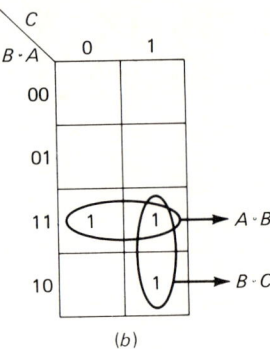

Karnaugh reduction is simpler than algebraic reduction. However, this example also demonstrates one reason that algebraic manipulation is still necessary. The Karnaugh reduction obtained by ORing can be further simplified to

$$L = B \cdot (A + C)$$

The advantage of this last reduction is discussed in Sec. 1.8.

An interesting feature of Karnaugh map reduction is illustrated in the next example. Situations of this type were intentionally avoided when algebraic reduction techniques were discussed.

Example 1.17 Write the basic and simplified versions of the Boolean equation represented by the Karnaugh map shown in Fig. 1.8a.

Solution The basic equation obtained by ORing all squares with 1s is

$$L = \bar{A} \cdot \bar{B} \cdot C + A \cdot B \cdot \bar{C} + \bar{A} \cdot B \cdot \bar{C} + A \cdot \bar{B} \cdot \bar{C} + A \cdot \bar{B} \cdot C$$

Figure 1.8
Maps for Example 1.17.
(*a*) Original map; (*b*) combined adjacencies; (*c*) alternate adjacencies.

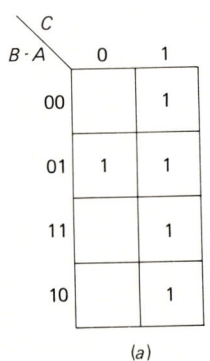
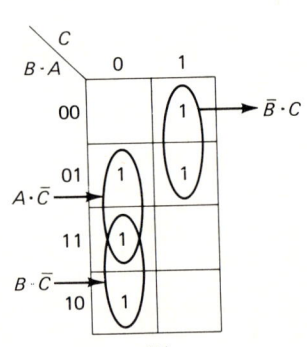
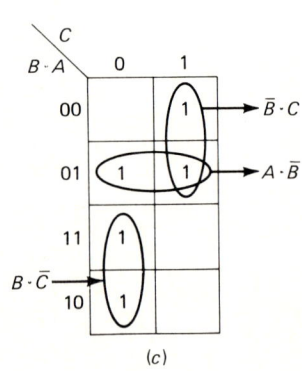

Combined adjacencies are shown in Fig. 1.8b. The reduced equation is

$$L = A \cdot \bar{C} + \bar{B} \cdot C + B \cdot \bar{C}$$

An alternate but equally valid set of adjacencies is shown in Fig. 1.8c. In this case the reduced equation is

$$L = A \cdot \bar{B} + \bar{B} \cdot C + B \cdot \bar{C}$$

These two versions differ by a single term, $A \cdot \bar{C}$ versus $A \cdot \bar{B}$. Although these terms are not equivalent, the entire equations are. Equivalency can be demonstrated by comparing truth tables. Thus Example 1.17 demonstrates that more than one "optimum" reduction may exist.

As with two-variable maps, a three-variable map may not contain any adjacencies. The meaning is the same, and no reduction is possible. However, with a three-variable map, situations occur in which 1s which appear unrelated are actually adjacent. For example, consider the Karnaugh map shown in Fig. 1.9a. The squares with 1s are $C \cdot \bar{B} \cdot \bar{A}$ and $C \cdot B \cdot \bar{A}$. In this case only B changes state, which means B can be eliminated, and the result is $\bar{A} \cdot C$. However, Fig. 1.9a does not appear to indicate adjacencies. If Karnaugh maps are to remain useful, our concept of adjacencies must be modified. The necessary modification is shown in Fig. 1.9b.

Geographic maps are two-dimensional representations of the three-dimensional planet called earth. A Karnaugh map can also be considered as three-dimensional. In this version the cylinder is continuous and the "edges" are adjacent. Figure 1.9c indicates the adjacencies on the "unrolled map." Corresponding corners of a three-variable map are adjacent.

On a Karnaugh map adjacencies can only be reduced if the number of adjacencies is a power of 2. Several situations have been described in which adjacencies were paired. The next higher power is 2^2, and Fig. 1.10 shows a map with four adjacent 1s. According to principle 4 for two-variable Karnaugh maps, adjacent 1s can be considered as two pairs of adjacencies with two

Figure 1.9 Edge adjacencies. (a) Basic map; (b) cylindrical version; (c) corresponding adjacencies.

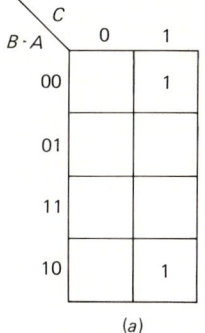
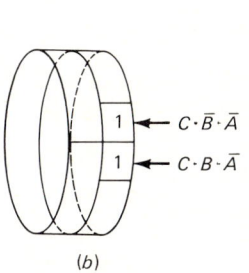
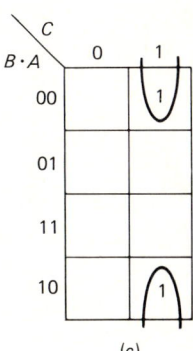

Figure 1.10 Four adjacent 1s.

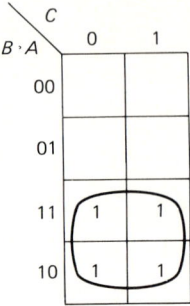

squares in each, which can then be reduced by Boolean manipulation. However larger groups result in greater simplification. Four adjacent 1s reduce to the single variable which does not change state.

It may not be immediately apparent which quantity in a four-adjacent-square situation remains constant. This is readily determined by listing the values of each adjacent square. For example in Fig. 1.10 the adjacent squares are

$$\bar{C} \cdot B \cdot A$$
$$\bar{C} \cdot B \cdot \bar{A}$$
$$C \cdot B \cdot A$$
$$C \cdot B \cdot \bar{A}$$

The only constant term is B, and the simplified equation is

$$L = B$$

Four adjacent squares can also occupy an entire column. Another four-adjacent-square situation occurs when the top two squares are combined with the bottom two squares. This is an edge-connected four-adjacent-square configuration. Regardless of which four-adjacent-square configuration occurs, the reduced version is a single variable.

Larger adjacencies result in greater simplification. The goal in simplifying a map is to enclose as many larger adjacencies as possible. Using a common square in more than one adjacency is one common technique for achieving greater reduction.

1.7 FOUR-VARIABLE MAPS

Advantages of Karnaugh maps are more apparent at the four-variable level. Since there are 16 possible combinations of four variables, a four-variable Karnaugh map must contain 16 squares. The same numbering sequence used on the combined variable side of a three-variable map is used on both sides of a four-variable map. This preserves the one-variable change between adjacent squares. Figure 1.11 shows a typical four-variable map. This map contains six 1s, and adjacencies are indicated.

Figure 1.11
Four-variable map.

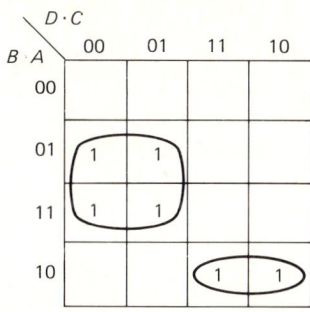

Example 1.18 Write the basic and simplified Boolean equations for the map shown in Fig. 1.11.

Solution The basic equation ORs squares with 1s. It is helpful to write complex Boolean products in descending alphabetical order:

$$L = \bar{D} \cdot \bar{C} \cdot \bar{B} \cdot A + \bar{D} \cdot \bar{C} \cdot B \cdot A + \bar{D} \cdot C \cdot \bar{B} \cdot A + \bar{D} \cdot C \cdot B \cdot A + D \cdot C \cdot B \cdot \bar{A} + D \cdot \bar{C} \cdot B \cdot \bar{A}$$

Again the variables which do not change are retained.
1. For the two-square adjacency

$$D \cdot \left. \begin{array}{c} C \\ \bar{C} \end{array} \right| \cdot B \cdot \bar{A}$$

C is the only variable which changes. This adjacency reduces to $D \cdot B \cdot \bar{A}$.

2. For the four-square adjacency

$$\bar{D} \cdot \left. \begin{array}{c} \bar{C} \\ \bar{C} \\ C \\ C \end{array} \right| \cdot \left. \begin{array}{c} \bar{B} \\ B \\ \bar{B} \\ B \end{array} \right| \cdot A$$

In this case two variables change state. This adjacency reduces to $\bar{D} \cdot A$.

3. The reduced equation is obtained by ORing the simplified terms. In this case

$$L = \bar{D} \cdot A + D \cdot B \cdot \bar{A}$$

This example demonstrates two features of a four-variable Karnaugh map: two adjacent squares reduce to a three-variable quantity and four adjacent squares reduce to a two-variable quantity. It is also possible that a single square can be part of several adjacencies on a four-variable map.

Figure 1.12
Map for Example 1.19.

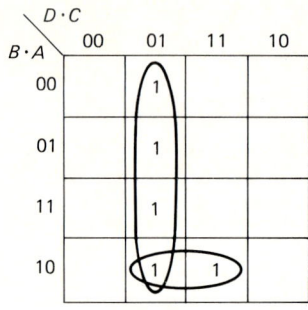

Example 1.19 Use a Karnaugh map to simplify

$$L = \bar{D}\cdot\bar{C}\cdot\bar{B}\cdot\bar{A} + \bar{D}\cdot C\cdot\bar{B}\cdot A + \bar{D}\cdot C\cdot B\cdot A + \bar{D}\cdot C\cdot B\cdot\bar{A} + D\cdot C\cdot B\cdot\bar{A}$$

Solution The map with adjacencies is shown in Fig. 1.12.

1. The four-square adjacency reduces to $\bar{D}\cdot C$.
2. The two-square adjacency reduces to $C\cdot B\cdot\bar{A}$.
3. Thus the best Karnaugh reduction is

$$L = \bar{D}\cdot C + C\cdot B\cdot\bar{A}$$

Edge adjacencies can also occur on four-variable Karnaugh maps. When three-variable Karnaugh maps were discussed, edge adjacencies were discussed in terms of visualizing the map as a horizontal cylinder. Since four-variable maps contain two variables on each side, it is necessary to think in terms of both horizontal and vertical adjacencies. Fig. 1.13 illustrates both types.

In this case the basic equation is

$$L = \bar{D}\cdot\bar{C}\cdot\bar{B}\cdot A + D\cdot\bar{C}\cdot B\cdot A + D\cdot C\cdot\bar{B}\cdot\bar{A} + D\cdot C\cdot B\cdot\bar{A}$$

Figure 1.13
Edge adjacency pairs. (*a*) Map; (*b*) horizontal cyclinder; (*c*) vertical cylinder.

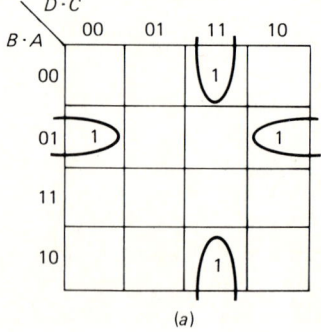
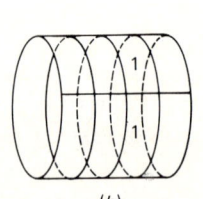
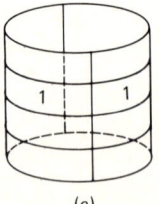

(*a*)   (*b*)   (*c*)

and the reduced equation is
$$L = \bar{C} \cdot \bar{B} \cdot A + D \cdot C \cdot \bar{A}$$

Reduction is possible when adjacencies occur in two- or four-square groupings. Six adjacent squares cannot be reduced, since adjacencies must be in powers of 2. Thus the next higher reduction occurs at 2^3, or 8, adjacent squares.

Example 1.20 Simplify the Boolean equation

$$L = \bar{D} \cdot \bar{C} \cdot \bar{B} \cdot \bar{A} + \bar{D} \cdot \bar{C} \cdot \bar{B} \cdot A + \bar{D} \cdot \bar{C} \cdot B \cdot A + \bar{D} \cdot \bar{C} \cdot B \cdot \bar{A} + \bar{D} \cdot C \cdot \bar{B} \cdot A + \bar{D} \cdot C \cdot B \cdot A +$$
$$D \cdot C \cdot \bar{B} \cdot A + D \cdot C \cdot B \cdot A + D \cdot \bar{C} \cdot \bar{B} \cdot \bar{A} + D \cdot \bar{C} \cdot \bar{B} \cdot A + D \cdot B \cdot \bar{C} \cdot A + D \cdot \bar{C} \cdot B \cdot \bar{A}$$

Solution The Karnaugh map is shown in Fig. 1.14; it has two overlapping sets of eight adjacent squares.

When the horizontal adjacencies are listed as below

$$\bar{D} \cdot \bar{C} \cdot \bar{B} \cdot A$$
$$\bar{D} \cdot \bar{C} \cdot B \cdot A$$
$$\bar{D} \cdot C \cdot \bar{B} \cdot A$$
$$\bar{D} \cdot C \cdot B \cdot A$$
$$D \cdot C \cdot \bar{B} \cdot A$$
$$D \cdot C \cdot B \cdot A$$
$$D \cdot \bar{C} \cdot \bar{B} \cdot A$$
$$D \cdot \bar{C} \cdot B \cdot A$$

only one quantity does not change. This set of eight adjacencies reduces to A. Similarly, when the edge-connected set is listed, the only quantity which does not change is $\bar{C}$. The simplified equation is

$$L = A + \bar{C}$$

Figure 1.14 Map for Example 1.20.

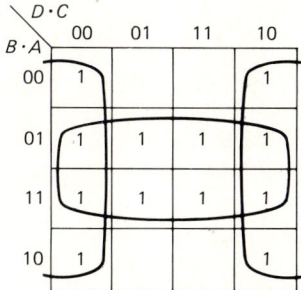

TABLE 1.20
Tabulator Truth Table

Grade	DCBA form	Tabulator output
0	0000	0
1	0001	0
2	0010	0
3	0011	0
4	0100	0
5	0101	0
6	0110	1
7	0111	1
8	1000	1
9	1001	1
10	1010	1

Thus, on the four-variable map of this example, eight adjacent squares reduce to a single variable. This example also emphasizes the need for combining as many 1s as possible. Selecting smaller groupings results in a more complicated equation.

The Karnaugh maps which have been discussed consist either of squares which contain 1s or blank squares which represent 0s. In some circumstances a third possibility exists. Consider the following situation.

A digital electronics instructor gives weekly 10-question true/false tests. The passing grade is 6. To keep track of how many students pass each exam, the instructor builds a test tabulator. After exams are graded, each score is entered into the test tabulator. If the grade is 6 to 10, the tabulator count increases by 1; if the grade is below 6, the tabulator retains the previous count. Thus, after all grades have been entered, the tabulator displays the total number of passing students.

This situation, although somewhat artificial, defines the requirements for a digital circuit. The next step is converting the requirements into a truth table. As shown in Table 1.20, the truth table lists the grades, the $D \cdot C \cdot B \cdot A$ equivalent of the test, and conditions for which outputs result in 1 or 0. The next design is writing the Boolean equation from the truth table. This is obtained by ORing all variable combinations which result in a 1

$$L = \bar{D} \cdot C \cdot B \cdot \bar{A} + \bar{D} \cdot C \cdot B \cdot A + D \cdot \bar{C} \cdot \bar{B} \cdot \bar{A} + D \cdot \bar{C} \cdot \bar{B} \cdot A + D \cdot \bar{C} \cdot B \cdot \bar{A}$$

An attempt at reduction is the next step, and Fig. 1.15 shows the Karnaugh map. These adjacencies reduce the original equation to

$$L = \bar{D} \cdot C \cdot B + D \cdot \bar{C} \cdot \bar{B} + D \cdot \bar{C} \cdot \bar{A}$$

which is a considerable improvement.

Figure 1.15 Test tabulator map.

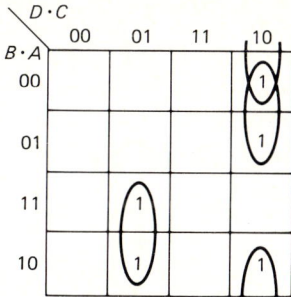

However greater reduction is possible if the truth table is "completed." There are 16 possible combinations of 4 variables. For the test tabulator, 0 to 5 must generate a 0 and 6 to 10 must generate a 1. The numbers 11 to 15 cannot occur; thus we don't care what is generated for 11 to 15. Table 1.21 shows the truth table for 0 to 15 in terms of 0, 1, and d for "don't care" situations.

The don't cares are impossible situations. Therefore d can be replaced with 0 or 1. As shown in Fig. 1.16, replacing ds with 1s yields larger map adjacencies. The simplified equation for this map is

$$L = D + C \cdot B$$

which is much better. Treating impossible conditions as 1s results in simpler equations.

TABLE 1.21 Expanded Calculator Truth Table

Decimal	DCBA sequence	L
0	0000	0
1	0001	0
2	0010	0
3	0011	0
4	0100	0
5	0101	0
6	0110	1
7	0111	1
8	1000	1
9	1001	1
10	1010	1
11	1011	d
12	1100	d
13	1101	d
14	1110	d
15	1111	d

Figure 1.16
Improved tabulator map.

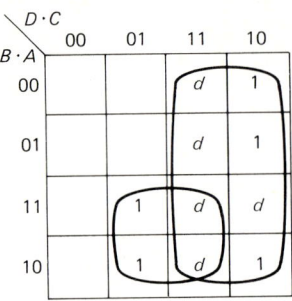

In principle Karnaugh maps can be extended to five or more variables. However, it is difficult to work with such large maps. Too many variables make for a complicated map and the advantages of mapping equations vanish. Computer programs based on Karnaugh methods simplify equations with five or more variables—thus, in effect a computer is used to simplify the design of another computer.

1.8 ELECTRONIC SIMPLIFICATION

Since computer logic is performed electronically, the purpose of reducing equations is to obtain a simpler circuit. In rare cases reductions result in the two simplest Boolean equations

$$L = 0 \quad \text{and} \quad L = 1$$

From an electrical standpoint $L = 0$ represents a circuit which is always off. Such a circuit does not perform any function. Therefore $L = 0$ does not have to be built. On the other hand, $L = 1$ represents a circuit which is always on and must be built. But $L = 1$ is a jumper wire between the appropriate terminals. Except for $L = 0$ and $L = 1$, Boolean equations require real electronic components.

Basic electronic components perform basic Boolean operations. Circuits which perform NOT, OR, and AND are called *gates* specifically, NOT gates, OR gates, and AND gates. A gate is a circuit with one output and one or more inputs. While circuits such as power supplies, amplifiers, and mixers have single outputs and one or more inputs, they are not considered gates. Power supplies, amplifiers, etc., process voltages which vary continuously over a range of voltage levels. Gates, on the other hand, have only two possible voltages, the electronic equivalents of 1 and 0.

The simplest gate is the NOT gate shown in Fig. 1.17a. It uses the standard amplifier symbol with a circle, or "bubble," at the output terminal. Schematics use bubble notation to indicate inversion. Some computer circuits require a variable and its invert at the same time. The circuit shown in Fig. 1.17b is an effective solution. Input A passes undisturbed along a wire, while $\bar{A}$ is obtained from the inverter. Another NOT gate application is shown in Fig.

Figure 1.17
Inverter circuits. (a) Symbol; (b) opposite outputs; (c) cascading stages.

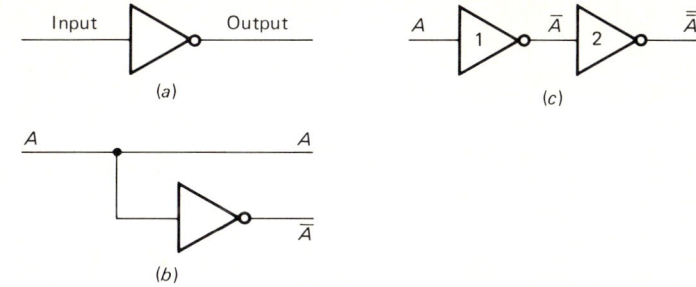

1.17c, which shows a circuit consisting of two inverters connected in cascade. Since the input is A, the output of inverter 1 is $\bar{A}$; this means the output of inverter 2 is $\bar{\bar{A}}$. Thus this circuit is the electronic equivalent of Boolean theorem (1.3a).

Other gates have at least two inputs. A 2-input OR-gate symbol is shown in Fig. 1.18a. This circuit outputs the electronic equivalent of 1 if at least one input is the electronic equivalent of 1. Figure 1.18b shows a 3-input OR gate, and gates with additional inputs are drawn in the same manner. AND is the other basic gate. The electronic version of AND is a circuit which outputs the electronic equivalent of 1 only when all inputs are the electronic equivalent of 1. Figure 1.19 shows the AND-gate symbol.

Any Boolean theorem or equation can be built using only NOT, OR, and AND gates. For example, the equation for Andy or Bruce going to the movies was

$$M = \bar{A} \cdot B + A \cdot \bar{B}$$

Since this equation also represents a basic computer circuit, it is a useful introduction to converting equations into circuits. The inputs are A and B and the output is M. Figure 1.20a is the block diagram and Fig. 1.20b is the circuit configuration.

$\bar{A}$ and $\bar{B}$ are derived from the inputs with inverters; then $\bar{A} \cdot B$ and $A \cdot \bar{B}$ are obtained by ANDing; and finally the ANDed quantities are ORed to obtain M. This particular equation requires 5 gates: 2 inverters, 2 AND gates, and 1 OR gate. This is an example of a circuit which cannot be simplified, and the inverse equation requires the same number of components. However when equations can be simplified, circuits contain fewer gates.

Figure 1.18
OR gates. (a) 2-input; (b) 3-input.

Figure 1.19
AND gates. (a) 2-Input; (b) 3-input.

Figure 1.20
$M = \bar{A} \cdot B + A \cdot \bar{B}$. (*a*) Block diagram; (*b*) logic diagram.

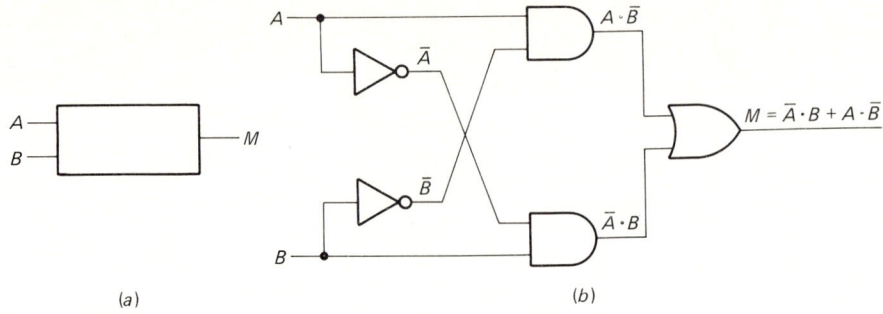

Example 1.21 Compare logic diagrams for the equations developed in Example 1.16.

Solution The original equation is $L = \bar{A} \cdot B \cdot C + A \cdot B \cdot C + A \cdot B \cdot \bar{C}$, and a Karnaugh map reduced this to $L = A \cdot B + B \cdot C$. Finally, Boolean algebra resulted in $L = B \cdot (A + C)$. These three logic diagrams are shown in Fig. 1.21.

The results are summarized as follows:

Original	Karnaugh	Boolean
Three 3-input AND gates	Two 2-input AND gates	One 2-input AND gate
Two inverters	One 2-input OR gate	One 2-input OR gate
One 3-input OR gate		

Improvements are substantial. Karnaugh reduction cuts the number of gates in half and reduces the types and complexity of gates. Similarly, Boolean reduction results in one-third fewer gates than Karnaugh reduction. The

Figure 1.21
$L = \bar{A} \cdot B \cdot C + A \cdot B \cdot C + A \cdot B \cdot \bar{C}$.
(*a*) Original equation; (*b*) Karnaugh reduction; (*c*) Boolean improvement.

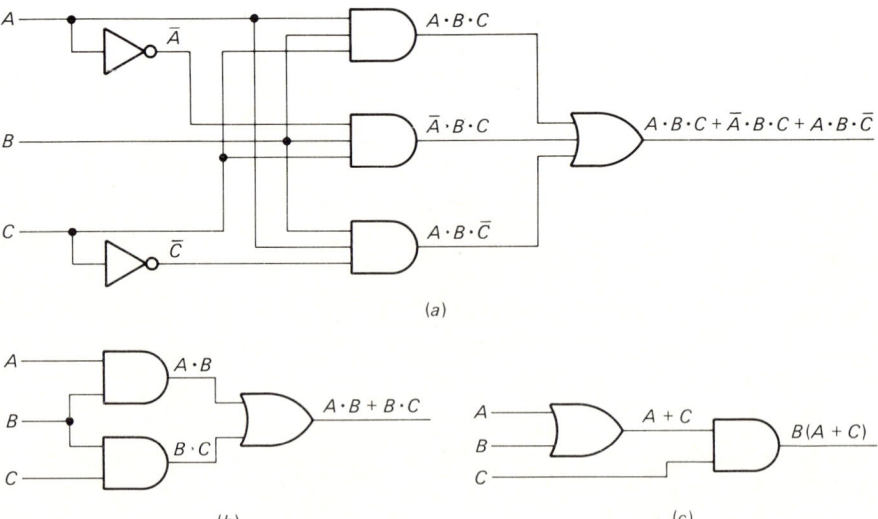

number of gates and gate complexity determine circuit cost. To a first approximation gate cost is determined by the number of signal leads. For example, a 2-input OR gate costs the same as a 2-input AND gate. While 2-input gates have three signal leads, an inverter has two signal leads. Since an inverter has one-third fewer leads, it costs about one-third less than a 2-input OR gate or AND gate; if the cost of a 2-input gate is taken as 1.0 unit, an inverter costs 0.67 unit.

Similarly, a 3-input gate is more expensive than a 2-input gate, since it has 4 signal leads, which is one-third more than a 2-input gate has. The 3-input gate therefore costs about one-third more than a 2-input gate; if the 2-input gate costs 1.0 unit, then the 3-input gate costs 1.33 units. Thus when a Boolean equation can be reduced from three to two variables, cost is reduced by at least one-third.

Example 1.22 Compare the relative costs of implementing circuits from the previous example.

Solution The original version requires six gates.

Gate type	Relative cost
Two inverters	2 × 0.67 = 1.34 units
Three 3-input ANDs	3 × 1.33 = 3.99 units
One 3-input OR	1 × 1.33 = 1.33 units
Six gates for a total cost of	6.66 units

The Karnaugh version requires three gates.

Gate type	Relative cost
Two 2-input ANDs	2 × 1.0 = 2.0 units
One 2-input OR	1 × 1.0 = 1.0 unit
Three gates for a total cost of	3.0 units

The Boolean improvement requires two gates.

Gate type	Relative cost
One 2-input AND gate	1 × 1.0 = 1.0 unit
One 2-input OR gate	1 × 1.0 = 1.0 unit
Two gates for a total cost of	2.0 units

Figure 1.22 NAND gates. (*a*) Two-gate versions; (*b*) single-gate equivalent.

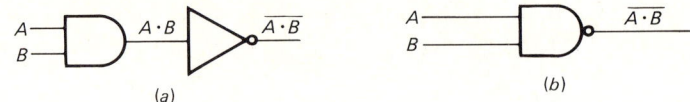

If the simplest version is taken as the reference, the Karnaugh reduction is 50 percent more expensive and the original version is 333 percent more expensive.

This cost analysis only considers gate cost. Besides using fewer gates, this final version requires less wiring. Simpler wiring decreases artwork required in laying out a printed circuit board and also reduces the number of holes to be drilled in the printed circuit board. A simpler circuit board is less expensive and results in further cost reduction.

Reduced cost is an important advantage of implementing simpler circuits. Another advantage is improved reliability. Assuming that all gates are equally prone to failure, a circuit with fewer components has fewer components that can fail.

The extended inverted functions NAND and NOR, which were introduced with De Morgan's theorem, are also implemented with gates. Figure 1.22 shows two methods of obtaining NAND. With two gates the output of an AND gate is connected to an inverter (Fig. 1.22*a*). The single-gate NAND shown in Fig. 1.22*b* also exists—in fact, it is the basic circuit from which all electronic gates are derived. The NAND gate uses fewer components than the AND gate and costs about 5 percent less.

Example 1.23 Compare the gate cost of

$$L = \bar{A} + \bar{B} + \bar{C}$$

with that of its De Morgan equivalent.

Solution The circuit for $L = \bar{A} + \bar{B} + \bar{C}$ is shown in Fig. 1.23*a*. The cost is

Gate type	Relative cost
Three inverters	3 × 0.67 = 2.0 units
One 3-input OR	1 × 1.33 = 1.33 units
Four gates for a total cost of	3.33 units

The De Morgan equivalent is

$$L = \overline{A \cdot B \cdot C}$$

and this circuit is shown in Fig. 1.23*b*. It uses one 3-input NAND gate for a

Figure 1.23
$L = \bar{A} + \bar{B} + \bar{C}$. (a) Original circuit; (b) De Morgan equivalent.

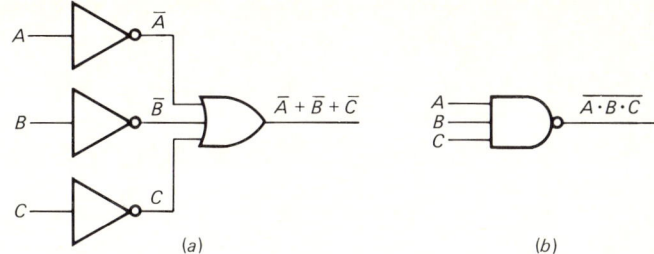

relative cost of $1 \times 1.33 \times 0.95 = 1.26$ units. In this case, the original version is 264 percent more expensive than the De Morgan equivalent.

NOR gates also can be constructed by using an OR gate connected to an inverter. However, the single-gate NOR also exists, and Fig. 1.24 shows the 2- and 3-input symbols.

SUMMARY

1. Boolean theorems are based on quantities which either do or do not exist. The basic Boolean relationhsips are NOT, OR, AND.

2. A Boolean equation is written by ORing all truth table combinations which result in a 1. The inverse equation is written by ORing all combinations which result in a 0.

3. Boolean theorems are used to simplify Boolean equations. De Morgan's theorem reduces terms containing extended inverts.

4. Minterm and maxterm forms of a Boolean equation are the canonical forms. The minterm is the sum-of-products form, and the maxterm is the product-of-sums form.

5. Karnaugh maps are a pictorial technique for Boolean reduction. Their principle is a change of state of only one variable between adjacent squares.

6. Three-variable Karnaugh maps are helpful in simplifying Boolean equations. However, the most efficient reduction may still require algebraic manipulation.

7. Four-variable Karnaugh maps are the most efficient method of reducing four-variable equations. In special cases the don't care condition leads to further simplification.

8. Circuits which implement Boolean equations use gates. Simpler and fewer gates reduce computer cost and complexity.

Figure 1.24 NOR gates.

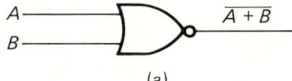
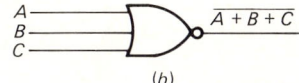

PROBLEMS

1.1 Determine if Boolean theorem (1.5a) is valid by using a truth table.

Answer: Valid.

1.2 Use a truth table to test the validity of $A = \bar{A} \cdot A + \bar{A}$.

1.3 Using a truth table determine if $A = \bar{A} \cdot A + A$ is valid.

Answer: Valid.

1.4 Use a truth table to test the validity of $A \cdot B = A \cdot \bar{B} + \bar{A} \cdot B$.

1.5 Using a truth table determine if $(A + B) \cdot (B + C) = A + B \cdot C$.

Answer: Not valid.

1.6 Andy and Bruce of Example 1.1 have learned to get along. They now go to the movies alone or with each other. Write the equation for going to the movies.

1.7 Mary wants to purchase a bicycle. The bicycle must have brakes (B). She will buy a bike which has either a hand brake (H) or a foot brake (F). No bicycle has both types. Write the Boolean equation for buying a bicycle.

Answer: $B = F \cdot \bar{H} + \bar{F} \cdot H$.

1.8 One step in a manufacturing process depends on the flow rate (F), pressure (P), and temperature (T) of the material. Safety considerations require that an alarm A be sounded before the process becomes dangerous. Dangerous conditions occur when flow and pressure are low and also when pressure and temperature are high. Write the Boolean equation for sounding the alarm.

1.9 Simplify the Boolean equation $L = A(B \cdot A + B \cdot \bar{A})$.

Answer: $L = A \cdot B$.

1.10 Simplify the Boolean equations (a) $L = E \cdot 1 \cdot \bar{E}$; (b) $M = A + 0 + A + 0$.

1.11 Simplify $D = A \cdot B \cdot C + \bar{A} \cdot B \cdot C + A \cdot B \cdot \bar{C} + \bar{A} \cdot B \cdot \bar{C}$.

Answer: $D = B$.

1.12 Simplify $L = A \cdot B + A \cdot \bar{B} \cdot C + \bar{A} \cdot B \cdot \bar{C}$.

1.13 Simplify $L = \bar{A} \cdot \bar{B} \cdot C \cdot D + A \cdot \bar{B} \cdot C \cdot D + \bar{A} \cdot \bar{B} \cdot \bar{C} \cdot D + A \cdot \bar{B} \cdot C \cdot D$.

Answer: $L = \bar{B} \cdot D$.

1.14 Simplify $L = A \cdot \bar{B} \cdot C \cdot \bar{D} + A \cdot \bar{B} \cdot C \cdot D + \bar{A} \cdot B \cdot C \cdot D + \bar{A} \cdot \bar{B} \cdot C \cdot D$.

1.15 Simplify $L = B + A \cdot B \cdot C \cdot D + \bar{A} \cdot B \cdot C \cdot D + \bar{A} \cdot B \cdot \bar{C} \cdot D + A \cdot B \cdot C \cdot \bar{D}$.

Answer: $L = B$.

1.16 Simplify $L = \bar{B} \cdot C \cdot D + A \cdot \bar{B} \cdot \bar{C} \cdot D + A \cdot \bar{B} \cdot \bar{C} \cdot \bar{D} + \bar{A} \cdot \bar{B} \cdot C \cdot D$.

1.17 Write $L = A \cdot B \cdot C \cdot D$ in a form which completely eliminates ANDing.

Answer: $L = \overline{\bar{A} + \bar{B} + \bar{C} + \bar{D}}$.

1.18 Write $L = \bar{A} \cdot \bar{B} \cdot C$ so that an extended invert exists above A, B, and C.

1.19 Simplify $L = (\overline{A \cdot B \cdot C} + A \cdot \bar{B} \cdot C)C$ and show original and simplified circuits.

Answer: $L = C(\bar{A} + \bar{B})$.

1.20 A Boolean equation is given as $L = \bar{A} \cdot \bar{B} + \bar{A} \cdot B$. Find the maxterm form.

1.21 Write the minterm form of $L = (A + B + C) \cdot (\bar{A} + B + C) \cdot (A + B + \bar{C})$.

Answer: $L = \bar{A} \cdot B \cdot \bar{C} + \bar{A} \cdot B \cdot C + A \cdot \bar{B} \cdot C + A \cdot B \cdot \bar{C} + A \cdot B \cdot C$.

1.22 Write $L = A + B \cdot C$ in (a) minterm and (b) maxterm form.

1.23 Write the basic and simplified Boolean equations for the Karnaugh map shown.

B \ A	0	1
0		1
1		1

Answer: $L = A \cdot \bar{B} + A \cdot B = A$.

1.24 Write the basic and simplified Boolean equations for the Karnaugh map shown.

B \ A	0	1
0		1
1	1	1

1.25 Write the simplest Boolean equation for the Karnaugh map shown.

B \ A	0	1
0	1	
1		1

Answer: $L = \bar{A} \cdot \bar{B} + A \cdot B$.

1.26 Write the basic and simplest Boolean equations for the Karnaugh map shown.

A\B	0	1
0	1	1
1	1	1

1.27 Use a Karnaugh map to simplify $L = \bar{A}\cdot\bar{B}\cdot\bar{C} + A\cdot\bar{B}\cdot\bar{C} + \bar{A}\cdot B\cdot\bar{C}$.

Answer: $L = \bar{B}\cdot\bar{C} + \bar{C}\cdot\bar{A}$.

1.28 Simplify $L = \bar{A}\cdot B\cdot C + A\cdot\bar{B}\cdot C + A\cdot B\cdot\bar{C}$ using a Karnaugh map.

1.29 Using a Karnaugh map simplify $L = \bar{A}\cdot\bar{B}\cdot\bar{C} + \bar{A}\cdot\bar{B}\cdot C + \bar{A}\cdot B\cdot\bar{C} + A\cdot B\cdot\bar{C}$.

Answer: $L = \bar{A}$.

1.30 Simplify $L = \bar{C}\cdot B\cdot\bar{A} + \bar{C}\cdot B\cdot A + C\cdot\bar{B}\cdot\bar{A} + C\cdot B\cdot\bar{A} + C\cdot B\cdot A$.

1.31 Use a Karnaugh map to simplify
$L = D\cdot\bar{C}\cdot\bar{B}\cdot\bar{A} + D\cdot\bar{C}\cdot\bar{B}\cdot A + D\cdot\bar{C}\cdot B\cdot\bar{A} + D\cdot\bar{C}\cdot B\cdot A$.

Answer: $L = D\cdot\bar{C}$.

1.32 Simplify $L = \bar{D}\cdot C\cdot\bar{B}\cdot A + D\cdot C\cdot\bar{B}\cdot\bar{A} + D\cdot C\cdot B\cdot A + D\cdot\bar{C}\cdot\bar{B}\cdot A$ using a Karnaugh map.

1.33 Use a map to simplify
$L = \bar{D}\cdot\bar{C}\cdot B\cdot\bar{A} + \bar{D}\cdot C\cdot B\cdot\bar{A} + D\cdot C\cdot B\cdot\bar{A} + D\cdot\bar{C}\cdot B\cdot\bar{A} + D\cdot\bar{C}\cdot B\cdot\bar{A}$.

Answer: $L = B\cdot\bar{A} + D\cdot\bar{C}\cdot\bar{A}$.

1.34 Find the simplest Karnaugh reduction.

B·A \ D·C	00	01	11	10
00	1			
01	1	d		
11	1	d		
10	1			

1.35 Compare the gate cost of implementing $L = A(\bar{A} + B)$ with its simplified equivalent.

Answer: Original is 267 percent as expensive.

1.36 Show the logic diagram for the inverse equation for Fig. 1.20 and compare the relative costs.

1.37 Compare the gate cost of the Karnaugh reduction versions of the test tabulator described in Sec. 1.7.

Answer: First reduction is 367 percent as expensive as second reduction.

2
NUMBER SYSTEMS

2.0 INTRODUCTION

Computers are useful because they permit tedious pencil and paper computations to be performed rapidly by using electronic circuits. The output of a computer may be data or a set of instructions to control an automatic process. In any event computers accept decimal data and output decimal data. We are so familiar with decimal numbers that we take them for granted. However, there is nothing "natural" about the decimal number system.

In the previous chapter logic circuits only used 1s and 0s. The same condition applies to arithmetic circuits. Since logic and arithmetic circuits operate with only 1s and 0s, an understanding of computer operation requires some knowledge of a number system which consists entirely of 1s and 0s. Such a number system is called *binary*, and this chapter introduces basic features of binary arithmetic.

2.1 BINARY NUMBERS

The answer to any question which begins with "How many...?" requires the ability to count. How many sheep, bushels of grain, or days between full moons are a few examples of counting in ancient times. Counting is the first mathematical operation which was ever required. These days we take counting for granted. A brief review of the basics of counting is helpful because it is possible to take too much for granted.

Counting uses single symbols to represent numbers. Dots, vertical lines, or slashes may do the job. Symbols for ropes, flowers, and letters of the alphabet have all been used. Any agreed upon symbol can represent a number, and many number systems have been developed. As civilizations grew, the need to count larger quantities also grew. Number systems with many unrelated symbols became unwieldy.

A number system with many unrelated symbols, which barely survives today, is that of Roman numbers. In the Roman system 1 is represented by I, 10 is X, 100 is C, and 1000 is M. It is easy for us, with the tremendous advan-

44 Computer Circuit Concepts

TABLE 2.1 Roman Numbers

Roman	Decimal	Roman	Decimal
I	1	VIII	8
II	2	IX	9
III	3	X	10
IV	4	L	50
V	5	C	100
VI	6	D	500
VII	7	M	1000

tage of hindsight, to point out the relationship between 1, 10, 100, and 1000. We "know" related numbers should have related symbols. There is nothing wrong with using letters of the alphabet to represent numbers. The problem with Roman numbers is that it has too many unrelated symbols. Table 2.1 compares the Roman number system with its decimal equivalents.

It is tedious to write Roman numbers. For example, 1988 is MCMLXXXVIII. Performing addition, subtraction, and other arithmetic operations with Roman numbers is possible but not very convenient. Also no symbol for zero exists in Roman numbers. The concept of zero as a number to represent the quantity *none* is vital to modern mathematics. In fairness, it should be mentioned that this unwieldy number system did not prevent the Roman Empire from conquering most of Europe, North Africa, and the near eastern part of Asia.

Even before the Roman Empire, more practical number systems were in use. More than 2000 years ago the Hindus used a positional number system and possibly knew about zero. About 800 years ago Arab traders brought Hindu numbers back with them. Arabic symbols were substituted and gradually developed into the symbols which are used today. This system is now called *decimal*.

Position is the principal advantage of the decimal system. There are few symbols, and the position of each symbol within the number indicates value. Decimal numbers may seem "natural" because we have 10 fingers. Counting on fingers is still used during childhood. The term *decimal* comes from the Latin word for "ten" and the term *digit* from the Latin word for "finger." Figure 2.1*a* shows the conventional method of counting on fingers from 1 through 10.

Figure 2.1*b* shows true decimal counting. In a positional number system, we start with the lowest number and count up to the highest number before moving on to the next position. Therefore the symbols 0, 1, 2, ..., 9 must be used before a two-position number is counted. Positional value and 0 are closely related. When the lowest position is filled, we place a 0 in this position and begin again with a 1 in the second position, obtaining 10, 11, 12, ..., 19. Actually 10 is a derived number. It uses the same symbol as 1 but in a differ-

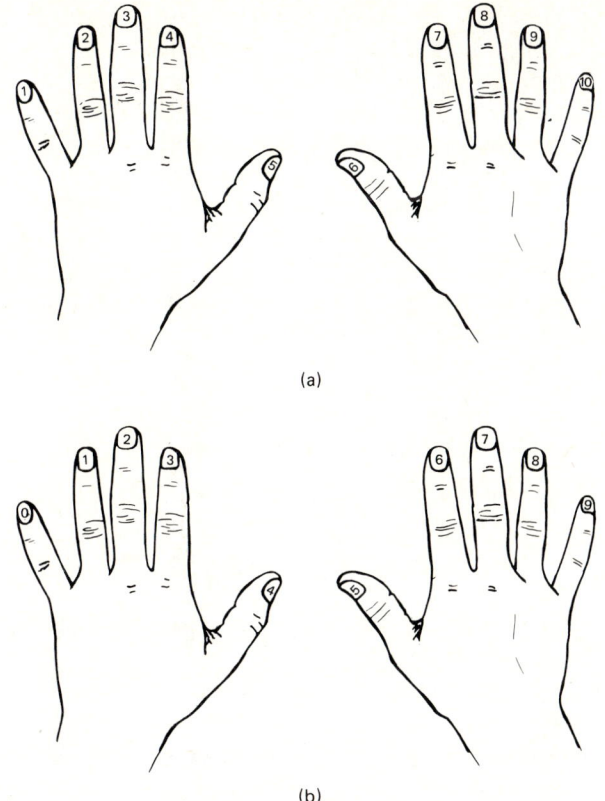

Figure 2.1 Finger counting. (*a*) Conventional; (*b*) true decimal.

ent position. Placing a 0 and moving the 1 symbol over gives the size of the number. This is the connection between 1, 10, 100, 1000, etc. Such a relationship does not exist in a nonpositional number system.

Continuing counting, we reach 19 and again place a 0 in the low position and begin again with a 2 in the second position, forming 20, 21, 22, ..., 29, etc., until we reach 90, ..., 97, 98, 99. At this point we place 0s in the two lowest number positions and start over again with 100, 101, 102, ...

The concept of first using all the symbols and then placing a 0 and starting over again with the same symbols may seem "too obvious" to be worth mentioning. However, it is needed to understand counting in the positional number systems used in computers.

Position automatically shows the value, or weight, of the number; thus 421, 214, and 124 all use the same symbols but have different meanings. The smallest three-position number is greater than the largest two-position number. Position has no value in Roman numbers: IX, which uses two symbols, is greater than VIII, with three symbols. In the decimal system each position is 10 times more significant than the previous position. A 2 in the third position is 10 times more important than a 2 in the second position and 100 times more important than a 2 in the first position. There are units, 10s, 100s, ..., etc., positions.

Consider the 3-digit number 279. It contains three different powers of 10, the smallest digit being in the highest digit position. The 7, which is larger than 2, is not as significant because it is in a lower position, and 9, which is the largest of these 3 digits, is the least significant. Each higher position represents the number multiplied by 10 to the next higher power, and powers begin with the lowest decimal number, which is 0; thus 10^0 is 1 and is the units position. Similarly, 10^1 is the 10s position, 10^2 is the 100s position, etc. We understand 279 as:

$$\begin{aligned} 9 \times 10^0 &= 9 \times 1 = 9 \\ + 7 \times 10^1 &= 7 \times 10 = 70 \\ + 2 \times 10^2 &= 2 \times 100 = 200 \\ \hline & 279 \end{aligned}$$

In this case, 9 is the *least significant digit* (LSD) and 2 is the *most significant digit* (MSD).

The base, or *radix*, of a system is the number of symbols which the system contains. The decimal system has a base of 10. Besides indicating how many symbols are used, the base is also the factor by which each successive position increases. The base is one number higher than the highest number in the units position. In the decimal system the base is 10, while the highest number is 9.

However, 10 is not the only possible base for a positional number system. Before Hindus used base 10, Babylonians used a positional number system with base 60. Vestiges of base 60 still exist. In measuring time we use 60 seconds to a minute and 60 minutes to an hour. Another surviving example of base 60 is used in measuring angles; 360° describes a circle.

Since 10 and 60 are practical bases, it should not be surprising to discover number systems with other bases. Any number except 0 can be a base. For example, a base 1 system would have only one symbol. If we wanted to use 1 as the symbol for a base 1 system, the decimal number 3 would be *///*. Six could be *//////*, or perhaps *////*/. In base 2 there are two symbols, 1 and 0, which have the same meaning as in decimal. A two-symbol number system is particularly useful for numbers processed by electronic circuits.

A different electronic condition must represent each possible number. Electronic representation of decimal numbers requires 10 different electronic states, but with a base 2 system only two electronic conditions are required. Two obvious electronic conditions are *on* and *off*. We use the on condition to represent 1 and the off condition to represent 0.

The base 2 system is called *binary*. The two-state nature of the binary is very compatible with the on/off nature of a circuit. In fact "digital" computers are actually binary computers. In a computer, decimal data are converted to binary and all calculations are performed in binary.

Since binary is a positional system, counting in binary follows the same procedure as in decimal. We count in sequence until the least significant position reaches the largest number. Binary begins with 0 and ends with 1. The least significant binary position fills up rather quickly:

0
1

To count higher we place a 1 in the next significant position and begin counting again in the least significant position.

0
1
10
11

With two positions filled, the next step is placing 0s in both positions and counting again from the first position:

0
1
10
11
100
101
110
111

Continuing this procedure allows counting to any number, just as in decimal. The term *digit* represents a number in decimal. Similarly, *bit*, from *binary digit*, represents a number in the binary system. For example 101 is a 3-bit number in which the *least significant bit* (LSB) is a 1 and the *most significant bit* (MSB) is also a 1.

In decimal the positional weights are units, tens, hundreds, etc. because the base is 10. In binary the base is 2 and the positional weights are units, 2s, 4s, 8s, 16s, etc. The LSB is in the 2^0 position. The next higher bit position is 2^1, and subsequent positions are, in increasing order, $2^2, 2^3, 2^4, \ldots$. Knowing that binary is a base 2 positional number system provides enough information to convert a binary number into its decimal equivalent.

48   Computer Circuit Concepts

Example 2.1   Find the decimal equivalent of the binary number 1101.

Solution   Start with the LSB, which is the 2^0 position, and work upwards.

$$
\begin{aligned}
1\ 1\ 0\ 1 \rightarrow \text{LSB} &= 1 \times 2^0 = 1 \times 1 = 1 \\
&= 0 \times 2^1 = 0 \times 2 = 0 \\
&= 1 \times 2^2 = 1 \times 4 = 4 \\
\text{MSB} &= 1 \times 2^3 = 1 \times 8 = 8 \\
& \overline{} \\
& 13
\end{aligned}
$$

Therefore 1101 in binary = 13 in decimal.

An alternate method of converting from binary to decimal begins with the MSB and works down to the LSB. This method automatically adjusts for the correct power of 2. In any binary number the MSB is always 1. Thus the starting number in converting binary to decimal is always 1. The next lower order bit can be either 0 or 1. If this bit is 0, then *double* the first number, which becomes 2. But if this bit is a 1, first double the number and then add 1, making it 2 + 1, or 3. Doubling and adding 1 is called *dabbling*. Continue to either double or dabble for each bit as required. The decimal equivalent results from operation on the LSB. This method is called *double-dabble*.

Example 2.2   Using double-dabble, find the decimal equivalent of the binary number 10101.

Solution   Starting with the 1 in the MSB position

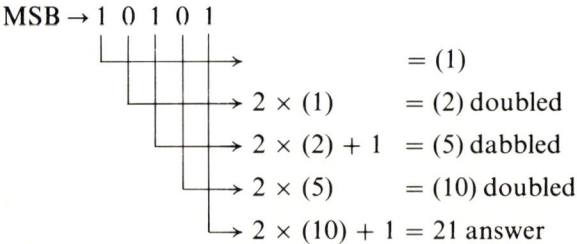

Therefore 10101 in binary = 21 in decimal.

The first 10 decimal and binary numbers are compared in Table 2.2.
    The first two numbers are the same in both systems, but binary numbers are usually longer than decimal. Increased length is a result of using base 2 instead of base 10. Binary numbers are typically three to four times as long as their decimal equivalents.

**TABLE 2.2
Decimal/Binary
Equivalents**

Decimal	Binary
0	0
1	1
2	10
3	11
4	100
5	101
6	110
7	111
8	1000
9	1001

Figure 2.2 shows two possible pocket calculators. The decimal calculator shown in Fig. 2.2a has an 8-position display, keys for the numbers 0 through 9, a decimal point, four basic arithmetic symbol keys, and in some cases an equal sign. The equivalent binary calculator is shown in Fig. 2.2b. This machine has a 21-position display instead of the 8-position display used in

Figure 2.2
Pocket calculators. (*a*) Decimal; (*b*) binary.

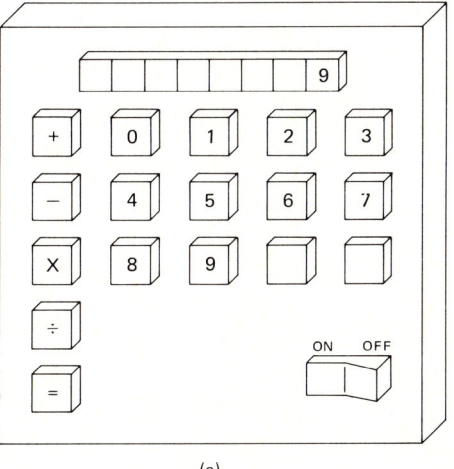

(a)

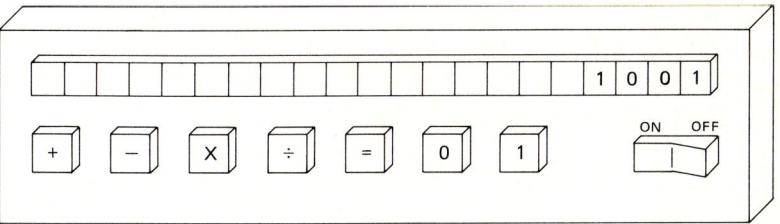

(b)

decimal. However the binary calculator only needs two number keys. Having been raised to think decimal, it is difficult to think binary. However, computers "think" binary.

An interesting feature of binary numbers is the ease of determining whether a number is odd or even. Except for 2^0, which is 1, every power of 2 is an even number. Therefore, if the LSB is 1 a binary number is odd, and if the LSB is 0 the binary number is even. Each number system has its own advantages and disadvantages.

Confusion can arise in determining if a number is binary or decimal. If any position is higher than 1, the number cannot be binary. However if the number contains only 1s and 0s, it may be binary or decimal. For example, 100, 101, 1100, etc., may be either binary or decimal. In such cases, using the base as a subscript removes any ambiguity: 100_{10} is 100 in base 10, while 100_2 is a binary number and is equivalent to the decimal number 4.

Besides converting from binary to decimal, it is necessary to perform the reverse operation. Several methods exist, and one of the easier methods is the reverse of double-dabble. The decimal number is divided by 2. If the quotient is an exact fit, the remainder is 0, and if the quotient is not an exact fit, the remainder can only be 1. Division of each successive quotient by 2 is continued until the quotient is 0. The sequence of remainders, in reverse order, is the binary equivalent.

Example 2.3 Convert 23_{10} into binary.

Solution

$23 \div 2 = 11$ and the remainder is 1

$11 \div 2 = 5$ and the remainder is 1

$5 \div 2 = 2$ and the remainder is 1

$2 \div 2 = 1$ and the remainder is 0

$1 \div 2 = 0$ and the remainder is 1

The last quotient is 0. The answer is the sequence of remainders in reverse order. Therefore $23_{10} = 10111_2$.

Usually this conversion is set up as a chain division. Each successive quotient is divided by 2 and the remainders appear next to the quotients. For 23_{10} we obtain

```
2) 23   remainder
2) 11      1
2)  5      1
2)  2      1
2)  1      0
    0      1
```

This result can be checked by converting from binary back to decimal.

2.2 OCTAL NUMBERS

Three positional bases have been discussed, 60, 10, and 2. Base 60 is practical but is mainly of historic interest. Base 10 is most familiar because we have grown up with it. Base 2 is "natural" for electronic computers because the most obvious electronic states are on and off. Other number bases are practical, and computers make use of several. In the 1950s vacuum tube computers operated in base 5, and results were displayed in a manner similar to the abacus. Some modern computers use base 8 to display internal conditions. Base 10 is known as decimal, and base 2 is known as binary. Similarly, base 8 is known as *octal*.

In any number system, the base is one number higher than the highest number in the units position. Therefore the highest number in octal is 7. The octal numbers are 0, 1, 2, ..., 7. To count higher than 7, a 0 is placed in the units position and counting continues with 10, 11, 12, ..., 17. After 17, 0 is again placed in the units position and the next numbers are 20, 21, ..., 27. This counting method is common to all positional number systems. Fill up a position, replace with 0, and start counting again.

Decimal and octal equivalents of the first 10 numbers are shown in Table 2.3. The first eight numbers, 0 through 7, are the same in both decimal and

TABLE 2.3 Decimal/Octal Equivalents

Decimal	Octal
0	0
1	1
2	2
3	3
4	4
5	5
6	6
7	7
8	10
9	11

octal. This can lead to confusion, as in binary, and subscripts clarify the situation. Statements such as 10_8 is equal to 8_{10} and 10_{10} is equal to 12_8 may seem strange at first, but experience cures strangeness.

Conversion from octal to decimal is similar to conversion from binary to decimal. The weight of the least significant position is the zero power of the base. Each successive position increases the weight by one power of the base. The only difference between converting binary to decimal and converting octal to decimal is weight power.

Example 2.4   Find the decimal equivalent of 103_8.

Solution   The least significant position has the weight of 8^0. Higher positions increase by one power of 8.

$$\begin{aligned} 103 \rightarrow 3 \times 8^0 &= 3 \times 1 = 3 \\ 0 \times 8^1 &= 0 \times 8 = 0 \\ 1 \times 8^2 &= 1 \times 64 = 64 \\ & \overline{67} \end{aligned}$$

Therefore $103_8 = 67_{10}$.

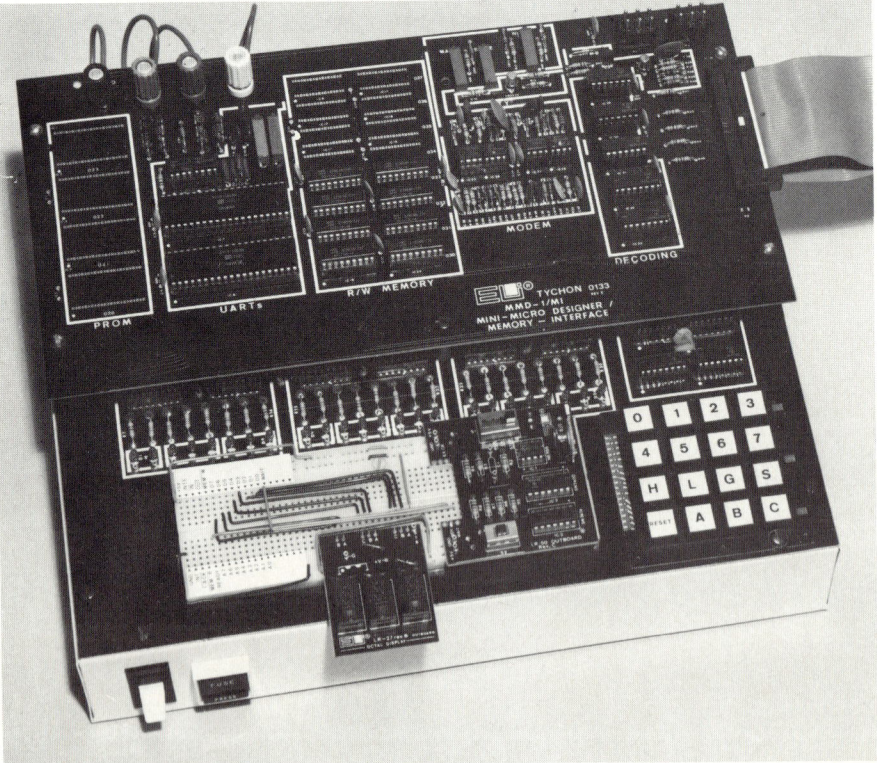

Figure 2.3 Octal keyboard computer. (*Courtesy E & L Instruments Corp.*)

In this example, a three-position octal number has a two-position decimal equivalent. In other cases octal and decimal equivalents will have the same number of positions. This is a considerable improvement over binary to decimal conversion. Octal and decimal positions are closer because there are almost as many octal as decimal numbers.

Two pocket calculators have been shown in Fig. 2.2. The decimal keyboard contains, among other keys, all the decimal numbers, while the binary keyboard contains both binary numbers. Because of the relative length of binary numbers, a binary keyboard calculator is not very practical. Since octal numbers and decimal numbers have about the same lengths, octal keyboards are practical. Figure 2.3 shows a computer with an octal keyboard.

It is also necessary to consider decimal to octal conversion. Decimal to binary conversion is obtained from chain division by 2 and decimal to octal conversion by chain division by 8.

Example 2.5 Find the octal equivalent of 175_{10}.

Solution Using chain division by 8

$$
\begin{array}{r|l}
8)\,175 & \text{remainder} \\
\hline
8)\ \ 21 & 7 \\
8)\ \ \ \ 2 & 5 \\
\ \ \ \ \ \ 0 & 2
\end{array}
$$

Therefore $175_{10} = 257_8$.

This result can be verified by converting 257_8 back to decimal

$$
\begin{aligned}
257 \rightarrow 7 \times 8^0 &= 7 \times 1 = 7 \\
\rightarrow 5 \times 8^1 &= 5 \times 8 = 40 \\
\rightarrow 2 \times 8^2 &= 2 \times 64 = 128 \\
\hline
&\qquad\qquad\quad 175 \checkmark
\end{aligned}
$$

Thus far binary to decimal and octal to decimal conversions have only considered integers. However, computer keyboards accept fractions in positional form. A decimal point separates the integer and fractional portions of a decimal number, for example, 1/4 is 0.25. Similarly, an *octal point* separates the integer and fractional portions of an octal number.

Converting octal fractions into decimal fractions is a continuation of positional weighting. In all positional systems, weights decrease by one power for each position to the right. In octal the weights are ... 8^3, 8^2, 8^1, 8^0. The least significant integer position is immediately to the left of the octal point. Positional weights continue to decrease by one power of 8 for each position past the octal point. Thus, since -1 is the number which is 1 less than zero, the first number past the octal point has a weight of 8^{-1}; the next number past the octal point has a weight of $8^{-2},\ldots$, etc.

Example 2.6 Find the decimal equivalent of 0.37_8.

Solution On the fractional side of the octal point the first position has a weight of 8^{-1}

$$0.37$$
$$3 \times 8^{-1} = 3 \times 0.125 = 0.375$$
$$7 \times 8^{-2} = 7 \times 0.015625 = 0.109375$$
$$\overline{0.484375}$$

Therefore $0.37_8 = 0.484375_{10}$.

This example demonstrates that converting an octal fraction into its decimal equivalent is a natural continuation of decreasing the weight by one power for each position moving towards the right. However, converting a decimal fraction to octal is not as straightforward; it is related to but not the same as converting decimal integers into octal integers. Integer conversion of decimal to octal numbers requires chain division and writing remainders in reverse order.

With fractions, the weighting is by negative powers, $8^{-1}, 8^{-2}, 8^{-3}$, etc. But 8^{-1} is the same as $1/8$, 8^{-2} is the same as $(1/8)^2$, etc. Negative exponents in the numerator are the same as positive exponents in the denominator. Therefore, conversion of a decimal fraction into octal uses chain multiplication instead of a chain division. A decimal fraction is multiplied by 8, and the integer part of the product is the first figure of the octal equivalent. Next, the fractional part of the product is multiplied by 8, and the integer part of this second product is the second significant figure, etc.

Example 2.7 Find the octal equivalent of 0.484375_{10}.

Solution Multiply the decimal fraction by 8

$$0.484375 \times 8 = 3.875$$

The first figure of the octal equivalent is 3. Multiply the fractional part of this multiplication by 8.

$$0.875 \times 8 = 7.0$$

The second figure of the octal equivalent is 7. The fractional part is 0 and no further multiplication is required.

$$0.484375 = 0.37_8$$

This example verifies Example 2.6, in which the reverse conversion was performed. Conversion of a decimal fraction into octal can also be set up in "cookbook" fashion

$$\begin{array}{ll} & \downarrow \\ 0.484375 \times 8 = & 3.875 \\ 0.865 \quad \times 8 = & 7.0 \\ 0 \quad \times 8 = & 0 \end{array}$$

in which case the octal equivalent is the sequence of integers beneath the arrow.

Conversion of binary fractions into decimal follows the same procedure. Naturally, powers of 2 are used.

Example 2.8 Find the decimal equivalent of 0.101_2.

Solution Beginning at binary point the positional weights are 2^{-1}, 2^{-2}, and 2^{-3}.

$$\begin{array}{l} 0.101 \\ \quad \rightarrow 1 \times 2^{-1} = 1 \times 0.5 \quad = 0.5 \\ \quad \rightarrow 0 \times 2^{-2} = 0 \times 0.25 \ = 0.0 \\ \quad \rightarrow 1 \times 2^{-3} = 1 \times 0.125 = 0.125 \\ \hline \qquad\qquad\qquad\qquad\qquad\quad\ 0.625 \end{array}$$

$$0.101_2 = 0.625_{10}$$

Conversion of decimal fractions into binary follows the same multiplication procedure used in Example 2.7. The necessary difference is multiplication by 2 instead of by 8.

Example 2.9 Find the binary equivalent of 0.8125_{10}.

Solution Multiply the decimal fraction by 2 and continue multiplying the resulting fractional portion by 2.

$$\begin{array}{ll} & \downarrow \\ 0.8125 \times 2 = & 1.625 \\ 0.625 \ \ \times 2 = & 1.25 \\ 0.25 \quad\ \times 2 = & 0.5 \\ 0.5 \quad\ \ \times 2 = & 1.0 \end{array}$$

The remainder is 0 and no further multiplication can alter the result.

$$0.9125_{10} = 0.1101_2$$

TABLE 2.4 Number Equivalents

Decimal	Octal	Binary
0	0	0
1	1	1
2	2	10
3	3	11
4	4	100
5	5	101
6	6	110
7	7	111
8	10	1000
9	11	1001

Table 2.4 compares the first 10 integers in decimal, octal, and binary. Placing 0s ahead of the most significant number position has no effect on the number. Regardless of base, 100.0 is the same as 0100.0—in other words, leading 0s do not alter a number. This observation is put to clever use for conversion between binary and octal.

Table 2.5 compares the first eight numbers in octal and binary. Leading 0s make each binary number 3 bits long. The first eight octal numbers are all possible numbers in the units position. The next higher octal number requires two positions. We note that 7_8 is 111_2 and 111_2 is the highest possible 3-bit number. There is a 3-bit number which corresponds to each octal number. Binary to octal conversion only require grouping the binary number into 3-bit segments.

Example 2.10 Find the octal equivalent of 110010.101_2.

Solution Begin at the octal point and group into 3-bit segments. Then write the octal equivalents of each group.

$$\underbrace{110}\underbrace{010}.\underbrace{101}$$
$$6 \quad 2 \ . \ 5$$
$$110010.101_2 = 62.5_8$$

A binary number does not always contain bits which group into sets of 3

TABLE 2.5 3-Bit Equivalent

Octal	0	1	2	3	4	5	6	7
Binary	000	001	010	011	100	101	110	111

bits. However, it is perfectly valid to add leading 0s to the integer and following 0s to the fraction. 10101.1_2 can be written as 010101.100_2. This does not change the value but does group into 3-bit segments.

The ready conversion between octal and binary is a convenience. It is very easy for humans to transpose a 1 or a 0 in a binary number, particularly when binary numbers are 8 bits or longer. Since conversion from binary to octal is simpler than conversion from binary to decimal, many computers convert binary to octal for display. While we are likely to err in reading 11010101_2, it is relatively easy to read 325_8.

2.3 HEXADECIMAL NUMBERS

Converting decimal to binary is complicated because decimal and binary are unrelated bases. On the other hand conversion between octal and binary is relatively simple because 8 is a power of 2. Conversion is essentially complete when an octal number is written in 3-bit equivalents. Since octal is a convenient binary conversion, a higher power should be even better. The power of 2 after 8 is 16. Base 16 is *hexadecimal* and is frequently called *hex*.

Each base requires a single symbol for each number. When a base is less than 10, enough symbols are available from the decimal system. When a base is higher than 10, additional symbols are required. Nothing is magic about the symbols 0 through 9; other symbols can represent numbers. The decimal system works just as well if letter symbols are substituted for numbers, for example, if A is 0, B is 1, C is 2, and J is 9. In this alphabetical position system 91_{10} is JB, 103_{10} is BAD, etc. Nothing is wrong with using letters to represent numbers.

There is a very old joke about a gambler who wants to show off his youngster's counting ability. The child stands up and in a loud and clear voice announces, "Two, three, four, ..., nine, ten, Jack, Queen, King." The child has demonstrated that any agreed upon set of symbols can represent numbers. In hex the agreed upon symbols are 0 through 9 followed by A through F for the last six numbers. The first 16 numbers in hex and decimal are shown in Table 2.6.

Counting in hex continues in the same manner as in any other positional system. After the highest number in the units position is reached, 0 is placed in the units position and counting begins again. The hex numbers after F are 10, 11, 12, ..., 19, 1A, 1B, 1C, ..., 1F, 20, 21, etc. The subscript H indicates a hex number. For example, $9_H = 9_{10}$, $10_H = 16_{10}$, $1A_H = 26_{10}$, etc.

Hex numbers are even more compact than decimal numbers. This makes

TABLE 2.6
Hex/Decimal Equivalents

Hex	0	1	2	3	4	5	6	7	8	9	A	B	C	D	E	F
Decimal	0	1	2	3	4	5	6	7	8	9	10	11	12	13	14	15

Figure 2.4 Hex keyboard computer. (*Courtesy Intel Corp.*)

hex an extremely practical base, and a hex computer with keyboard is shown in Fig. 2.4.

All positional conversions follow the same pattern. Hex conversion is not difficult.

Example 2.11 Find the decimal equivalent of $40B_H$.

Solution Weighting in hex is by powers of 16

$$
\begin{aligned}
40B \rightarrow B \times 16^0 &= 11 \times 1 = 11 \\
\hookrightarrow 0 \times 16^1 &= 0 \times 16 = 0 \\
\hookrightarrow 4 \times 16^2 &= 4 \times 256 = 1024 \\
\hline
&1035
\end{aligned}
$$

$$40B_H = 1035_{10}$$

In this case, the decimal number requires one more position than the hex number. Hex numbers are compact. In fact, any base greater than 10 uses fewer positions than decimal.

Example 2.12 Find the decimal equivalent of $0.B8_H$.

Solution The first position to the right of the hex point has a weight of 16^{-1}, the next 16^{-2}, etc.

$$0.B8$$

$$B \times 16^{-1} = 11 \times 0.0625 = 0.6875$$
$$8 \times 16^{-2} = 8 \times 0.00390625 = 0.03125$$
$$\phantom{8 \times 16^{-2} = 8 \times 0.00390625 =\ } \overline{0.71875}$$

$$0.B8_H = 0.71875_{10}$$

Conversion of decimal integers into hex follows the same chain division process used to convert decimal into binary and octal. In converting from decimal to hex, division by 16 is required.

Example 2.13 Find the hex equivalent of 1035_{10}.

Solution Chain division is used. Remainders higher than 9 have alphabetic equivalents.

	decimal remainder	hex remainder
16) 1035		
16) 64	11	B
16) 4	0	0
0	4	4

$$1035_{10} = 40B_H$$

Similarly conversion of a decimal fraction into hex use chain multiplication by 16.

Example 2.14 Find the hex equivalent of 0.71875_{10}.

Solution Chain multiplication is by 16. Integers higher than 9 have alphabetic equivalents.

$$0.71875 \times 16 = 11.5 = B.5$$
$$0.5 \times 16 = 8.0 = 8.0$$
$$0.71875_{10} = 0.B8_H$$

TABLE 2.7
0.1_{10} **to Hex Conversion**

Hex	Decimal equivalent	Error, %
0.1	0.625	37.5
0.19	0.09765625	2.34
0.199	0.0998535156	0.146
0.1999	0.0999908447	0.0092

Fractional conversion into binary, octal, or hex into decimal always results in an exact number. An exact equivalent does not always occur when the reverse conversion is performed. There is no guarantee that converting a decimal fraction in any base 2 fraction will result in a "perfect fit."

Example 2.15   Find the hex equivalent of 0.1_{10}.

Solution   Chain multiplication by 16 is used.

$$0.1 \times 16 = 1.6$$
$$0.6 \times 16 = 9.6$$
$$0.6 \times 16 = 9.6$$
$$0.1_{10} = 0.199\ldots_H$$

In this case, the hex equivalent continues indefinitely. We can write as many 9s as we have patience for but a perfect fit will never occur. Each additional 9 makes the hex value a closer approximation of the decimal fraction. Conversion stops when an acceptable tolerance is achieved.

Suppose an error of less than 0.1 percent, which is less than 1 part in 1000, is acceptable. The hex fraction must be reconverted to decimal and compared with the original decimal fraction as each successive value is obtained. The results are presented in Table 2.7. Error decreases as the number of positions in the hex equivalent increases. For 0.1_{10} a four-position hex number is required to reduce the error to less than 0.1 percent.

Additional positions are required to obtain the same accuracy when 0.1_{10} is converted into octal. The results are presented in Table 2.8. In this case, a five-position octal fraction is required to reduce the error to less than 0.1 percent.

TABLE 2.8
0.1_{10} **to Octal Conversion**

Octal	Decimal equivalent	Error, %
0.06	0.09375	6.25
0.063	0.099609375	0.391
0.0631	0.0998635156	0.146
0.06314	0.0999755859	0.024

TABLE 2.9
0.1_{10} to Binary Conversion

Binary	Decimal equivalent	Error, %
0.0001	0.0625	37.5
0.00011	0.09375	6.25
0.00011001	0.09765625	2.34
0.000110011	0.99609375	0.39
0.000110011001	0.998535156	0.146
0.0001100110011	0.999755859	0.024

Since binary is the least compact of all number systems, many more positions are needed to obtain the same accuracy. The results are presented in Table 2.9. A 13-bit binary fraction is required to reduce the error to less than 0.1 percent. In summary, we can get close, but many decimal fractions do not have an exact equivalent in binary-based systems.

The conversions require a good deal of computation. Fortunately, there is no need to perform fractional conversion on a regular basis. Whenever we enter 0.1_{10} or any other fraction on a keyboard, circuits perform rapid conversion. In fact more time is needed to punch the keyboard than for the conversion.

As shown, a small difference may exist between a number and its computer equivalent in a computer's number system. Our pencil and paper calculations were based on an error of 1 part in 1000. Pocket calculators work with errors of less than 1 part in 100 million. Such errors are insignificant. An error of 1 part in 100 million is equivalent to that of a watch which loses less than 1 second in a year.

The previous section described a simple conversion method between octal and binary based on grouping into 3-bit segments. This method can be extended to hex by grouping into 4-bit segments. The 4-bit equivalents of hex numbers are given in Table 2.10, which shows that every possible 4-bit combination corresponds to a hex number. Table 2.5 showed the corresponding relationship of 3-bit combinations to octal numbers.

Example 2.16 Find the hex equivalent of 10011101.00111110_2.

Solution Group into 4-bit segments. Then use Table 2.10 to find the hex equivalents.

1001 1101 . 0011 1110
 9   D   . 3   E

Therefore $11011101.00111110_2 = 9D.3E_H$.

TABLE 2.10 Hex/Binary Equivalents

Hex	0	1	2	3	4	5	6	7	8	9	A	B	C	D	E	F
Binary	0000	0001	0010	0011	0100	0101	0110	0111	1000	1001	1010	1011	1100	1101	1110	1111

TABLE 2.11
Hex/Octal/Binary Equivalent

Hex	Octal	Binary
0	0	0
1	1	1
2	2	10
3	3	11
4	4	100
5	5	101
6	6	110
7	7	111
8	10	1000
9	11	1001
A	12	1010
B	13	1011
C	14	1100
D	15	1101
E	16	1110
F	17	1111

Only octal/hex conversions have not been discussed. The easiest method begins with conversion to binary. Then bits are grouped into either 3- or 4-bit segments. Table 2.11 shows the first 16 numbers in hex, octal, and binary. Leading 0s can be added as required.

Because of the compatability of binary numbers with on/off electronic states it is likely that computers will always operate in binary. The first pocket calculators were introduced in the early 1970s. These calculators used 4 bits, and 4-bit calculators still exist. Improvement in manufacturing techniques resulted in integrated circuits which could process 8 bits at a time instead of only 4 bits. Later 16-bit integrated circuits were introduced, and now 32- and 64-bit circuits exist.

2.4 ADDITION

After counting, the most basic arithmetic operation is adding. Addition is actually an extension of counting. By counting to 4 and then 3 more, 3 is added to 4. In adding two numbers together to obtain a sum, we no longer consider the steps. For us, addition has become an "automatic" operation. But when addition was first encountered in elementary school, many stumbling blocks had to be overcome before addition could be performed comfortably. Counting in an unfamiliar base causes problems until familiarity is achieved. Similarly, learning to add in an unfamiliar base presents problems at the outset. A brief review of decimal addition smoothes an understanding of addition in other bases.

Addition requires aligning numbers according to weighted position. In adding a 3-digit number to a 2-digit number, the units columns of both numbers are lined up. Positioning appropriate columns recognizes the weighted nature of the decimal system. After numbers are correctly positioned, column by column addition is performed. Adding two numbers in the same column together can result in a single column sum or a sum with *carry*. A carry results when the sum is greater than the highest number in the base. In decimal any sum higher than 9 results in a carry. A *carry* is the addition of 1 to the next higher significant digit. While 4 + 5 does not generate a carry, 4 + 6 does. In this case, 1 is carried into the 10s column. Increasing the weight of the 10s column by 1 means that 10 has been added to the 10s column. Similarly, 60 + 40 results in a carry of 1 into the 100s column. This means the weight of the 100s column is increased by 100. The weight of each higher position increases, but addition is performed column by column as if adding in the units column. When all the columns have been added, the sum has the correct positional weights.

The ability to add requires correct positioning as well as learning sums of all possible combinations of single-digit numbers. When both requirements have been satisfied and experience gained, addition becomes an "automatic" operation. An addition table such as Table 2.12, presents sums by row and column.

Table 2.12 contains all 100 possible combinations obtained by adding two digits. Primes (') in front of a number indicate a carry into the next higher column. In Table 2.12 the row is x, the column is y, and the sum $z = x + y$ is the intersection of the appropriate x and y values. For example, the intersection of $x = 5$ with $y = 6$ is shown as '1, which means 1 with a carry of 1. Although we are familiar with the decimal addition table, this review simplifies use of addition tables in other bases.

TABLE 2.12 Decimal Addition, $x + y = z$

y\x	0	1	2	3	4	5	6	7	8	9
0	0	1	2	3	4	5	6	7	8	9
1	1	2	3	4	5	6	7	8	9	'0
2	2	3	4	5	6	7	8	9	'0	'1
3	3	4	5	6	7	8	9	'0	'1	'2
4	4	5	6	7	8	9	'0	'1	'2	'3
5	5	6	7	8	9	'0	'1	'2	'3	'4
6	6	7	8	9	'0	'1	'2	'3	'4	'5
7	7	8	9	'0	'1	'2	'3	'4	'5	'6
8	8	9	'0	'1	'2	'3	'4	'5	'6	'7
9	9	'0	'1	'2	'3	'4	'5	'6	'7	'8

**TABLE 2.13
Binary Addition,
$x + y = z$**

y \ x	0	1
0	0	1
1	1	'0

When two numbers are added, the top number is the *addend* and the bottom number is the *augend*.

    Addend
+ Augend
———
    Sum

Addition is *commutative*. This means that numbers can be added in either order, but the result is the same.

Addition can be extended to more than two numbers at a time. However, computers are limited to adding two numbers at a time and our goal is to understand computers. Since there are only two numbers in binary, the binary addition table is much smaller than that for decimal. Table 2.13 is the complete binary addition table and has the same form as the decimal addition table. In both cases each sum is 1 greater than the sum directly above it. Of course, the reduced size of the binary addition table is balanced by the extended length of binary numbers.

Example 2.17   Find the sum of 1110_2 and 1101_2.

Solution   Arrange the numbers under each other in positional order

```
  1 1 1 0
+ 1 1 0 1
---------
```

Add the numbers one position at a time using the binary addition table. Begin with the LSB

```
  1 1 1 0
  1 1 0 1
---------
        1   There is no carry
```

Proceed to the next higher significant bit:

```
  1 1 1 0
  1 1 0 1
---------
      1 1   There is no carry
```

Continue to the next higher significant bit:

```
'1 1 1 0
 1 1 0 1
─────────
   1 1 1   There is a carry
```

The remaining addition is $1_2 + 1_2 + 1_2$. This condition is not included in the binary addition table, just as $9 + 9 + 9$ is not included in the decimal addition table. However $9 + 9 + 9$ can be added by obtaining a partial sum with the first two 9s and then adding the last 9. Similarly from the binary addition table, $1_2 + 1_2 = {'0}$, and adding 1_2 to $'0$ we obtain $'1_2$. That is $1_2 + 1_2 + 1_2 = 11_2$. This makes sense, since 11_2 is 3_{10}. Therefore the final result is

```
 '1 1 1 0
  1 1 0 1
─────────
1 1 0 1 1
```

As long as we position numbers correctly, addition tables are valid for any length. This applies to both integers and fractions. The binary point is positioned in the same manner as the decimal point.

With the appropriate addition table, addition can be performed in any base without memorizing sums. Table 2.14 is the complete octal addition table. This table has the same form as decimal and binary addition tables. Again, each sum is 1 greater than the sum directly above. A binary addition table has 4 sums; in octal there are 64 sums; and in decimal there are 100. Higher base systems have more numbers and more combinations.

TABLE 2.14 Octal Addition, $x + y = z$

y \ x	0	1	2	3	4	5	6	7
0	0	1	2	3	4	5	6	7
1	1	2	3	4	5	6	7	'0
2	2	3	4	5	6	7	'0	'1
3	3	4	5	6	7	'0	'1	'2
4	4	5	6	7	'0	'1	'2	'3
5	5	6	7	'0	'1	'2	'3	'4
6	6	7	'0	'1	'2	'3	'4	'5
7	7	'0	'1	'2	'3	'4	'5	'6

66 Computer Circuit Concepts

Example 2.18 Find the sum of 106_8 and 713_8.

Solution Position the numbers by weight:

$$\begin{array}{r} 1\ 0\ 6 \\ +\ 7\ 1\ 3 \\ \hline \end{array}$$

Use the octal addition table, and start with the least significant position:

$$\begin{array}{r} 1\ '0\ 6 \\ 7\ 1\ 3 \\ \hline \end{array}$$

      1   There is a carry

Proceed to the next more significant position:

$$\begin{array}{r} 1\ '0\ 6 \\ 7\ 1\ 3 \\ \hline \end{array}$$

    2 1   There is no carry

Continue with the most significant position:

$$\begin{array}{r} '1\ '0\ 6 \\ 7\ 1\ 3 \\ \hline \end{array}$$

1 0 2 1   There is a carry

$106_8 + 713_8 = 1021_8$

Since hex has more numbers than decimal, the hex addition table, Table 2.15, is larger. Hex addition is performed in the same manner used for any other positional system. The augend is placed below the addend with the same positional weights in the same columns. Then numbers are added by using the addition table.

Example 2.19 Find the sum of 79_H and 92_H.

Solution Correctly position the numbers and add column by column:

$$\begin{array}{r} 7\ 9 \\ +\ 9\ 2 \\ \hline \end{array}$$

      B   There is no carry

TABLE 2.15
Hex Addition, $x + y = z$

y \ x	0	1	2	3	4	5	6	7	8	9	A	B	C	D	E	F
0	0	1	2	3	4	5	6	7	8	9	A	B	C	D	E	F
1	1	2	3	4	5	6	7	8	9	A	B	C	D	E	F	'0
2	2	3	4	5	6	7	8	9	A	B	C	D	E	F	'0	'1
3	3	4	5	6	7	8	9	A	B	C	D	E	F	'0	'1	'2
4	4	5	6	7	8	9	A	B	C	D	E	F	'0	'1	'2	'3
5	5	6	7	8	9	A	B	C	D	E	F	'0	'1	'2	'3	'4
6	6	7	8	9	A	B	C	D	E	F	'0	'1	'2	'3	'4	'5
7	7	8	9	A	B	C	D	E	F	'0	'1	'2	'3	'4	'5	'6
8	8	9	A	B	C	D	E	F	'0	'1	'2	'3	'4	'5	'6	'7
9	9	A	B	C	D	E	F	'0	'1	'2	'3	'4	'5	'6	'7	'8
A	A	B	C	D	E	F	'0	'1	'2	'3	'4	'5	'6	'7	'8	'9
B	B	C	D	E	F	'0	'1	'2	'3	'4	'5	'6	'7	'8	'9	'A
C	C	D	E	F	'0	'1	'2	'3	'4	'5	'6	'7	'8	'9	'A	'B
D	D	E	F	'0	'1	'2	'3	'4	'5	'6	'7	'8	'9	'A	'B	'C
E	E	F	'0	'1	'2	'3	'4	'5	'6	'7	'8	'9	'A	'B	'C	'D
F	F	'0	'1	'2	'3	'4	'5	'6	'7	'8	'9	'A	'B	'C	'D	'E

Proceed to the next higher significant position:

```
    7 9
  + 9 2
  -------
  1 0 B    There is a carry
```

Therefore $79_H + 92_H = 10B_H$.

Because letters as well as conventional numbers are involved, hex addition seems strange. However addition in any base seems strange, including decimal when it is first encountered.

2.5 SUBTRACTION

Usually subtraction is considered after addition. A brief review of decimal subtraction simplifies subtraction in other bases.

Addition is commutative but subtraction is not. The result of $6 + 4$ is the same as $4 + 6$, but the result of $6 - 4$ is not the same as $4 - 6$. In addition numbers are combined to form a sum, and in subtraction numbers are combined to form a *difference*. A *subtrahend* is deducted from a *minuend* to obtain the difference:

>   Minuend
>   − Subtrahend
>   ─────────────
>   Difference

Although addition and subtraction are opposite, there are common features. In order to perform either operation:

- Numbers must be aligned in corresponding columns according to weight.
- Numbers are combined starting with the least significant position.
- Numbers from one column can generate a result which affects the next higher significant figure, namely a carry for addition and a borrow for subtraction.

A carry occurs when the sum is more than the highest number in the base. A *borrow* occurs when the subtrahend is greater than the minuend. For example, to subtract 3 from 2 we borrow 10 from the next significant digit in the minuend. This converts the minuend 2 into a 12 and the difference is 9. The borrow has increased the minuend value by 10. The minuend must remain constant, and a borrow is compensated by reducing the digit from which the borrow was made.

Addition tables simplify adding in the various bases. Similarly, subtraction tables simplify subtracting in these same bases. In the decimal subtraction table shown in Table 2.16, a prime in front of a number represents a borrow. The form of the subtraction table is $z - x = y$, where z is the minuend, x is the subtrahend, and y is the difference. To use this table locate the minuend z and the subtrahend x in the same column. The difference is obtained by moving

TABLE 2.16 Decimal Subtraction Table, $z - x = y$

y \ x	0	1	2	3	4	5	6	7	8	9
0	0	1	2	3	4	5	6	7	8	9
1	1	2	3	4	5	6	7	8	9	'0
2	2	3	4	5	6	7	8	9	'0	'1
3	3	4	5	6	7	8	9	'0	'1	'2
4	4	5	6	7	8	9	'0	'1	'2	'3
5	5	6	7	8	9	'0	'1	'2	'3	'4
6	6	7	8	9	'0	'1	'2	'3	'4	'5
7	7	8	9	'0	'1	'2	'3	'4	'5	'6
8	8	9	'0	'1	'2	'3	'4	'5	'6	'7
9	9	'0	'1	'2	'3	'4	'5	'6	'7	'8

across the minuend row to the value of y, which is the difference. For example, to subtract 6 from 8 line up the z value of 8 and the x value of 6 in the same column as shown in Table 2.16. Then move across from the z value of 8 to the y value of 2. The other example illustrated in Table 2.16 is 3 − 7. In this case the subtrahend is greater than the minuend and a borrow is required. Therefore ′3 and 7 are located in the same column. By moving across from ′3 the difference is found to be 6. Care must be taken to decrease the next higher digit in the minuend to compensate for the borrow.

Observe that Table 2.16 for decimal subtraction is exactly the same as Table 2.12 for decimal addition; thus the same table can be used to perform decimal addition and decimal subtraction. When using the table for addition, work from the outside towards the inside, and for subtraction work from the inside towards the outside. A prime represents a carry for addition and a borrow for subtraction. Dual use of a single table for addition and subtraction also applies to other bases. In decimal a carry and a borrow are equal to 10, the base of the system. Similarly, in binary carry and borrow have a value of 2, in octal of 8, and in hex of 16.

Example 2.20   Subtract 1101_2 from 11011_2.

Solution   Position the subtrahend below the minuend by corresponding weights. Leave a blank row for the adjusted minuend.

```
       Minuend    1 1 0 1 1
Adjusted minuend
    Subtrahend      1 1 0 1
                   ─────────
    Difference
```

Starting with the LSB, perform subtraction bit by bit, using Table 2.13 as a subtraction table. The binary addition/subtraction table is small enough to show. For this case no borrow is required in the LSB.

```
       Minuend    1 1 0 1 1
Adjusted minuend          1
    Subtrahend    1 1 0 1
                 ─────────
    Difference           0
```

Continue to the next higher significant bit. For this case no borrow is required.

```
       Minuend    1 1 0 1 1
Adjusted minuend         1 1
    Subtrahend      1 1 0 1
                   ─────────
    Difference          1 0
```

In the next bit the minuend is less than the subtrahend and a borrow is required. Since a borrow from 0 is not possible, the borrow must be continued until a nonzero number is reached. Of course in binary the only nonzero number is 1.

Minuend	1	1	0	1	1
Adjusted minuend	0	'0	'0	1	1
Subtrahend		1	1	0	1
Difference		1	1	1	0

Therefore $11011_2 - 1101_2 = 1110_2$.

Octal subtraction is performed in the same manner as decimal and binary subtraction by using Table 2.14 as a subtraction table.

Example 2.21 Subtract 71.3_8 from 102.1_8.

Solution Position the subtrahend below the minuend by corresponding weights, and leave a blank row for the adjusted minuend.

Minuend	1	0	2	. 1
Adjusted minuend				
Subtrahend		7	1	. 3

Starting with the least significant position perform subtraction. In this case a borrow is required. This reduces the next higher significant figure of the minuend from 2 to 1.

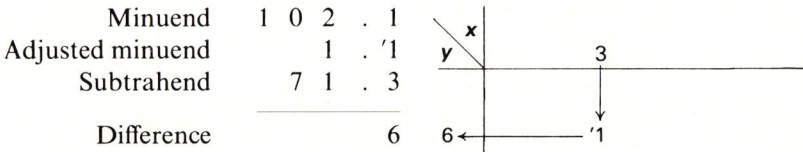

Minuend	1	0	2	. 1
Adjusted minuend			1	. '1
Subtrahend		7	1	. 3
Difference				6

Next bring down the octal point and perform subtraction for the next higher significant position. In this case no borrow is required.

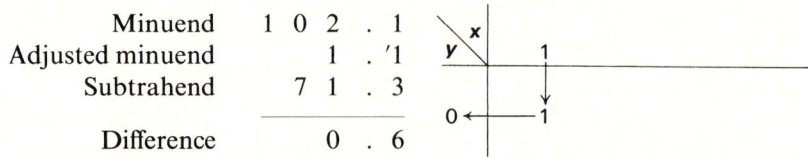

Minuend	1	0	2	. 1
Adjusted minuend			1	. '1
Subtrahend		7	1	. 3
Difference			0	. 6

Proceed to the next higher significant figure. A borrow is required. In this case, the next higher position is 0, and the borrow must be continued. The most significant figure of this minuend reduces to 0.

Minuend	1 0 2 . 1
Adjusted minuend	0 ′0 1 . 1
Subtrahend	7 1 . 3
Difference	1 0 . 6

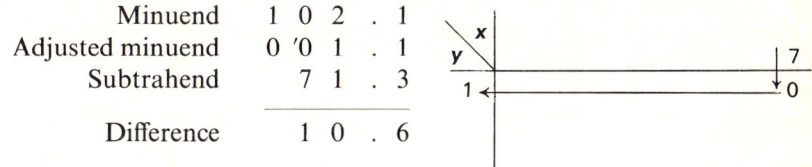

Therefore $102.1_8 - 71.3_8 = 10.6_8$.

The remaining base of interest is hex, and the procedure for hex subtraction is the same as for decimal, binary, and octal. Table 2.15 is used as a hex subtraction table.

Example 2.22 Subtract $9AC7_H$ from $F8B6_H$.

Solution Position the subtrahend below the minuend by corresponding weights, and perform subtraction starting with the least significant position.

Minuend	F 8 B 6
Adjusted minuend	E ′7 ′A ′6
Subtrahend	9 A C 7
Difference	5 D E F

Therefore $F8B6_H - 9AC7_H = 5DEF_H$.

2.6 MULTIPLICATION AND DIVISION

In grade school four fundamental mathematical operations are introduced, namely addition, subtraction, multiplication, and division. More advanced procedures, such as raising a number to a power, taking the square root of a number, and using the binomial theorem, are extensions of the four fundamental operations. Therefore any computer which performs addition, subtraction, multiplication, and division has the capability of performing all mathematical operations.

While some applications require a knowledge of multiplication and division in all bases, our requirements are satisfied by understanding multiplication and division in binary. Binary multiplication and division are easy to understand after a brief review of decimal multiplication and division.

In multiplication, the *multiplicand* times the *multiplier* yields the *product*.

    Multiplicand
$\times$ Multiplier
    Product

Multiplication is commutative. It doesn't matter whether 6 is multiplied by 5 or 5 is multiplied by 6—the results are the same. Table 2.17 shows the 100

TABLE 2.17
Decimal Multiplication Table, $x \cdot y = z$

y \ x	0	1	2	3	4	5	6	7	8	9
0	0	0	0	0	0	0	0	0	0	0
1	0	1	2	3	4	5	6	7	8	9
2	0	2	4	6	8	10	12	14	16	18
3	0	3	6	9	12	15	18	21	24	27
4	0	4	8	12	16	20	24	28	32	36
5	0	5	10	15	20	25	30	35	40	45
6	0	6	12	18	24	30	36	42	48	54
7	0	7	14	21	28	35	42	49	56	63
8	0	8	16	24	32	40	48	56	64	72
9	0	9	18	27	36	45	54	63	72	81

products of the decimal system. In each column, the increase from one product to the next is equal to the x value of the column. In the $x = 6$ column, the numbers are 0, 6, 12, 18, ..., an increase of 6 for each successive value. Thus, multiplication can be considered as repetitive addition. Multiplication involves calculating the partial product for each multiplier digit. To account for weight, each decimal partial product is shifted one position to the left. Calculating partial products can require addition because a carry may be generated during individual multiplications. The sum of partial products is the total product.

Binary multiplication is performed with a binary multiplication table, shown as Table 2.18. Since there are only four products, binary multiplication is easy. Shifting one position to the left for each partial product allows for the increased weight of each multiplier bit.

Example 2.23 Multiply 1101_2 by 1011_2.

Solution Follow the same procedure as for decimal multiplication.

```
                                1 1 0 1
                                1 0 1 1
                              ─────────
Units partial product           1 1 0 1
2s partial product            1 1 0 1
4s partial product          0 0 0 0
8s partial product        1 1 0 1
                          ─────────────
Total product             1 0 0 0 1 1 1 1
```

**TABLE 2.18
Binary Multiplication
Table, $x \cdot y = z$**

y \ x	0	1
0	0	0
1	0	1

In any number system, the maximum number of product bits is the sum of the number of bits in the multiplicand and multiplier. Actually, binary multiplication is even easier than using the binary multiplication table. Since any number times 0 is 0 and any number times 1 is the number itself, there are only two possibilities in binary multiplication:

- When the multiplier is 0, the entire partial product is 0.
- When the multiplier is 1, the partial product is the multiplicand.

Division is the reverse of multiplication. In division the quotient is the result of dividing the dividend by the divisor.

$$\frac{\text{Dividend}}{\text{Divisor}} = \text{Quotient}$$

Just as subtraction is performed by using an addition table, decimal division is performed using the multiplication table shown as Table 2.17. This is accomplished by locating the dividend z and the divisor x in the same column. Then move across from this divisor to y which is the required quotient.

Long division is the most difficult of the four basic arithmetic operations, requiring trial-and-error methods. A trial quotient for the most significant portion of the dividend is tested. If the product of the trial quotient and the divisor is greater than that portion of the dividend, the following subtraction gives a negative difference. This means that the trial quotient is too large, and the next lower trial quotient is attempted. After a suitable trial quotient is obtained, the process is repeated on the remaining portions of the dividend until the last remainder is 0.

Binary division can be performed in the same manner as decimal division by using Table 2.18 as a binary division table. As in any other number system, division by 0 is not permitted. This reduces the number of binary division possibilities:

- When the dividend is 0 the quotient is 0: $0/1 = 0$.
- When the dividend is 1 the quotient is 1: $1/1 = 1$.

Example 2.24 Divide 100111_2 by 110_2.

Solution Set up the long division procedure. The first place where 1110 can go into the dividend is the 1001 portion.

$$
\begin{array}{r}
1 \\
110\overline{)100111} \\
110 \\
\hline
11
\end{array}
$$

The difference, 11_2, is positive. Therefore bring down the next significant bit.

$$
\begin{array}{r}
11 \\
110\overline{)100111} \\
110\!\downarrow \\
\hline
111 \\
110 \\
\hline
1
\end{array}
$$

The difference 1_2 is positive. Bring down the next bit.

$$
\begin{array}{r}
1\;1 \\
110\overline{)100111} \\
110 \\
\hline
111 \\
110 \\
\hline
11
\end{array}
$$

In this case the dividend 11 is less than the divisor. Bring down a 0 from the right-hand side of the binary point.

$$
\begin{array}{r}
110.1 \\
110\overline{)100111.0} \\
110 \\
\hline
111 \\
110 \\
\hline
110 \\
110 \\
\hline
0
\end{array}
$$

The remainder is 0, indicating that division is complete. Thus 100111/110 = 110.1. Decimal conversion yields 39/6 = 6.5, which is correct.

The methods which have been presented to perform addition, subtraction, and multiplication in binary are the "pencil and paper" equivalent of the same operations in decimal. Computers perform these operations in modified forms, as will be discussed in later chapters.

2.7 SIGNED NUMBERS

As a practical matter, negative numbers are just as important as positive. Real computers accept and process negative as well as positive numbers.

When working with pencil and paper, a choice exists for indicating positive numbers—either nothing is done or a plus sign is placed in front of the number. For example, 13 and +13 have the same meaning. However, there is no choice when indicating a negative number—a minus sign must be used to indicate it.

Inside a computer there are only 1s and 0s. No other symbols are available. Thus, 1s and 0s must also be used to represent the sign of a number. With pencil and paper the sign, if any, is placed in front of the number. In a computer the closest thing to placing something in front of a number is using the MSB to represent signs. The convention is:

- When the MSB is 0 the number is positive.

- When the MSB is 1 the number is negative.

Table 2.19 compares 4-bit numbers in terms of unsigned and signed values. For the same number of bits, the largest unsigned number is more than twice as large as the largest signed number. With 4 bits the largest unsigned number is 15, while the largest signed number is +7. This is a consequence of using bit position to represent sign. Also, the range of unsigned numbers, 0 through 15, is not quite the same as the range of signed numbers, −7 through +7. The range difference occurs because there are two ways to represent 0 in signed form. There is 0000, which is $+0_2$, and 1000, which is -0_2. A decrease in number size and a slight decrease in range are the penalties for obtaining negative as well as positive numbers.

Addition, subtraction, and other mathematical operations can be performed with signed numbers. Early computers used signed numbers, and the operation was similar to that using pencil and paper methods. The sign of a result was determined by the sign of the numbers used.

To avoid confusion in reading, a method of indicating number sign is necessary. In this text, the MSB is underlined for signed numbers. Of course, this convention is for clarity. No such distinction exists inside a computer.

76 Computer Circuit Concepts

TABLE 2.19
Unsigned/Signed Numbers

Binary	Unsigned	Signed
0000	0	+0
0001	1	+1
0010	2	+2
0011	3	+3
0100	4	+4
0101	5	+5
0110	6	+6
0111	7	+7
1000	8	−0
1001	9	−1
1010	10	−2
1011	11	−3
1100	12	−4
1101	13	−5
1110	14	−6
1111	15	−7

Example 2.25 Using 8-bit signed numbers add $+8_{10}$ to $+13_{10}$.

Solution In signed binary $\quad +13_{10} = 0\underline{0001101}$
and $\quad\quad +8_{10} = 0\underline{0001000}$

When both numbers are positive, signed addition is performed by adding magnitude bits and retaining the common sign bit for the result:

$$0\underline{0001101}$$
$$0\underline{0001000}$$
$$\overline{00010101}$$

Besides demonstrating signed addition, this example illustrates a fundamental limitation of any real computer. In this case the largest positive number is $0\underline{1111111}$, or $+127_{10}$. Any larger number results in an *overflow*, that is, a number exceeding machine capacity. Similarly, the largest signed 16-bit number is $0\underline{111111111111111}$, or $+32767_{10}$. Any real computer is subject to overflow. Typically, computers have a largest number in either the 10^{39} or 10^{99} range. While either of these limits is extremely large for practical applications, no number limitation exists with pencil and paper.

Example 2.26 Find the sum of 10110111 and 10101111.

Solution Since both signs are the same, add magnitude bits and retain the common sign bit.

$$\underline{1}0110111$$
$$\underline{1}0101111$$
$$\overline{\underline{1}1100110}$$

In this example the sum approaches overflow, and circuits to detect overflow are required.

Rules for signed binary addition are the same as those for signed decimal. The result of adding two positive numbers is a positive number. Similarly the result of adding two negative numbers is a negative number. It is also necessary to add numbers with unlike signs. The simpler case occurs when addend magnitude is greater than augend magnitude; as with signed decimal numbers, the difference between two binary numbers is determined and the addend sign is retained.

When the signs are opposite and the addend magnitude is less than the augend magnitude, the situation is more complicated. With pencil and paper it is easy to use the commutative property of addition. After all $+10 + (-17)$ is the same as $-17 + 10$. However, testing magnitude and interchanging numbers require many complex circuits. An alternative solution might permit operation with an addend magnitude which is smaller than the augend. Unfortunately, such an approach requires additional rules and additional computer circuits.

Another set of conditions is required for subtraction. It is one thing to say subtraction is performed by changing the sign of the subtrahend and following the rules for addition. However, it is another thing to build such circuits.

An improvement in working with signed numbers involves the concept of *complements*. The complement of a number is the difference between the number and the highest number in the base. For example in decimal the complement of 3 is 6, and the complement of 7 is 2. In decimal, the process is called finding the 9s complement.

Example 2.27 Find the 9s complement of 143.

Solution To find the complement of a 3-digit number, subtract from the 3-digit number which is all 9s

$$\begin{array}{r} 999 \\ -143 \\ \hline 856 \end{array}$$

Therefore the 9s complement of 143 is 856.

An interesting feature of 9s complement is that subtraction can be performed by adding numbers.

Example 2.28 Subtract 143 from 801.

Solution Instead of subtracting in the conventional manner, add the 9s complement of 143 to 801.

$$\begin{array}{rl} 801 & \text{minuend} \\ 856 & \text{9s complement of 143} \\ \hline 1657 & \end{array}$$

Next, perform end-around carry. This is accomplished by adding the last carry to the LSD.

$$\begin{array}{r} 801 \\ 856 \\ \hline (1)657 \\ \hookrightarrow 1 \\ \hline 658 \end{array}$$

$$801 - 143 = 658$$

The same result is obtained by conventional subtraction. In principle 9s complement subtraction requires only addition. This is useful if subtraction can actually be performed by adding, as there is no need to design and build subtraction circuits and addition and subtraction can be performed in the same circuit. Actually, 9s complement subtraction is more involved because subtraction is required to obtain the 9s complement. Subtracting with 9s complements is interesting but not practical.

However a unique feature of binary complements provides considerable improvement in performing binary subtraction. Decimal complements are called 9s complements. Similarly, binary complements are called 1s complements. The sense is the same. In decimal the complemented number is subtracted from a string of 9s. In binary the number is subtracted from a string of 1s.

Example 2.29 Find the 1s complement of 1101.

Solution The 1s complement is obtained by subtracting from a number which is at least 4 bits long and consists entirely of 1s.

$$\begin{array}{r} 1111 \\ -1101 \\ \hline 0010 \end{array}$$

Therefore the 4-bit 1s complement of 1101 is 0010.

All 1s complements can be found in the same manner. Table 2.20 shows the results for 4-bit numbers. Observe that a binary complement can be obtained by reversing 1s and 0s. Subtraction is not required to find the 1s complement. Circuits for reversing 1s and 0s are simpler than subtraction circuits.

Moreover, 1s complement numbers permit addition and subtraction in the same circuit. Positive numbers in 1s complement are indicated in the same way as are ordinary signed numbers; thus 0 in the MSB indicates a positive number. A negative 1s complement number is obtained by reversing all bits, including the MSB. For example, $+5$ is 0101 and -5 is 1010. Table 2.21 compares 4-bit unsigned, signed, and 1s complement values. There are still two ways to write 0 in 1s complement. However, 1s complement arithmetic is simpler to implement than signed numbers.

TABLE 2.20 4-Bit Complements

Number	Complement
0000	1111
0001	1110
0010	1101
0011	1100
0100	1011
0101	1010
0110	1001
0111	1000
1000	0111
1001	0110
1010	0101
1011	0100
1100	0011
1110	0001
1111	0000

TABLE 2.21
Number Comparison

Binary	Unsigned	Value Signed	1s Complement
0000	0	+0	+0
0001	1	+1	+1
0010	2	+2	+2
0011	3	+3	+3
0100	4	+4	+4
0101	5	+5	+5
0110	6	+6	+6
0111	7	+7	+7
1000	8	−0	−7
1001	9	−1	−6
1010	10	−2	−5
1011	11	−3	−4
1100	12	−4	−3
1101	13	−5	−2
1110	14	−6	−1
1111	15	−7	−0

2.8 COMPLEMENTARY ARITHMETIC

The same three cases must be considered in 1s complement addition: both numbers positive, both numbers negative, and opposite signs. Since positive 1s complement numbers are the same as ordinary signed numbers, addition is the same. Adding two negative 1s, complement negative numbers requires end-around carry.

Example 2.30 Using 8-bit 1s complements, add -8_{10} to -13_{10}.

Solution In 1s complement a negative number is obtained by reversing all bits, including the sign bit.

$$+13_{10} = \underline{0}000\ 1101 \qquad \text{Therefore } -13_{10} = \underline{1}111\ 0010$$
$$+8_{10} = \underline{0}000\ 1000 \qquad \text{Therefore } -8_{10} = \underline{1}111\ 01111$$

Add these negative numbers, including the sign bits:

$$\underline{1}111\ 0010$$
$$\underline{1}111\ 0111$$
$$\overline{1\underline{1}110\ 1001}$$

A carry-out is generated. Since negative 1s complement numbers begin with $\underline{1}$, a carry-out will always occur when two negative numbers are added. Perform an end-around carry:

$$\underbrace{①\underline{1}111\ 1001}_{\longrightarrow 1}$$

$$\underline{1}110\ 1010 \quad \text{1s complement sum}$$

To determine if this sum is correct, "decode" this negative number by reversing the magnitude bits.

$$\underline{1}110\ 1010 \quad \text{1s complement sum}$$
$$\underline{1}001\ 0101 \quad \text{decoded sum}$$

The decoded sum converts to -21_{10} and is correct.

The next two examples illustrate 1s complement addition with unlike signs. When unlike signs were discussed for signed numbers, the more difficult situation was not covered. In 1s complement there is no difficult situation.

Example 2.31 Using 1s complements add $+13_{10}$ to -8_{10}.

Solution Determine the 1s complements

$$+13_{10} = \underline{0}000\ 1101 \text{ and } -8_{10} = \underline{1}111\ 0111$$

Since sign bits are included in addition, it doesn't matter whether a larger or smaller number is the addend. This is a significant improvement over ordinary signed addition.

$$\begin{array}{r} \underline{0}000\ 1101 \\ +\underline{1}111\ 0111 \\ \hline 1\underline{0}000\ 0100 \end{array}$$

Again a carry-out is generated and an end-around carry is required.

$$\underbrace{①\underline{0}000\ 0100}_{\longrightarrow 1}$$

$$\underline{0}000\ 0101$$

The sign is correct, and the decimal equivalent, $+5_{10}$, is also correct.

The next example illustrates addition when the negative number is larger.

Example 2.32   Using 1s complements add -13_{10} to 8_{10}.

Solution   The 1s complement equivalents of these numbers were determined in previous problems.

$$-13_{10} = \underline{1}111\ 0010 \quad \text{and} \quad +8_{10} = \underline{0}000\ 1000$$

Add these numbers, including the sign bits.

```
 1111  0010
 0000  1000
 ──────────
 1111  1010
```

In this case, no carry-out is generated. The MSB of the sum is $\underline{1}$, so the result is negative. We must decode the magnitude bits to determine the decimal equivalent.

$$\underline{1}111\ 1010 \quad \text{1s complement sum}$$
$$\underline{1}000\ 0101 \quad \text{decoded}$$

The decimal equivalent is -5_{10}.

Subtraction in 1s complement is the equivalent of decimal subtraction. When decimal subtraction is performed, the sign of the subtrahend is reversed and the subtrahend is added to the minuend. The 1s complement equivalent is addition of the subtrahend's 1s complement to the minuend.

Rules for 1s complement addition and subtraction can be summarized as follows:

- *Addition*   Add the 1s complement numbers, including the sign bits. When a carry-out occurs, perform an end-around carry. Sum sign and magnitude are correct.

- *Subtraction*   Form the 1s complement of the subtrahend and add it to the minuend. If carry-out occurs, perform end-around carry. Sign and magnitude of the difference are correct.

Processing sign bits is the difference between signed and 1s complement arithmetic. In signed arithmetic, the sign bits are tested to determine the sign of the result. In 1s complement arithmetic, sign bits are part of the arithmetic process. Since no sign test is required, fewer circuits are required to perform 1s complement arithmetic.

In the early days 1s complement circuits replaced signed arithmetic

TABLE 2.22 Equivalents Compared

Binary	Decimal equivalent			
	Unsigned	Signed	1s Complement	2s Complement
0000	0	+0	+0	+0
0001	1	+1	+1	+1
0010	2	+2	+2	+2
0011	3	+3	+3	+3
0100	4	+4	+4	+4
0101	5	+5	+5	+5
0110	6	+6	+6	+6
0111	7	+7	+7	+7
1000	8	−0	−7	−8
1001	9	−1	−6	−7
1010	10	−2	−5	−6
1011	11	−3	−4	−5
1100	12	−4	−3	−4
1101	13	−5	−2	−3
1110	14	−6	−1	−2
1111	15	−7	−0	−1

circuits. More recently an even better method called *2s complement* has replaced 1s complement. In 2s complement arithmetic, positive numbers are the same as in 1s complement. A negative 2s complement is obtained by adding 1 to the 1s complement negative number.

Suppose -3_{10} is required as a 4-bit 2s complement number. Start with $+3_{10}$, which is 0011. To obtain -3_{10} in 1s complement reverse all bits: thus -3_{10} is 1100. Adding 1 to the 1s complement yields -3_{10} in 2s complement: thus, -3_{10} in 2s complement is 1101. Table 2.22 compares 4-bit values.

The only difference between 1s and 2s complement numbers is that negative 2s complements differ by 1 from the corresponding 1s complements. This seemingly trivial difference results in simpler circuits. There is only one way to write 0 in 2s complement; this eliminates the need for circuits to recognize two forms of 0. With only one version of 0, the range of 2s complement numbers is exactly the same as the range of unsigned numbers. There are 15 different signed and 1s complement 4-bit numbers, but there are 16 different unsigned and 2s complement numbers. However, greater ease in performing arithmetic calculations in 2s complement is the real advantage.

There are "brute force" methods to obtain 2s complement numbers, but an easy method, similar to the computer circuit technique, is available. Starting with the LSB, each bit of the signed positive version is recopied, including the first 1. All remaining bits are reversed.

84 Computer Circuit Concepts

Example 2.33   Express -24_{10} as an 8-bit 2s complement number.

Solution

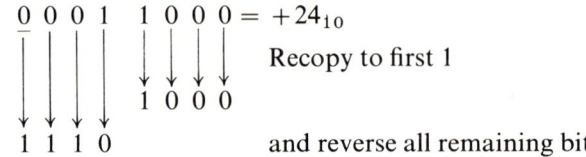

Thus, $-24_{10} = \underline{1}100\ 1110$ in 2s complement.

The opposite procedure is used for the decimal equivalent.

Example 2.34   Find the decimal equivalent of the 2s complement number $\underline{1}011\ 1010$.

Solution   Use the recopy and reverse method.

$\underline{1}011\ 1010$ Number

$\underline{1}100\ 0110$ Recopy from LSB including first 1.

Reverse remaining bits except sign bit.

The same cases must be investigated as previously. The first is addition of two positive numbers but since positive numbers are the same in both 1s and 2s complement, there is no need to repeat this procedure. The next case is addition of two negative numbers.

Example 2.35   Use 2s complement arithmetic to add -15_{10} to -9_{10}.

Solution   Using the recopy-and-reverse procedure

$+15 = \underline{0}000\ 1111$ thus $-15_{10} = \underline{1}111\ 0001$ in 2s complement

$+9 = \underline{0}000\ 1001$ thus $-9_{10} = \underline{1}111\ 0111$ in 2s complement

Then add the 2s complement numbers, including sign bits.

$-15_{10} = \underline{1}111\ 0001$
$-9_{10} = \underline{1}111\ 0111$
─────────────────
$\phantom{-9_{10} = }①\ \underline{1}110\ 1000$

Ignore the carry-out.
Thus the 2s complement sum is $\underline{1}110\ 1000$.
To determine the decimal equivalent, recopy-and-reverse is used, but the sign bit is retained.

$\underline{1}110\ 1000$ 2s complementary sum

$\underline{1}001\ 1000$ Recopy-and-reverse procedure except for sign bit

The decimal equivalent of $\underline{1}001\ 1000$ is -24_{10}, which is correct.

Next consider 2s complement addition of opposite signed numbers when the larger number is positive.

Example 2.36 Using 2s complement add $+24_{10}$ to -4_{10}.

Solution First obtain the 2s complement equivalents.

$$+24_{10} = \underline{0}001\ 1000$$
$$+4_{10} = \underline{0}000\ 0100$$

Therefore $-4_{10} = \underline{1}111\ 1100$

Now add $+24_{10}$ to -4_{10}.

$$\underline{0}001\ 1000 = +24_{10}$$
$$\underline{1}111\ 1100 = -\ 4_{10}$$

$\underline{\textcircled{1}}0001\ 0100 = $ 2s complement sum

Again ignore the carry-out. The decimal equivalent of this 2s complement sum is $+20_{10}$, which is correct.

Addition in 2s complement of opposite-sign numbers when the larger number is negative also results in a carry-out. Again, the carry-out is ignored. Subtraction in 2s complement is performed by 2s-complementing the subtrahend and adding to the minuend.

Example 2.37 Using 2s complement, subtract $+35_{10}$ from -12_{10}.

Solution For the minuend

$$+12_{10} = \underline{0}000\ 1100; \quad \text{thus} -12_{10} = \underline{1}111\ 0100$$

For the subtrahend

$$+35_{10} = \underline{0}010\ 0011$$

To subtract, the subtrahend must first be 2s-complemented and then added to the minuend:

$$\text{minuend} = \underline{1}111\ 0100$$
$$\text{2s complement of subtrahend} = \underline{1}101\ 1101$$

2s complement difference $\textcircled{1}\ \underline{1}101\ 0001$

Again ignore the carry-out. Therefore the 2s complement difference is 1101 0001. Since this is a negative number, it must be interpreted as before.

    1101 0001   2s complement difference

    1010 1111   Recopy-and-reverse procedure except for MSB

The decimal equivalent is -47_{10}, which is correct.

The rules for 2s complement arithmetic are:

- *Addition* Add the 2s complement numbers, including the sign bits. Ignore any carry-out. Both the sign and magnitude of the sum will be correct.
- *Subtraction* Form the 2s complement of the subtrahend and add it to the minuend. Ignore any carry-out. Both the sign and magnitude of the difference will be correct.

Since 2s complement arithmetic ignores carry-outs, there is no need to perform end-around carry or to detect a carry-out. Results are achieved with fewer circuits.

SUMMARY

1. Because of the two-state nature of binary numbers, electronic "digital" computers are really binary computers. Conversion methods from binary to decimal and vice-versa are required.
2. Octal is an extension of binary. Therefore, conversion from octal to binary and vice versa is straightforward.
3. Hexadecimal, or base 16, is a further extension of binary with easy conversion to binary. Hexadecimal numbers are more compact than decimal.
4. The addition procedure is similar in all bases. Addition in any base is simplified by use of addition tables.
5. Subtraction is the reverse of addition and may be simplified by using addition tables in reverse. Subtraction is not commutative.
6. Binary multiplication is similar to decimal multiplication, and binary division is similar to decimal division.
7. The MSB is used for number sign. An MSB of 0 indicates a positive number and an MSB of 1 a negative number.
8. Fewer circuits are required for complementary arithmetic than for signed arithmetic. In particular, 2s complement arithmetic has fewer circuits than 1s complement.

PROBLEMS

2.1 Find the decimal equivalent of 10011_2.

Answer: 19_{10}.

2.2 Find the decimal equivalent of 11101_2 using double-dabble.

2.3 Find the binary equivalent of 25_{10}.

Answer: 11001_2.

2.4 Convert 100_{10} into binary.

2.5 Find the decimal equivalent of 144_8.

Answer: 100_{10}.

2.6 Find the decimal equivalent of 12.5_8.

2.7 Find the decimal equivalent of 101.11_2.

Answer: 5.75_{10}.

2.8 Find the octal equivalent of 11101.1101_2.

2.9 Find the binary equivalent of 130.5_8.

Answer: 1011000.101_2.

2.10 Find the decimal equivalent of 79_H.

2.11 Find the decimal equivalent of $A1.8_H$.

Answer: 176.5_{10}.

2.12 Find the hex equivalent of 0.2_{10} which is within 1 percent of the correct value.

2.13 Find the octal equivalent of 0.2_{10} which is within 1 percent of the correct value.

Answer: 0.0146_8.

2.14 Find the binary equivalent of 0.2_{10} which is within 1 percent of the correct value.

2.15 Convert 96.48_H to binary.

Answer: 10010110.01001_2.

2.16 As a check find the decimal equivalents of the hex and binary numbers in Prob. 2.15.

2.17 Convert $A03.92_H$ into octal.

Answer: 5003.444_8.

2.18 Find the decimal equivalents of the hex and octal numbers in Prob. 2.17.

2.19 The highest two-position hex number is FF. Find its equivalent in (a) binary, (b) octal, (c) decimal.

Answer: (a) 11111111, (b) 377, (c) 255.

2.20 Find the sum of 11001.1_2 and 101.1_2.

2.21 Verify the result of Prob. 2.20 by converting both numbers and the result into decimal.

Answer: $25.5 + 5.5 = 31$.

2.22 Find the sum of $110011_2 + 1111_2$.

2.23 Add 56_8 to 22_8.

Answer: 100_8.

2.24 Verify the result of Prob. 2.23 by converting both numbers and the result into decimal.

2.25 Find the sum of $15.6_8 + 62.2_8$.

Answer: 100.0_8.

2.26 Find the sum of $1F3_H$ and $30D_H$.

2.27 Find the sum of 18.7_H and 7.9_H.

Answer: 20.0_H.

2.28 Subtract 1010_2 from 1101_2.

2.29 Subtract 101.1_2 from 11111_2.

Answer: 11001.1_2.

2.30 Verify the results of Prob. 2.29 by converting the numbers to decimal.

2.31 Perform $11000100_2 - 100101_2$.

Answer: 10011111_2.

2.32 Subtract 89_H from 352_H.

2.33 Find the decimal value of the answer to Prob. 2.32.

Answer: 713_{10}.

2.34 Subtract $3BC_H$ from $5D0_H$.

2.35 Multiply 1111_2 by 1101_2.

Answer: 11000011_2.

2.36 Verify the result of Prob. 2.35 by converting to decimal.

2.37 Multiply 1001_2 by 1100_2.

Answer: 1101100_2.

2.38 $101000_2/1000_2 = ?$

2.39 Divide 1101110_2 by 1010_2.

Answer: 1011_2.

2.40 Verify the result of Prob. 2.39 by converting to decimal.

2.41 Find the decimal equivalent of $\underline{1}0101010$ assuming: (*a*) the number is signed; (*b*) the number is a 1s complement.

Answer: (*a*) -42; (*b*) -53.

2.42 Find the decimal equivalent of $\underline{0}0101010$ assuming: (*a*) the number is signed; (*b*) the number is a 2s complement.

2.43 Using 8-bit numbers add $+21_{10}$ to -21_{10} in 1s complement.

Answer: $\underline{1}111\ 1111 = -0_{10}$.

2.44 Using 8-bit numbers add $+21_{10}$ to -21_{10} in 2s complement.

2.45 Assuming that $\underline{0}000\ 0111$ and $\underline{1}111\ 1101$ are 2s complement numbers perform addition and convert to decimal as a check.

Answer: $\underline{0}000\ 0100 = +4_{10}$.

2.46 Add the 2s complement numbers $\underline{1}111\ 1111$ and $\underline{1}101\ 1000$ to find the sum.

2.47 As a check perform the equivalent operation in decimal.

Answer: $-1 + (-40) = -41$.

2.48 Add the 2s complement numbers $\underline{1}100\ 1110$ and $\underline{0}110\ 1010$. Verify the result by converting the problem to decimal.

3
GATES

3.0 INTRODUCTION

Boolean and arithmetic operations were introduced in the first two chapters. Since computers implement logic and arithmetic with pulses, the next step is the investigation of circuits for processing pulses.

Gates are the most basic computer switches, and their electrical specifications are considered in this chapter. Practical methods of implementing AND, OR, and NOT are described and then the more versatile NAND and NOR gates are introduced. Three-state gates, the computer equivalent of open circuits, are also discussed.

Logic and 3-state gates are grouped into families with compatible characteristics. Gate families are compared, and the relative advantages and disadvantages of each family are discussed.

3.1 THE SWITCH

A digital computer performs calculations using a separate quantity to represent each number. Digital computers are modern devices, but the origins of digital computing go back to ancient times. It is extremely difficult to say precisely when or where the first digital computer was used.

The *abacus* is a digital computer based on the position of beads on a rod. Beads represent numbers and the rod represents place value. By switching beads up and down the rod, an abacus performs addition, subtraction, multiplication, and division. Answers are displayed along the cross bar; for example, the answer shown in Fig. 3.1 is 1732. Numerical computation has been performed on the abacus for at least 5000 years. In ancient Egypt grooves drawn in the sand were used instead of rods and small stones instead of beads. Since sand was readily available, a bag of stones was in effect a portable digital computer. The Semitic word for sand is *abg*, from which the Greeks derived the word *abakos*, meaning a calculating table.

The shepherd shown in Fig. 3.2 used a digital computer which may be older than the abacus. In this case the digital computer is a jar of pebbles. As

Figure 3.1
A recent Chinese abacus.

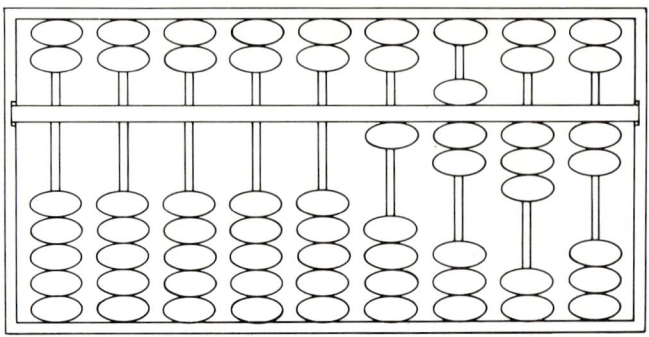

the shepherd took his flock out to pasture, he would switch one pebble into the jar for each sheep going out. Upon returning, the shepherd switched one pebble out of the jar. If at the end of this operation pebbles remained in the jar, the shepherd knew exactly how many sheep to look for. The ancient shepherd may not have been very good with arithmetic, but a jar of pebbles performed the correct digital computation. The Latin word for pebble is *calculus*.

In more recent times pebbles, stones, and beads have been replaced with modern equivalents. Gears, relays, vacuum tubes, and recently transistors and integrated circuits (ICs) have been used to perform digital computation. Switching of pebbles in and out of a jar or beads up and down a rod are

Figure 3.2
Bringing in the sheep.

methods of representing a specific number. An automobile odometer uses position of gear teeth to represent the number of miles traveled. Similarly, a relay contact, either open or closed, represents a number. The same function can be performed by vacuum tubes, which are either conducting or not. The newest electronic switching devices are transistors packaged into ICs.

Each of these switching devices use digital numbers to perform computation. Numbers are expressed as a sequence of digits and intermediate states do not exist. There are no half pebbles, half-open relay contacts, or partially conducting transistors—digital computation requires specific states. A payroll number, a social security number, and a telephone number are digital numbers. Digital numbers are unique and there is no partial credit for coming close. A digital number is either exactly correct or it is totally incorrect. There is nothing useful about dialing every digit in a telephone number correctly except for one. Digital numbers are and must be specific.

About 30 or 40 years ago desk-top calculators used gear position to represent a number. In more modern versions, the electrical condition of a circuit represents a number. The most practical electrical circuit conditions are on and off. Computer evolution from pebbles and beads to transistors and ICs is a history of switch improvement. Regardless of switching device, all computers operate by switching from a condition which represents one number to another condition which represents a different number. The basic computational unit is the switch.

An ideal switch has infinite *off* resistance and zero *on* resistance, passes infinite current, and operates in zero time. Real switches come close enough to these ideal conditions to be useful. For example, a wall switch requires good electrical isolation. When the switch is open, the entire voltage drop should be across the open contacts and the current should be zero. An open wall switch should have infinite resistance. When the wall switch is closed, maximum current should flow and the switch should have zero resistance. As a practical matter *off* resistance is determined by the plastic which supports the switch contacts and *on* resistance is determined by the metallic contacts. Figure 3.3 represents a wall switch. With modern materials, the *off* resistance of a wall switch is about 10^{10} Ω, and *on* resistance is about 0.02 Ω. Wall switches readily pass 10 A. Electromechanical relays have the same resistance and current carrying capabilities as wall switches.

Another form of electrical switch is the transistor. On/off characteristics are determined by base-emitter voltage. The collector of the *npn* transistor

Figure 3.3
Wall switch. (*a*) Open circuit; (*b*) closed circuit.

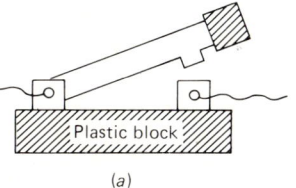
(a)

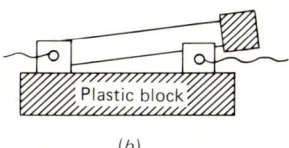
(b)

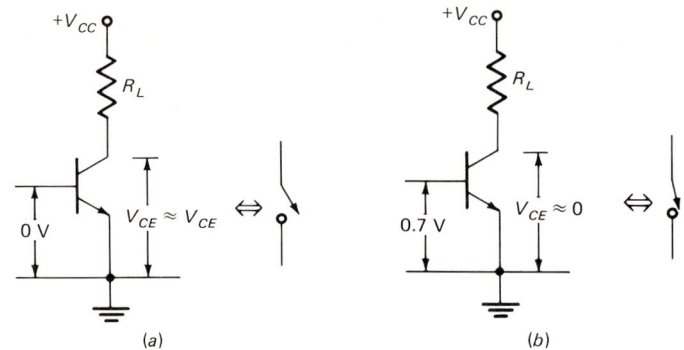

Figure 3.4 Basic transistor switch. (*a*) Base emitter reverse-biased; (*b*) base emitter forward-biased.

shown in Fig. 3.4 is connected to a positive dc supply. Except for leakage current, the transistor cannot conduct when the base-emitter junction is reverse-biased. Since no collector current flows, a reverse-biased transistor is equivalent to an open switch. When the base-emitter junction is heavily forward-biased, the transistor saturates and collector current is maximum. A saturated transistor is equivalent to a closed switch. Transistor *off* and *on* resistance are determined by Ohm's law in the cutoff and saturation conditions.

Example 3.1 A transistor with a 1-μA leakage current is connected to a 5-V power supply. When the transistor is saturated, V_{CE} is 50 mV and the saturation current is 10 mA. Determine: (*a*) the transistor *off* resistance; (*b*) the transistor *on* resistance.

Solution

(*a*) $R_{off} = \dfrac{V_{CE}}{I_{leakage}} \simeq \dfrac{V_{CC}}{I_{leakage}}$

$= \dfrac{5}{1 \times 10^{-6}} = 5 \times 10^6 \, \Omega$

(*b*) $R_{on} = \dfrac{V_{CE}}{I_{sat}}$

$= \dfrac{50 \times 10^{-3}}{10 \times 10^{-3}} = 5 \, \Omega$

These calculations show that transistor *off* and *on* resistance are poorer than wall switch or relay values. Transistors have much lower *off* resistance and much higher *on* resistance. However, any voltmeter can determine when a transistor is off or on: if V_{CE} is approximately the same as the power supply voltage, the transistor is off; if V_{CE} is approximately zero, the transistor is on. Actually, a voltmeter is not even required, as a light-emitting diode (LED) in series with a current-limiting resistor can determine whether a transistor is on

TABLE 3.1
Switch Characteristics

Characteristic	Relay	Transistor
Off resistance	$10^{10}\,\Omega$	$10^{6}\,\Omega$
On resistance	$0.02\,\Omega$	$5\,\Omega$
Current	10 A	10 mA
Switching speed	0.03 s	1×10^{-9} s

or off. Such LED circuits are the basis of logic probes used in troubleshooting computer circuits. The logic probe is attached to a point in the computer. If the LED is on, that point is at power supply voltage; if the LED is off, that point is at 0 V.

In terms of *off* resistance, *on* resistance, and current handling capability, wall switches and relays are better than transistors. Then why consider using transistors as switches? It is true that transistors are physically small, but the most important characteristic for computing is switching speed. Transistors readily switch from on to off or vice versa in nanoseconds (ns). Sometimes one nanosecond (1 ns) is called a *light-foot* because in this time light travels 0.984 ft, or approximately 1 ft. Thus transistor switch speeds are comparable with that of a ray of light traveling across the transistor. Table 3.1 compares switching characteristics of relays and transistors. Vacuum tubes, used in computers after relays and before transistors, are almost as fast as transistors but are physically larger and dissipate much more power.

Switch speed determines time required for a computer to perform calculations. This characteristic is included in Table 3.1. Since transistors are 30 million times faster than relays, a transistor computer is 30 million times faster than a relay computer. Speed, small size, and low power consumption make transistors and integrated circuits the most attractive components for performing digital computation.

3.2 AND Gates

Since digital numbers have precise values, circuits which represent and control numbers must be precise. Computer circuits provide outputs which are either open or closed; open/closed must thus take the place of true/false, yes/no, and, of course, 1/0. This is the reason for discussing Boolean algebra and binary numbers in the first two chapters. All circuits operate on this basis. There are only two possibilities. For computer purposes the question, "Is it raining?" cannot be answered with "It looks like rain" or "It is drizzling." If it looks like rain then it is not raining, and if it is drizzling then it is raining.

Similarly, are John and Mary on their way to school? If John is on his way to school and Mary is not or vice versa, then John and Mary are not on their way to school. Either both are going to school or else they are not. Questions must be asked so that true and false are the only possible answers. This does

Figure 3.5
Two-switch circuit.

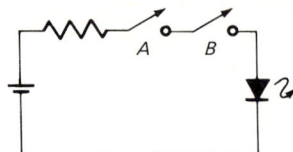

not require reorganizing the entire universe, but it means that computer questions must be carefully phrased. Computer circuits can have only two possibilities, and of the two possibilities only one can be correct. For this reason an on/off switch is a natural choice as the basic computer element. Computers are binary, or two-state, machines.

Gates, introduced in the first chapter, perform control functions. It is the gates which organize, sequence, and process data. Gates are combinations of switches. Input and output signals to a gate have only two possible states, 1 or 0.

Figure 3.5 shows a circuit containing a battery, resistor, two switches, and an LED. All components are connected in series. Clearly, the LED will only glow when switches A and B are both closed. However, this circuit requires further attention. With two switches, there are four possible combinations of switch conditions. Both switches can be off. Either switch can be on while the other is off. Both switches can be on. These four switch combinations along with the LED conditions are shown in Table 3.2a. Off and on are not the only terms which can be used to describe switch conditions. We could, for example, use *open* instead of *off* and *closed* instead of *on*. It is also possible to assign *low* (L) to represent *off* and *high* (H) to represent *on*. Low and high refer to switch and lamp voltages. Equivalent conditions in terms of L and H are shown in Table 3.2b.

Other terms can be used to describe switch conditions. Two interesting possibilities are true/false and 1/0. In true-false terminology *true* could represent *on*, in which case *false* would represent *off*. For the 1/0 case if 1 represents *on*, then 0 represents *off*. Of course we have arrived at the truth table again. When the truth table was introduced, it was a Boolean concept. The switch condition of the circuit shown in Fig. 3.5 is the electrical equivalent of a truth table. The expression *truth table* does not have any deep philosophical meaning. A truth table lists all possible states which a circuit can experience.

TABLE 3.2
Switch Combinations

(a) On/Off			(b) H/L		
Switch A	Switch B	Lamp	Switch A	Switch B	Lamp
Off	Off	Off	L	L	L
Off	On	Off	L	H	L
On	Off	Of	H	L	L
On	On	On	H	H	H

TABLE 3.3
2-Input AND-Gate Truth Tables

(a) T/F

A	B	Lamp
F	F	F
F	T	F
T	F	F
T	T	T

(b) 1/0

A	B	Lamp
0	0	0
0	1	0
1	0	0
1	1	1

Table 3.3 shows the truth tables for the same conditions listed in Table 3.2. These tables describe a 2-input AND gate. An AND gate is a circuit whose output is true only when all of its inputs are true. AND is used here in its Boolean sense, each and every.

When 1/0 is used, all possible input combinations can be included by counting in binary. This is one of many reasons for discussing binary numbers in Chap. 2. As mentioned, ANDing is sometimes referred to as logical multiplication, and the Boolean equation for the two-switch-with-lamp circuit in Fig. 3.5 is

$$L = A \cdot B \tag{3.1}$$

This is the logic equation for a 2-input AND gate. Some texts omit the logical multiplication dot, in which case the equation becomes

$$L = AB \tag{3.2}$$

However, dots clarify Boolean operations and are used throughout this book. Logic equations and gate truth tables are equivalent: one form is a mathematical equation and the other is a list of conditions, but the meanings are identical. Symbols for gates were introduced in Chap. 1. The logic diagram for this two-switch circuit is the 2-input AND gate shown in Fig. 3.6a.

Boolean equations have other applications. A 3-input AND gate might be a burglar alarm consisting of a front door switch F, a back door switch B, and a window switch W connected in series. Every switch, as shown in Fig. 3.7a, must be closed to keep the relay activated. If any of the three switches open, the relay deactivates and voltage to operate the burglar alarm A exists. The logic diagram is shown in Fig. 3.7b.

An interesting computer application of an AND gate is a circuit which allows pulses to be transmitted only when another pulse is present at the same time. An AND gate has a high output only when all inputs are high. As

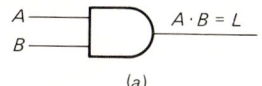 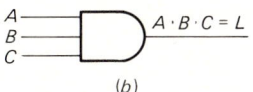

Figure 3.6 AND gate symbols. (a) 2-Input; (b) 3-input.

Figure 3.7
Burglar alarm. (*a*) Circuit; (*b*) logic diagram.

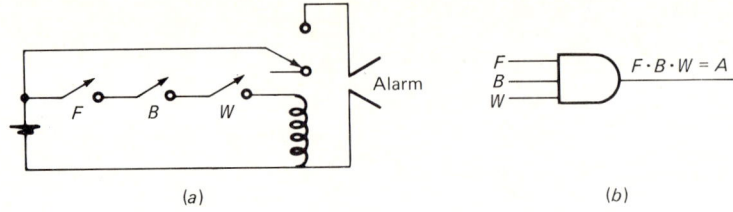

Figure 3.8
Coincidence circuit.

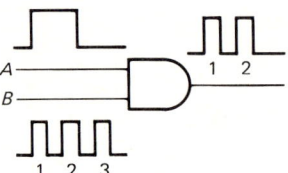

shown in Fig. 3.8, pulses 1 and 2 at the *B* input are high at the same time that the *A* input is high. Therefore these pulses appear at the output. When pulse 3 appears at the *B* input, *A* is low. Therefore pulse 3 cannot appear at the output. Such a circuit is called a *coincidence* circuit and is used, for example, to determine the time during which a number may be transferred from one computer circuit to another. In this circuit the pulse which determines time duration is connected to input *A* and the number goes to input *B*. The electrical equivalent of an AND gate is a circuit with switches in series.

3.3 OR Gates

Other gates are also important, and Table 3.4 is for an OR gate. The output is high if at least one input is high. Thus, for a 2-input OR gate the output will be high if either input or both inputs are high. Similarly, for a 3-input OR gate the output will be high if any one, any two, or all inputs are high. The term OR is used in the sense of *any*.

Comparison of OR- with AND-gate truth tables shows some similarity. Both gates have the same outputs when all inputs are low and when all inputs are high, but when inputs are not the same, OR and AND gates have different outputs. Some resemblance does not represent equivalence. Since OR and

TABLE 3.4
OR-Gate Truth Table

A	B	L
0	0	0
0	1	1
1	0	1
1	1	1

Figure 3.9
2-Input OR gate. (*a*) Switch version; (*b*) logic symbol.

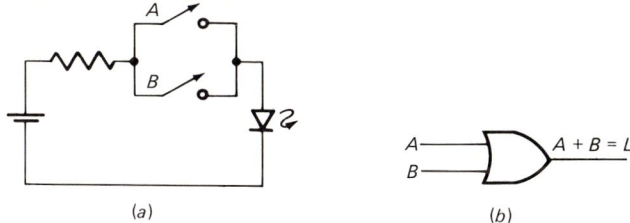

AND do not correspond for all combinations, they are completely different gates.

The similarity between the OR gate and arithmetic addition was discussed. For this reason an OR gate is said to perform logical addition. We have

$$L = A + B \tag{3.3}$$

as the Boolean equation for a 2-input OR gate, and similarly

$$L = A + B + C \tag{3.4}$$

for a 3-input OR gate, etc. Although logical addition using OR and logical multiplication using AND resemble ordinary addition and multiplication, the resemblance is not perfect. Therefore, OR and AND gates cannot be directly used to perform numerical addition and multiplication. Arithmetic circuits are discussed in later chapters.

A 2-input OR gate using switches is shown in Fig. 3.9*a*. When switches are used, OR is achieved by connecting switches in parallel. The 2-input OR-gate symbol is shown in Fig. 3.9*b*.

Truth tables, logic equations, and logic diagrams are different but equivalent ways of representing gate performance. Each is better for some applications and worse for others. Graphs are also helpful because computers use pulses. *Timing diagrams* are graphs of pulse amplitude versus time; an *on* pulse represents a logic 1 and an *off* pulse represents a logic 0. The timing diagram shows an oscilloscope display. Figure 3.10 compares a 2-input OR-

Figure 3.10
OR-gate conditions. (*a*) Truth table; (*b*) timing diagram.

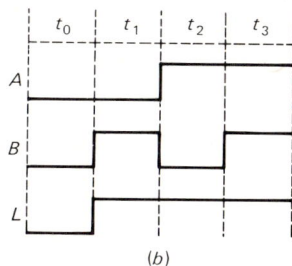

Figure 3.11
Diode OR gate.

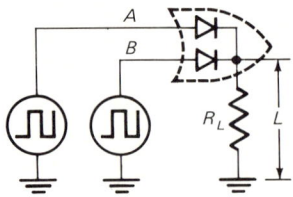

gate truth table with a timing diagram. The four methods of representing gate performance thus are:

- Logic equations
- Truth tables
- Timing diagrams
- Logic symbols

We can use as many or as few as are needed to clarify computer circuit performance. The logic symbol for an OR gate is the same regardless of which components perform the OR function. OR gates made from switches, diodes, or transistors use the same symbol.

A 2-input diode OR gate is shown in Fig. 3.11. As before, inputs are A and B and the output is L. Inputs are from the positive sides of the diodes to ground, and the output is from the negative diode junctions and ground. Typically, 5 V represents a logic 1 and 0 V a logic 0. Figure 3.12 shows all possible combinations. In Fig. 3.12a both inputs are at circuit ground, which is 0 V. This corresponds to $A = 0$, $B = 0$ of the truth table. In this case L is also at circuit ground. Figure 3.12b shows a 5-V pulse at A, but with B still at

Figure 3.12
Diode OR gate conditions. (a) $A = 0$, $B = 0$; (b) $A = 1$, $B = 0$; (c) $A = 1$, $B = 1$; (d) voltage truth table.

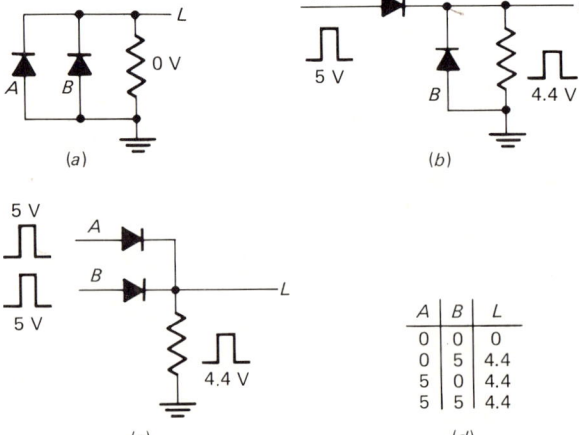

Figure 3.13 Cascaded OR gates.

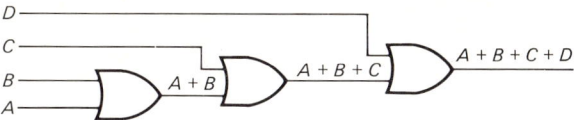

0 V. This is the $A = 1$, $B = 0$ truth table combination. The diode at terminal A is forward-biased and the diode at B is reverse-biased. For a silicon diode the forward-biased voltage drop across the diode is between 0.6 and 0.7 V. Therefore, the voltage at L will be a pulse which is about 4.4 V. If the conditions shown in Fig. 3.13b are interchanged, the logic states are $A = 0$, $B = 1$. The output pulse is still 4.4 V.

Figure 3.12c shows simultaneous 5-V pulses connected to A and B. The inputs are in parallel because both are from 5 V to ground; therefore the L is still 4.4 V. Figure 3.12d is the voltage truth table for this diode OR gate. A high output is not quite the same as a high input. However, there is no difficulty in distinguishing a high output from a low output. A 4.4-V output operates an LED circuit, but a 0-V output cannot.

However when diode OR gates are cascaded, as shown in Fig. 3.13, it becomes difficult to determine if an output is high or low. The output of the first OR gate is $A + B$. In turn this is ORed with C and $A + B + C$ is the output of the second OR gate. Similarly, the output of the third OR gate is $A + B + C + D$. If diode OR gates are used, the output voltage drops as the cascade becomes longer. The output of a 2-input diode OR gate is 4.4 V for a 5-V input. Similarly, the output of the second OR gate is 0.6 V lower. The output of the second OR gate shown in Fig. 3.13 is about 3.8 V, and the output of the third OR gate is about 3.2 V. If variations in diode characteristics and resistor tolerances are considered, the output voltage of each diode OR gate could be even lower. Cascading too many diode OR gates makes it difficult to determine whether an output is high or low.

Figure 3.14 shows two diodes connected as an AND gate. Compared with a diode OR gate, the connections are reversed, since here the inputs are the negative sides of the diodes and the output resistor goes to a positive 5-V power supply. Examination of a diode AND-gate circuit shows that it is compatible with the diode OR gate. In this sense *compatible* means high outputs are about 5 V and low outputs are about 0 V. Actually, the AND-gate low is about 0.6 V, and the high is very close to 5 V. These values assume

Figure 3.14 Diode AND gate.

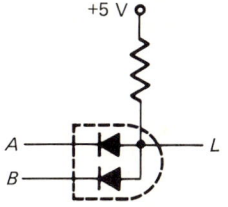

Figure 3.15
Decimal-to-binary encoder. (*a*) Number equivalents; (*b*) circuit.

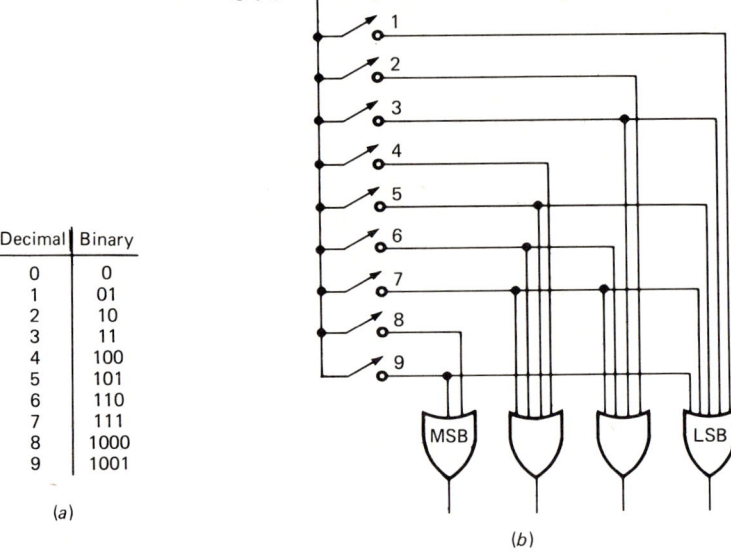

Decimal	Binary
0	0
1	01
2	10
3	11
4	100
5	101
6	110
7	111
8	1000
9	1001

(*a*)

(*b*)

a load connected at L which is much greater than the AND-gate resistor. If such is not the case, the high output drops in a manner similar to that in the cascaded OR gate.

OR gates can convert numbers from one base to another. Since humans use decimal and computers use binary, a vital part of any computer is an *encoder*, such as that shown in Fig. 3.15. Encoders convert decimal numbers to binary equivalents.

Figure 3.15*a* shows the numbers 0 through 9 in decimal together with binary equivalents. In Fig. 3.15*b*, one side of each decimal number switch is connected to a 5-V supply, and the other side of each number switch is connected to various OR gates. For example, decimal number switch 9 only connects to the OR gates of the MSB and LSB. Therefore, when switch 9 is activated, only these gates have high outputs. The two middle OR gates have low outputs since all these inputs are low. In this case the gate outputs indicate 1001, which is the binary equivalent of 9_{10}. Similarly, when 5_{10} is activated, 0101 appears at the OR-gate outputs, etc. This particular decimal-to-binary encoder has several practical problems. There is no way to distinguish the number 0 from the condition when no number is entered. Another problem is whether or not 5-input OR gates exist.

Another important part of any computer is the circuitry at the output, where binary numbers are converted back to decimal. A *decoder* converts from binary, or another number system, to decimal. One method of making a binary-to-decimal decoder uses AND gates and is included as a problem at the end of this chapter.

3.4 Inverters

The other basic Boolean gate is the *inverter*. As described, an inverter is a gate with an output which is opposite to the input. Since there are only two possible input conditions in binary, the meaning of opposite is clear:

- When an input is low, the output is high.
- When an input is high, the output is low.

The term *NOT gate* is used instead of *inverter*, and another term which means the same is *complement*. NOT, complement, and invert all mean opposite. As before, the symbol for invert is a bar over the particular quantity: $\bar{A}$ is the invert of A. The logic equation for an inverter is

$$L = \bar{A} \tag{3.5}$$

Figure 3.16 shows the truth table and timing diagram for an inverter.

Since an inverter has a single input, two input conditions are possible. OR and AND gates with two input leads have four possible conditions. This progression continues for gates with multiple inputs. In general

$$C = 2^N \tag{3.6}$$

where N is the number of input leads and C is the number of possible combinations. Consider an inverter. N is 1 and 2^1 is 2, which is consistent with inverter operation. For 2-input gate N is 2 and 2^2 is 4. This too is correct. Input combinations for a 2-input gate are 0, 0; 0, 1; 1, 0; and 1, 1.

Equation (3.6) is useful because the number of input combinations which must be investigated is specified. Consider a gate with four inputs. In this case N is 4 and C is 16, that is, 16 different input conditions exist for a 4-input gate. These possibilities can be listed in numerical order by counting in binary through the decimal equivalents of 0 through 15.

Figure 3.17a shows a possible inverter circuit built with a switch. In this version the open switch represents A and the closed switch represents $\bar{A}$. Using $\bar{A}$ as 5 V means A is 0 V. A gate of this type is practical but is not compatible with the previous logic levels. All the gates which have been presented use A as 5 V and $\bar{A}$ as 0 V. This is opposite to the circuit described in Fig. 3.17a.

In previously considered gates an open switch is logic 0 and a closed

Figure 3.16
Inverter. (*a*) Truth table; (*b*) timing diagram.

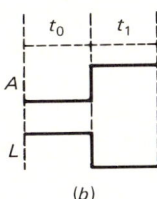

Figure 3.17 Inverter circuits. (*a*) Possible; (*b*) compatible.

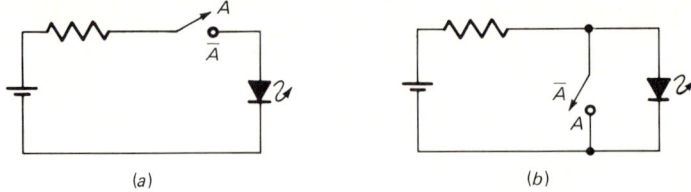

switch is logic 1. For a compatible inverter the LED should be on when the switch is open and off when the switch is closed. As shown in Fig. 3.17*b*, these requirements are achieved by placing the switch in parallel with the LED. When the switch is open, there is voltage across the LED, but when the switch is closed, a short circuit exists across the LED. Therefore the LED is on when the switch is open and off when the switch is closed. This version is compatible with previous gates.

Two amplifiers are shown in Fig. 3.18. Figure 3.18*a* is a noninverting amplifier. Figure 3.18*b* shows an inverting amplifier, of which the input and ouput are out of phase. Inversion is indicated by a circle, or *bubble*, at the output.

It is one thing to show a device symbol but quite another to construct the actual circuit. Understanding digital circuits requires manipulating logic symbols as well as knowledge of circuit operation. The conditions for obtaining noninverting and inverting amplifiers are a natural consequence of transistor action.

Figure 3.19*a* shows a noninverting amplifier. It is the *common collector* configuration, also known as the *emitter follower*. When base current increases, collector current increases and therefore emitter current increases. The output is taken across the load resistor R_L which is in the emitter leg. Thus an increase in base current will result in an increase at the output. In the common collector, input and output are in phase; therefore the common collector is a noninverting amplifier. This configuration is used in *buffer* circuits. A buffer increases output current and permits many devices to be driven from the same gate.

Figure 3.19*b* shows the *common emitter* configuration. In a common emitter, an increase in base current also results in an increase in collector current, which causes an increase in the voltage across the load resistor. However, the output is from collector to emitter. Kirchhoff's voltage law shows that the collector-to-emitter voltage is the difference between supply voltage and load resistor voltage. Therefore, in a common emitter an increase

Figure 3.18 Amplifier symbols. (*a*) Noninverting; (*b*) inverting.

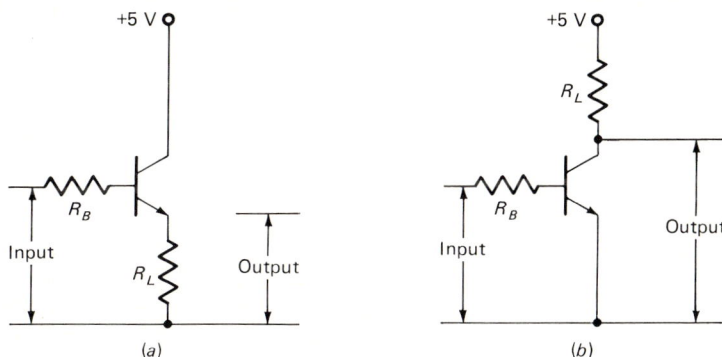

Figure 3.19
Basic transistor circuits.
(*a*) Common collector;
(*b*) common emitter.

in the load resistor voltage results in a decrease in the output voltage. Since an input increase results in an output decrease, the common emitter is an inverting amplifier.

Actually, there is no amplification of signals in a digital computer. Signals are at one of two levels. Whether the circuits shown in Fig. 3.19 are amplifiers or semiconductor switches depends on the values of β, R_B, and R_L. If these values are chosen for switching, then a 5-V signal at R_B saturates the transistor. When the transistor is saturated, the output of a common collector is approximately 5 V, and the common emitter output is very nearly 0 V. When the input to R_B is 0 V, the transistor is cut off. In this case the output of the common collector is nearly 0 V, and the output of the common emitter is approximately 5 V.

All Boolean expressions can be built from OR, AND, and NOT gates. Implementing a Boolean expression which involves ANDing two terms requires a 2-input AND gate, ANDing a three-term Boolean expression requires a 3-input AND gate, etc. The same situation applies to ORing.

Ordinarily, a single switch controls a single circuit, two switches control two circuits, three switches control three circuits, etc. If these conditions could not be improved, computers would be extremely large. For example, an important part of any computer is the memory section. At this stage the memory can be considered as a huge mailbox with thousands of cubbyholes. Each cubbyhole contains a letter and each cubbyhole door is controlled by a separate switch. Under these circumstances there are as many switches as there are cubbyholes. Fortunately, switch count is reduced with a circuit called an *address decoder*.

An address decoder uses NOT gates together with AND gates. The decrease in switches resulting from its use can be represented by an equation of the same form as Eq. (3.6). Two switches control four cubbyholes and three switches control eight cubbyholes. Address decoding is really efficient for nine or more switches; nine switches control 2^9, or 512, separate cubbyholes. An address decoder in which two signals at the input control four circuits at the

Figure 3.20
1-of-4 decoder.

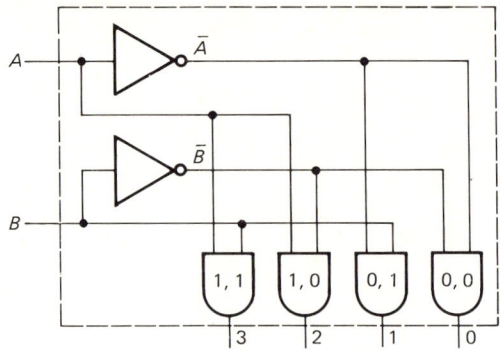

output is shown in Fig. 3.20. Both inputs are retained inside the circuit. Also, A and B are inverted, thus A, B, $\bar{A}$, and $\bar{B}$ are available. Under these circumstances four combinations exist: 0, 0; 0, 1; 1, 0; and 1, 1. The crux of wiring an address decoder is forcing each combination to operate only one of the AND gates.

A 2-input AND gate has a high output only when both inputs are high. Therefore when A and B are 0, 0, both inputs must be inverted. This is the case for the AND gate labeled 0, 0. None of the other AND gates receive two inverted inputs. If A and B are 0, 1, then A but not B is inverted to activate the AND gate labeled 0, 1. Also input signals of 1, 0 and 1, 1 have the correct AND connections. For each possible input combination, the inverted and noninverted combinations are appropriate. Such circuits are called 1-*of-N* *decoders*, and N is the number of possible combinations. This circuit is a 1-of-4 decoder. The same information can also be displayed as a truth table with two inputs and multiple outputs. In this case the inputs are A and B, while the outputs are the 4 AND gates. Truth tables with multiple outputs were introduced in Chap. 1.

Actually all the input combinations appear at all the AND-gate inputs at the same time, but the connections to each AND gate have been chosen to permit only the correct AND gate to function. AND gates and inverters are relatively simple circuits, yet the correct connection of ANDs with inverters, as just seen, significantly reduces the number of switches required to unlock the cubbyholes of a computer memory. As a matter of fact, virtually every circuit in a computer is built from the three basic gates. There are many sophisticated circuits in a computer, and an address decoder is one of the

Figure 3.21
NAND gate. (*a*) Components; (*b*) symbol; (*c*) truth table.

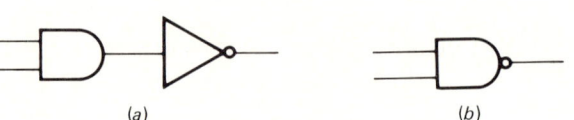

Figure 3.22
NOR gate. (*a*)
Components; (*b*) symbol;
(*c*) truth table.

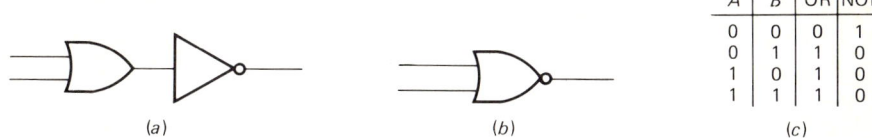

A	B	OR	NOR
0	0	0	1
0	1	1	0
1	0	1	0
1	1	1	0

simpler examples. Regardless of circuit complexity the basic building blocks are AND, OR, and NOT gates.

As described in Chap. 1, it is useful to combine an inverter with an AND gate into a NAND gate. The components of and symbol for a NAND gate are shown in Fig. 3.21. As before, a bubble indicates inversion. The truth table for a NAND gate is the invert of the truth table for an AND gate. De Morgan's theorem leads rather naturally to NAND and NOR gates. The symbol for a NOR gate is an OR gate with a bubble at the output, as shown in Fig. 3.22. A NOR-gate truth table is the invert of the OR-gate truth table.

Any Boolean expression or De Morgan equivalent can be built from the gate types which have been discussed, but greater reduction in gate types is possible. Either a NAND or NOR gate can be used as an inverter. Therefore a NAND can function as either a NAND or an inverter. Similarly a NOR can function as either a NOR or an inverter.

When De Morgan's theorem was presented, either OR or AND could be completely eliminated from a Boolean expression. Since NAND and NOR can function as NOT, an entire computer can be constructed by using a single gate type, either NAND or NOR. Such a computer, while feasible, is not practical—the total number of gates is much greater when using only a single gate type. However, the concept is important: regardless of the complexity, computer circuits are constructed from a few basic gates.

3.5 THREE-STATE GATES

Life in general, and engineering in particular, is a series of compromises in which advantages are maximized and disadvantages are minimized. Consider the two telephone systems shown in Fig. 3.23. In a party line connection several subscribers are connected on a single circuit to the central station. The

Figure 3.23
Telephone connection methods.

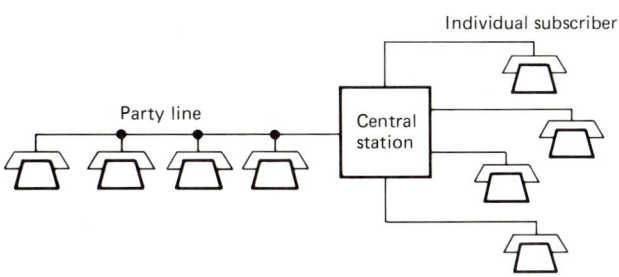

advantage to a party line subscriber is economic. Since fewer circuits are required, party line subscribers have lower telephone bills. The disadvantages of a party line are loss of privacy and a lower probability of being able to make a call.

The alternative to party lines is individual subscriber lines. A single subscriber enjoys privacy and has a greater probability of making or receiving a call. The disadvantage to a single subscriber is economic. Additional circuits as compared with a party line require more instrumentation, and individual subscribers pay a larger telephone bill. Assuming that telephone service is required, a potential subscriber weighs the advantages and disadvantages. On this basis either an individual or a party line is selected.

There are similarities between a telephone system and a digital computer. Telephones and computers contain circuits which must be interconnected and then disconnected. Prior to the 1970s interconnections between computer circuits were of the individual subscriber type, with each circuit individually wired to every other circuit and access guaranteed when required. However, individual interconnections required complicated control circuits and took up a lot of room.

More recently the party line approach has been adopted for interconnecting computer circuits. Complexity and cost are reduced. In terms of privacy, eavesdropping as such is not a problem inside a computer. However, situations often require one computer circuit to be connected to only one other circuit, and the remaining circuits must be disconnected.

Disconnected means no voltage is applied, and *connected* means voltage is applied. The precise manner of achieving a disconnection is critical. Is there, for example, a difference between 0 V and no voltage? Figure 3.24 shows two different zero-type conditions. In Fig. 3.24a current flows through the circuit. However, the output voltage is across a section of conductor and since conductors have negligible resistance, the output voltage is essentially zero. In Fig. 3.24b the output voltage is also zero. In this case zero voltage is measured across an open circuit. Switching was discussed in the first section of this chapter, and open and short circuits were presented as ideal cases. In reality, small voltages exist across the output terminals of both circuits.

The difference between these two circuits is the impedance level. When current flows, zero voltage is measured across a circuit section which is a good approximation of a short circuit. When no current flows, zero volts is

Figure 3.24
Zero conditions. (*a*) Zero volts; (*b*) no voltage.

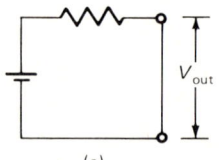
(a)

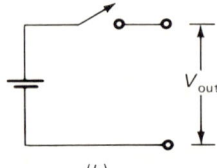
(b)

Figure 3.25
Buffer amplifier types.
(*a*) Conventional; (*b*) three-state.

measured across a circuit section which is a good approximation of an open circuit. Low-impedance and high-impedance zero voltages are both required in computer circuits. The short-circuit zero voltage is the logic 0 of a gate. The open-circuit zero voltage is required to disconnect one computer circuit from another.

Several high-impedance zero-voltage techniques are possible—one approach is removing power supply voltage from the disconnected circuit, while another variation is removing circuit ground. Since the early 1970s the *three-state gate* has been the most popular semiconductor method of obtaining high-impedance zero voltage.

Figure 3.25 shows two buffer amplifiers. The conventional noninverting buffer shown in Fig. 3.25*a* has been discussed. Figure 3.25*b* shows a buffer amplifier with an additional lead called *enable*. This circuit is one version of a 3-state gate. If a logic 1 is connected to the enable lead, a 3-state gate performs as a conventional buffer amplifier: a logic 0 at the input results in a logic 0 at the output, and a logic 1 at the input results in a logic 1 at the output. However, when a logic 0 is connected to the enable lead, there is no output regardless of the input. With logic 0 or logic 1 as an input, the output of a 3-state gate is an open circuit. As discussed, open circuit really means high impedance.

Three-state gates are used to connect or disconnect computer circuits. When connection is required, logic 1 is connected to the enable lead. In this case, whatever data exist at the input appear at the output. On the other hand, when disconnect is required, a logic 0 is connected to the enable lead. With logic 0 at the enable lead the output is an open circuit regardless of whether a 1 or a 0 exists at the input lead. Table 3.5 shows 3-state gate behavior.

A 3-state gate has three possible outputs, namely, open circuit, logic 0, and logic 1. However, it only has two logic levels, since open circuit is an output condition but is not a logic level. When a 3-state gate has an output, the output levels are the same binary values as for any other gate.

TABLE 3.5
3-State Gate Conditions

Enable	Input	Output
0	0	Open Circuit
0	1	Open circuit
1	0	0
1	1	1

Figure 3.26
Using 3-state gates.

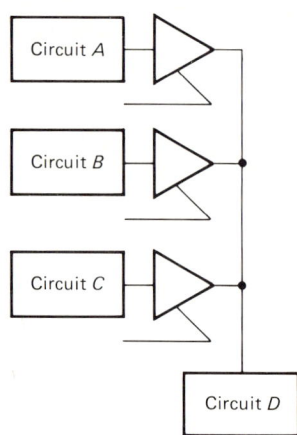

Figure 3.26 shows how 3-state gates are used in party line connections. Circuit *D* can receive data from circuit *A*, *B*, or *C*. When data must be transmitted from circuit *A* to circuit *D*, the 3-state gate connected to circuit *A* is enabled with logic 1 and remaining 3-state gates are disabled with logic 0. Similarly, to transmit data from any other circuits the appropriate 3-state gate is enabled and remaining gates are disabled. Circumstances may make it more convenient to enable with logic 0 and disable with logic 1; this form of 3-state gate also exists. Figure 3.27 shows two versions of a 3-state gate. In addition, there are 3-state gates which are inverters rather than buffer amplifiers.

Another method of combining data, which in fact is older than the 3-state gate, is the *open-collector* gate. Previous gates are complete within the integrated circuit chip; external components are not required. On the other hand, open-collector-gate ICs are almost but not quite complete. Open-collector circuits require an externally connected load resistor, called a *pull-up* resistor. When open-collector gates are used, many circuits can be connected to the same pull-up resistor.

Figure 3.28*a* shows three open-collector inverters connected to a common pull-up resistor. If the input to all three inverters is low, the output is high, but if the input to any inverter is high, the output is low. A single low at the output drives the entire output low. Figure 3.28*b* shows the truth table for a three-open-collector inverter circuit.

This circuit has the same truth table as a 3-input NOR gate. Open-collector

Figure 3.27
Three-state gate types.
(*a*) Enable with high: (*b*)
enable with low.

(a)   (b)

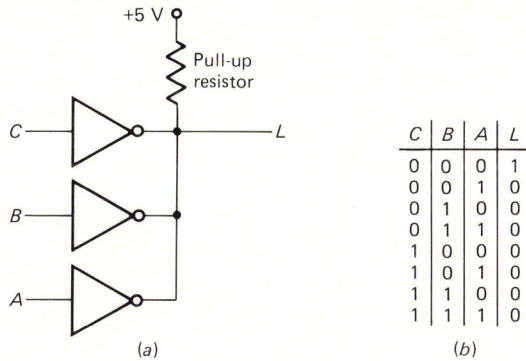

Figure 3.28 Connecting open-collector inverters. (*a*) Schematic; (*b*) truth table.

gates require wiring the pull-up resistor and are called *wired gates*. In this case, the resultant gate is called a wired NOR. Open-collector AND, NAND, and OR gates also exist, and outputs are connected to a common pull-up resistor. Figure 3.29*a* shows two open-collector NAND gates connected together. The Boolean equation for this circuit is

$$L = \overline{AB} \cdot \overline{CD} \tag{3.7}$$

Figure 3.29*b* shows the equivalent circuit using conventional gates. In this case an extra gate is required if conventional gates are used. With all open-collector gates, a single low output from any gate results in a low for the entire circuit.

In some cases open-collector circuits reduce gate count, and in other applications they simplify construction of multiple-input gates. Also, open-collector gates are less expensive than 3-state gates, but they are slower and more susceptible to noise.

3.6 BIPOLAR GATES

Computers used discrete transistors before ICs were developed. Individual transistors were connected to resistors, diodes, and capacitors to form circuits, which were mounted on plug-in circuit boards and connected to the computer main frame. In a discrete circuit, each transistor is a single

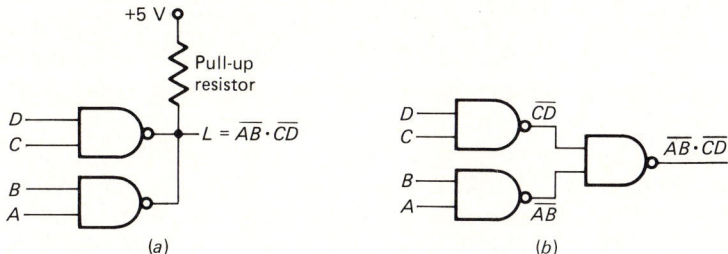

Figure 3.29 NAND gate circuits. (*a*) Open collector; (*b*) conventional equivalent.

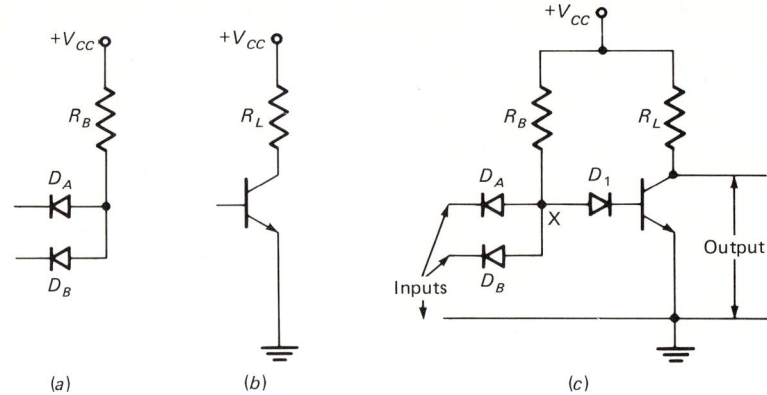

Figure 3.30 Basic DTL gate. (*a*) Diode AND gate; (*b*) inverter; (*c*) combined circuit.

silicon chip and components are soldered to the transistor leads. On the other hand, a single IC contains many transistors, diodes, and resistors on the same chip which are connected to perform the necessary functions. Entire circuits are packaged into single ICs. ICs are smaller, less expensive, faster, and dissipate less power than discrete-component equivalents. Digital ICs were introduced in 1963.

Resistor transistor logic, or RTL, was the first type of IC. RTL ICs combine resistors and transistors to obtain gates, a NOR gate being the basic RTL circuit. RTL ICs operate from a 3.5-V supply, and the difference between logic 1 and logic 0 is about 1 V.

This small difference is the main problem with RTL gates. Electrical noise is a problem in any circuit, but in digital circuits noise spikes are extraneous pulses and destroy the meaning of existing 1s and 0s. Except for noise, RTL characteristics are good; typical gates switch in 15 ns and dissipate 15 mW.

The next generation of ICs replaced input resistors with diodes, which decrease power dissipation and improve noise immunity. Since diodes and transistors are combined, the improved version is called *diode transistor logic* (DTL). A basic DTL gate is shown in Fig. 3.30. It consists of the diode AND gate previously shown in Fig. 3.14 combined with the inverter of Fig. 3.4.

Since D_1 and the base-emitter junction of the inverter are in series, approximately 1.4 V from point X to ground is required to forward-bias the transistor. Figure 3.31*a* shows D_A and D_B at logic 0. D_A and D_B are both forward-biased, and the voltage from X to ground is about 0.7 V. Since 1.4 V is required to turn the transistor on, the transistor is cut off when both diodes are at logic 0. When the transistor is off, no current flows through R_L, and therefore the output voltage is close to V_{CC}. The output is logic 1 when both inputs are logic 0.

If one input is at logic 0 while the other is high, there is still only 0.7 V from X to ground. The transistor remains cut off, and the output remains at logic 1. This condition is shown in Fig. 3.31*b*. The output is at logic 1 if one diode is

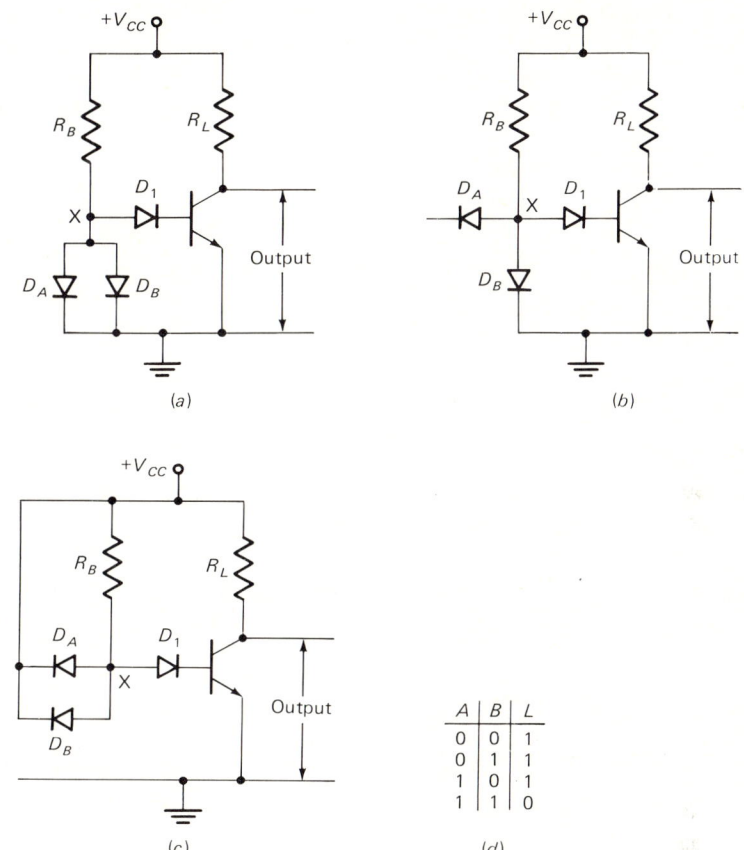

Figure 3.31
DTL gate conditions. (a) Both diodes low; (b) one diode low; (c) both diodes high; (d) truth table.

at logic 0. When both diodes are at logic 1, the output is logic 0. This condition is shown in Fig. 3.31c. D_A and D_B are both reverse-biased, and D_1 is forward-biased. This means there is 1.4 V between X and ground, and the transistor saturates. In a saturated transistor, the output voltage is close to 0 V. The output of this gate is at logic 0 only when both inputs are at logic 1. Figure 3.31d is the truth table for these conditions. This circuit is a 2-input DTL NAND gate, and the output voltage swing from logic 1 to logic 0 is about 5 V.

Owing to the series combination of base-emitter junction and D_1, any voltage lower than 1.4 V (Fig. 3.32a) cannot turn the transistor on. Therefore noise is not as serious. Another diode can be added in series with D_1 to further improve noise immunity, as shown in Fig. 3.32b. In this case three diodes, namely, D_1, D_2, and the transistor base-emitter junction, must conduct before the transistor can saturate, and the gate remains at logic 1 unless the voltage from X to ground is greater than 2.1 V. Since diodes dissipate less power than resistors, DTL gates dissipate less power than RTL

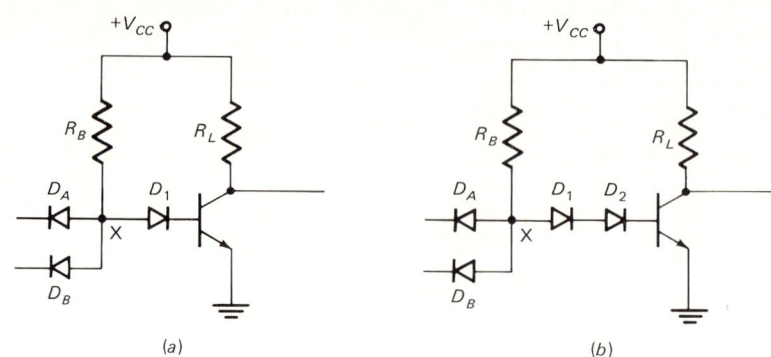

Figure 3.32
DTL NAND gates. (*a*) 1.4 V low; (*b*) 2.1 V low.

gates, specifically, about 10 mW versus 15 mW for RTL, but DTL gates switch in 30 ns, which is about twice as slow as RTL. In terms of trade-offs, improved noise immunity makes DTL more attractive than RTL.

The next improvement was the development of the multi-emitter transistor, which is a transistor containing two or more emitters. Q_1 in Fig. 3.33 is a two-emitter transistor and replaces D_A, D_B, and D_1 of the DTL NAND gate. On a silicon chip one transistor with two emitters takes up less room than three diodes. Multi-emitter gates are called *transistor-transistor logic* (TTL or T²L). If emitter A or emitter B is at ground, Q_1 is forward-biased. If both emitters are at +5 V, the base-emitter junction of Q_1 is cut off.

As shown in Fig. 3.34*a*, with at least one emitter at ground, the base-emitter junction of Q_1 is forward-biased, and Q_1 conducts into saturation. Saturation makes the voltage at the collector of Q_1 very low. Since the collector of Q_1 is the input of Q_2, Q_2 cuts off when Q_1 saturates. With Q_2 cut off, the output is high. The output will be high when emitter A, emitter B, or both are low.

Figure 3.34*b* shows both emitters of Q_1 connected to +5 V. The base-emitter junction of Q_1 is reverse-biased; therefore the base of Q_1 is at a higher voltage than the collector. This means the base-collector junction of Q_1 is forward-biased and current flows into the base of Q_2. Q_2 saturates and the output is about 0 V. The truth table is shown in Fig. 3.34*c*. This circuit is a 2-input TTL NAND gate.

Figure 3.33
Basic TTL NAND gate.

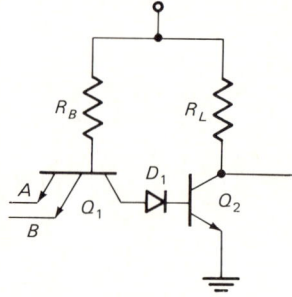

Figure 3.34 TTL NAND gate conditions. (*a*) At least one emitter low; (*b*) both emitters high; (*c*) truth table.

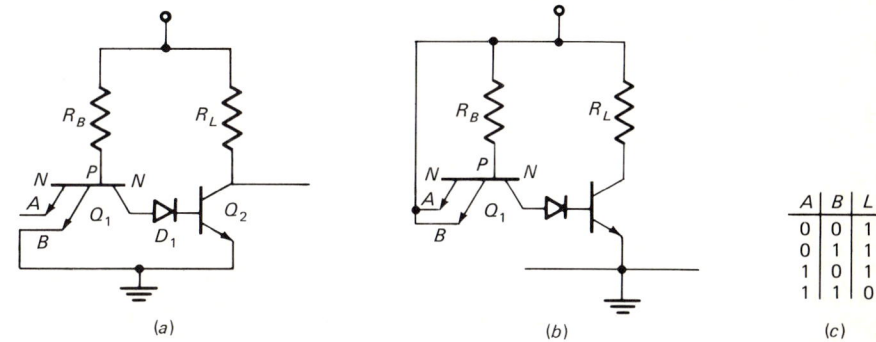

DTL and TTL NAND gates suffer from loading problems. Loading was introduced in Sec. 3.3. A high output is 5 V only if the load has a very high input impedance. But when the input impedance is the same as the gate output impedance, a logic high is much less than 5 V, and it becomes difficult to distinguish logic 1s from logic 0s.

Example 3.2 The Thevenin equivalent of a TTL NAND gate with a logic 1 output is approximated by the circuit in Fig. 3.35. Find the load voltage when the input impedance of the load is: (*a*) infinite; (*b*) 4 KΩ; (*c*) 400 Ω.

Solution This circuit is a voltage divider.

$$V_{load} = \frac{5 \times Z_{load}}{Z_{load} + 4000}$$

(*a*) For Z load = ∞

$$V_{load} = \frac{5 \times \infty}{\infty + 4000} = 5 \text{ V}$$

(*b*) For Z load = 4 KΩ

$$V_{load} = \frac{5 \times 4000}{4000 + 4000} = 2.5 \text{ V}$$

(*c*) For Z load = 400 Ω

$$V_{load} = \frac{5 \times 400}{400 + 4000} = 0.45 \text{ V}$$

Figure 3.35 Logic 1 TTL NAND equivalent circuit.

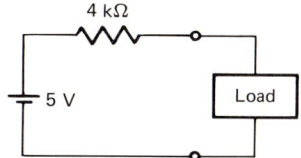

This example illustrates gate loading. When the load input impedance is infinite, a logic high is 5 V, but when the load input impedance is the same as the gate output impedance, the output is 2.5 V. This value is halfway between logic 1 and logic 0. It is difficult to interpret 2.5 V in terms of logic levels. The situation is even worse when the load has a much lower impedance than the gate. Thus, 400 Ω is equivalent to 10 loads of 4-kΩ input resistances in parallel. In this case what is supposed to be a 5-V logic high is only 0.45 V. Thus a TTL gate can possibly drive one other gate but no more.

Gate loading is improved by lowering gate output resistance. A DTL or TTL NAND gate has a high output impedance because the output stage is a common emitter. The output impedance of a common collector is much lower. Figure 3.36 shows the IC equivalent. Here Q_1 is still a multi-emitter input stage and Q_2, which was the output stage in the basic TTL NAND gate, is converted into a stage with two load resistors, R_2 and R_3. The output from collector resistor R_3 is out of phase with the input, and the output from emitter resistor R_2 is in phase. Q_2 is a phase splitter similar to those used in audio amplifiers. Transistor Q_4 is on top of Q_3. Because of its similarity to Indian carvings from the northwestern coast of the United States and Canada, this circuit is called a *totem pole*. A totem pole is the IC version of a discrete-transistor push-pull amplifier.

Q_1 saturates if either or both emitters of Q_1 are low. With Q_1 saturated, the input to Q_2 is low and Q_2 cuts off. When Q_2 is cut off, voltage at the collector of Q_2 is approximately V_{CC} and Q_4 conducts. Q_3 cannot conduct at the same time since there is no voltage across R_2 when Q_2 is cut off. With Q_4 conducting at saturation and Q_3 cut off, the output is taken from the emitter of Q_4 through D_1. The output is about 4.3 V, which is certainly a logic high. Thus when the output is a logic high, the output voltage is taken from an emitter, which means that the output has a common-collector configuration.

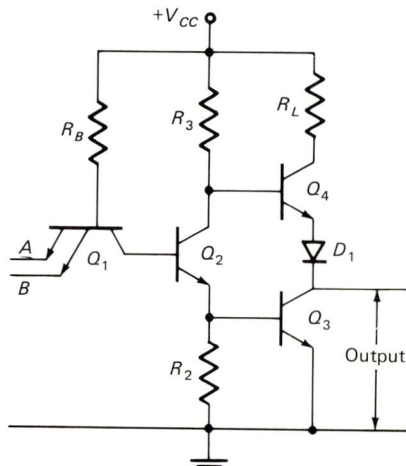

Figure 3.36 Totem-pole-output TTL NAND gate.

With totem pole output, a logic high gives a low output impedance. When both emitters of Q_1 are high, Q_1 cuts off. When Q_1 is cut off, Q_2 saturates. With Q_2 saturated, most of the supply voltage is across R_2, and Q_3 saturates. The voltage across R_3 is low enough to keep Q_4 cut off; since Q_3 is saturated, the output will be low. The truth table for a TTL NAND gate with totem pole is exactly the same as for one without totem pole.

Without totem pole, a logic 0 output is across a saturated transistor, which is also the case for a totem pole output. A saturated transistor has an output impedance of about 10 Ω. Without totem pole a logic 1 output is taken from a common emitter, which corresponds to a high output impedance, but with totem pole the common-collector output impedance is about 80 Ω. Thus, with totem pole, the output impedance is low for logic 1 and logic 0.

Example 3.3 Except for the lower output impedance, the Thevenin equivalent of a totem-pole TTL NAND gate is the same as shown in Fig. 3.35. Find the load voltage when the input impedance of the load is: (a) infinite; (b) 4 kΩ; (c) 400 Ω.

Solution As in Example 3.2, this circuit is a voltage divider. Therefore

$$V_{load} = \frac{4.3 \times Z_{load}}{Z_{load} + 80}$$

(a) For $Z_{load} = \infty$

$$V_{load} = \frac{4.3 \times \infty}{\infty + 80} = 4.3 \text{ V}$$

(b) For $Z_{load} = 4 \text{ k}\Omega$

$$V_{load} = \frac{4.3 \times 4000}{4000 + 80} = 4.22 \text{ V}$$

(c) For $Z_{load} = 400 \text{ }\Omega$

$$V_{load} = \frac{4.3 \times 400}{400 + 80} = 3.58 \text{ V}$$

Comparing Examples 3.2 and 3.3 demonstrates the advantage of low output impedance. Without totem pole a logic high can barely be connected to a single gate. With totem pole a logic high gate can realistically drive 10 gates.

An explanation is still needed for D_1 in the totem pole configuration. When totem pole output is low, Q_2 and Q_3 are saturated, and thus the collectors of Q_2 and Q_3 are low. Under these conditions it is possible for Q_4 to become forward-biased, but the voltage needed to forward-bias D_1 is not

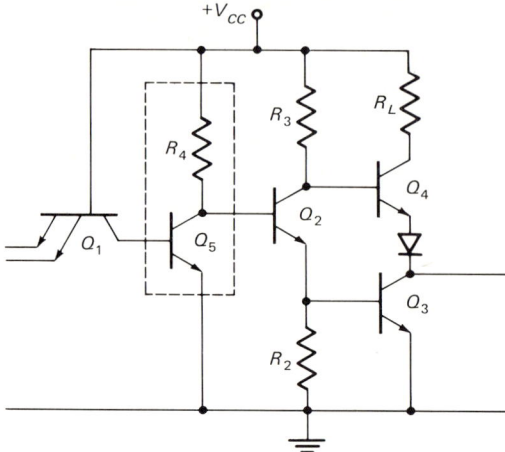

Figure 3.37 TTL AND gate.

available. However, there is insufficient voltage to forward-bias D_1 when Q_4 is low, and thus Q_4 cannot accidentally turn on when Q_3 saturates.

Another advantage of totem pole is faster switching. Switching delay is a result of capacitors, both transistor junction capacitance and stray wiring capacitance. These capacitances combine with circuit resistance to form RC networks. An RC circuit must charge and discharge before a transistor switches on and off; output-stage saturation resistance is especially important. Since totem-pole output decreases the output resistance, the output time constant is smaller.

Open-collector gates cannot have totem-pole outputs. Therefore shunt capacitance of the pull-up resistor, along with associated wiring capacitance, make open-collector gates slower. On the other hand, open-collector gates are better when multiple inputs are required. It is rare when a single device is superior in all respects.

A TTL NAND gate with totem-pole output is the basic gate for all TTL gates. Other TTL gates employ modifications of the basic NAND. As shown in Fig. 3.37, a TTL AND gate is formed by adding an inverter to the NAND. With the addition of the inverter stage Q_5 ahead of the phase splitter, the operation is reversed, and NAND converts to AND. Adding Q_5 after the totem pole would also cause inversion, but the advantage of totem pole would be destroyed.

A TTL NOR gate also uses a phase splitter and totem pole output. In a NOR gate two or more single emitter transistors are the input stage, and all input transistors are connected to the phase splitter. Component values result in saturation of the lower totem pole transistor unless all inputs are grounded. Similarly, an inverter in front of the phase splitter converts a NOR into an OR gate.

Low totem-pole-output resistance improves switching time, but further improvement is possible. Most of the delay in a transistor switch occurs when

Figure 3.38
Diode clamps. (*a*)
Germanium diode; (*b*)
Schottky diode; (*c*)
Schottky clamp symbol.

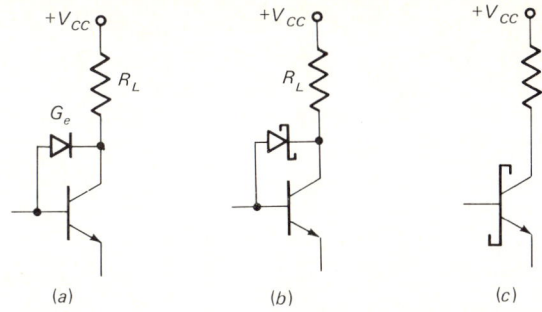

a transistor goes in and out of saturation. If a transistor reaches a low but unsaturated voltage, the time delay is much smaller. In Fig. 3.38*a* a germanium diode is connected between base and collector. The barrier voltage of a germanium diode is about 0.3 V, which is less than half the value for silicon. When a transistor is cut off, the diode has no effect, but when the transistor is switched on, voltage at the collector is clamped to the difference between base-emitter voltage and germanium diode voltage. Thus, the lowest collector-to-ground voltage is approximately 0.4 V, which is low enough for logic 0 but high enough to keep a transistor out of saturation. Preventing saturation by germanium diode clamping is a valid approach. Although it is not possible to deposit a germanium diode on a silicon IC, it is possible to make silicon diodes with the same barrier voltage as germanium. These diodes are called *Schottky*, and the symbol for a Schottky diode clamp is shown in Fig. 3.38*b*. The symbol for a transistor combined with a Schottky diode clamp is shown in Fig. 3.38*c*. A TTL gate with totem pole output using Schottky clamping is shown in Fig. 3.39. It is the same circuit as before except

Figure 3.39
Schottky-clamped TTL
NAND gate.

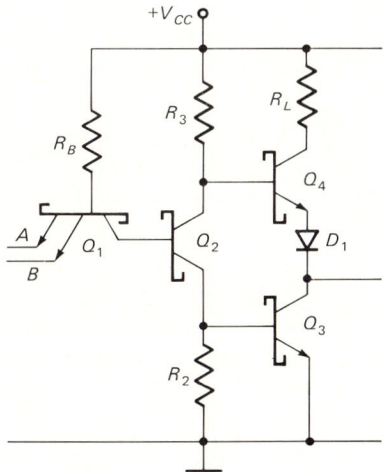

that clamped transistors are used. A Schottky gate is more than three times faster than an unclamped gate.

3.7 MOS GATES

Bipolar transistors inject current into a forward-biased diode and extract current from a reverse-biased diode. *Metal-oxide semiconductor field-effect transistors* (MOSFETs) operate on a different principle. The body of the MOSFET shown in Fig. 3.40a is *p*-doped and contains two regions of *n*-doped material. A layer of silicon dioxide, or glass, is grown over the MOSFET, and then a layer of aluminum is evaporated on top of the silicon dioxide. One *n*-doped region is connected to the body, and leads are attached to both *n*-doped regions and also to the aluminum layer. The *n*-doped region connected to the *p*-doped body is called the *source* (S) and the other *n*-doped region is called the *drain* (D). The aluminum layer is called a *gate* (G). Figure 3.40b shows a MOSFET symbol. Figure 3.40c is a simplified version.

With the source grounded, a positive voltage V_{DD} is connected to the drain. This reverse-biases the *n*-drain with respect to the *p*-body. With no voltage at the gate, only leakage current flows from drain to source. Zero voltage from gate to source, $V_{GS} = 0$, results in a cutoff MOSFET.

When V_{GS} is made positive, an electric field exists in the body of the MOSFET directly under the oxide layer between the drain and source. Since V_{GS} is positive, the induced electric field attracts electrons and repels holes. As V_{GS} increases, the electric field between drain and source attracts more electrons and repels more holes. This process continues until the drain-to-source *p*-doped region actually becomes a channel of *n*-material. When this *n*-channel is formed, drain current I_D flows from drain to source. Since conduction occurs in the formed *n*-channel, the device is called an *n-channel MOSFET*. The value of V_{GS} needed to initiate drain current is the threshold voltage (V_{th}). An *n*-channel only occurs when V_{GS} is greater than V_{th}.

MOSFET operation is different from a bipolar transistor in that no input MOSFET current exists but instead an input voltage controls an electric field, which in turn controls the output current. Since a MOSFET has no input current, the input resistance of a MOSFET is infinite. As a practical matter, the input resistance of a MOSFET, while not infinite, is extremely high. Typical MOSFET input resistance is in the 10^{12} Ω range.

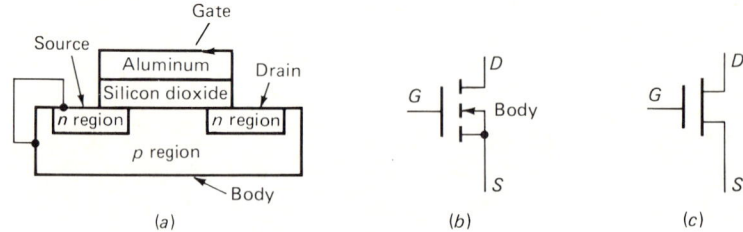

Figure 3.40 MOSFET device. (*a*) Structure; (*b*) symbol; (*c*) simplified symbol.

When V_{GS} varies, I_D varies accordingly. This output current variation, along with the extremely high input resistance, makes the MOSFET an almost perfect voltage amplifier. Unfortunately, the linear range of amplification is quite small, and MOSFETs are restricted to single-stage front-end amplifiers. In computers MOSFETs are used as switches. The small linear amplification range of a MOSFET is not a limitation.

A MOSFET only uses about one-fifth as much chip area as a bipolar transistor; thus, 5 MOSFETs occupy the same area as a single bipolar transistor. Obviously, more functions can be performed in the same space if MOSFETs replace bipolar transistors. Also, MOSFETs dissipate less power. On the other hand, MOSFETs are slower than bipolar transistors. MOSFET input is a metal layer–to–insulator semiconductor sandwich, and this is a capacitor. Input capacitance combined with high input resistance results in a long time constant circuit. The input capacitance has to be charged before a MOSFET switches on and discharged before a MOSFET switches off. Bipolar transistors are faster because input resistance is much smaller.

A MOSFET has a threshold voltage of about 1 V and can operate from a 5-V supply. Typical load resistances are several thousand ohms. With such values, cutoff voltage is about 4 V and saturation voltage is about 0.4 V. These values are in the same range as those of TTL circuits, and many MOSFET circuits can be connected directly to TTL circuits. However, there are MOSFETs with higher voltages and still others which require negative voltage. In these cases voltage translation circuits are required to interface with TTL.

IC resistors use more chip area than transistors. Since MOSFETs are small to begin with, it is practical to use them as resistors. This is accomplished by driving a MOSFET at constant bias. With V_{GS} at a fixed value, I_D is constant, and therefore V_{DS} is constant. Thus the ratio of V_{DS} to I_D is constant and represents a fixed resistor. In Fig. 3.41a the gate of Q_1 is directly connected to V_{DD}. Therefore Q_1 operates at constant bias and acts as a fixed resistor. The source of Q_1 is connected to the drain of Q_A. As shown in Fig. 3.41b, Q_1 is the load resistor for Q_A.

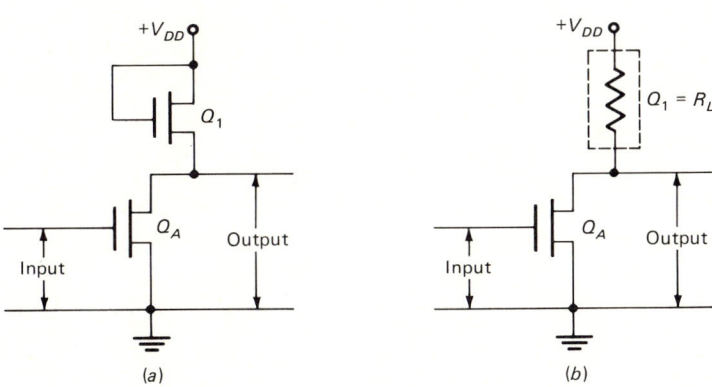

Figure 3.41 MOSFET inverter. (*a*) Actual circuit; (*b*) electrical equivalent.

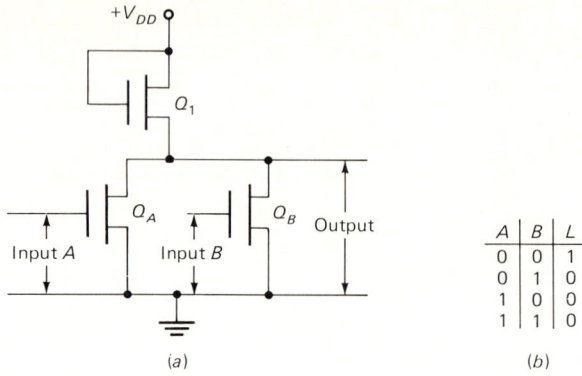

Figure 3.42
MOSFET NOR gate. (*a*) Circuit; (*b*) truth table.

V_{GS} of Q_A is the input of this circuit, and V_{DS} of Q_A is the output. When V_{GS} is less than threshold, Q_A is cut off and the output is logic 1. When V_{GS} is greater than threshold, Q_A saturates, and the output is logic 0. A low input results in a high output and vice versa. The circuit shown in Fig. 3.41*b* is the basic MOSFET inverter.

MOSFET gates do not require totem-pole outputs. Consider the circuit shown in Fig. 3.42. Q_A and Q_B are connected in parallel and the gates are inputs. With both Q_A and Q_B cut off, no current flows through Q_1, which is the load resistor. Since there is no voltage drop across Q_1, the output voltage is approximately V_{DD}. Thus, when A and B are both at logic 0, the output is at logic 1.

If either input is high while the other is low, the high MOSFET saturates. When one MOSFET saturates, Q_1 conducts and the output is approximately 0 V. Thus when either A or B is logic 1, the output is logic 0. Similarly, when both inputs are high, the output is low. The truth table for these conditions is shown in Fig. 3.42*b*. This circuit is a 2-input NOR.

In a different circuit, shown in Fig. 3.43*a*, the inputs are connected in series. When the inputs to A and B are both low, Q_A and Q_B are cut off. With both Q_A and Q_B cut off, Q_1 cannot conduct and the output is high. Thus, when both inputs are at logic 0, the output is a + logic 1. If either input is high while the other is low, the low still results in cutoff. With either Q_A or Q_B cut off, Q_1 cannot conduct, since the MOSFETs are connected in series. Thus if only A or only B is logic 1, the output is logic 1. Q_1 will only conduct when A and B are both high. With both Q_A and Q_B on, Q_1 saturates and the output is low. Therefore, the only time the circuit is at logic 0 occurs when both inputs are at logic 1. The truth table is shown in Fig. 3.43*b*, and this is a NAND gate.

Switches in series and parallel introduced this chapter. High input impedance permits MOSFETs to be connected in series and parallel to construct practical gates. However, logic symbols are independent of internal structure.

Because of the long time constant, MOSFETs are slower than bipolar

Figure 3.43
MOSFET NAND gate. (*a*) Circuit; (*b*) truth table.

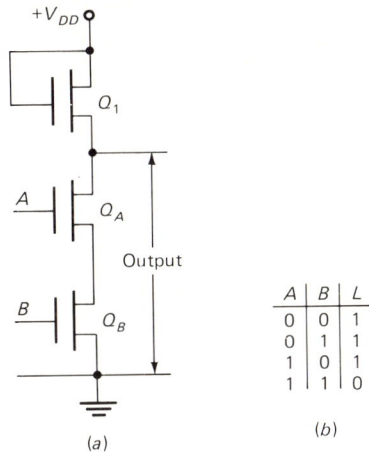

A	B	L
0	0	1
0	1	1
1	0	1
1	1	0

devices. However, the long time constant can be used to advantage in *dynamic* logic. Previous gates used *static* logic, in which the power supply is always on and gate outputs are available on a continuous basis. Figure 3.44 shows a dynamic gate.

As in the previous MOSFET inverter, Q_A is an inverter and Q_1 is the load

Figure 3.44
Dynamic MOSFET inverter. (*a*) Circuit; (*b*) waveforms.

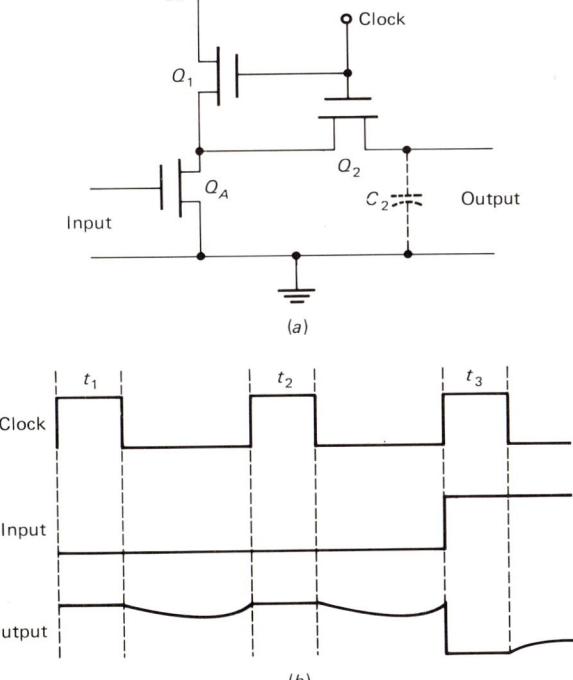

resistor, but in this case the gate of Q_1 is not connected to V_{DD}. Instead, the gates of Q_1 and Q_2 are connected to the terminal called *clock*. A clock is a train of periodic pulses. When the clock is low, Q_1 and Q_2 are cut off, and when the clock is high, Q_1 and Q_2 can conduct. Whether or not Q_1 conducts when the clock is high depends on the input to the inverter Q_A.

Consider the conditions shown in Fig. 3.44b. During clock pulse t_1 the input to Q_A is low. Although the clock is high during t_1, Q_A is cut off since the input is low. The output of Q_A is high and this high is passed through Q_2 since clock pulse t_1 has turned Q_2 on. Q_2 is the output stage. The schematic also shows the gate capacitance of Q_2. During the interval between t_1 and t_2, the clock is off. However, the output remains high during this interval because the long time constant of Q_2 maintains charge on the capacitance. During t_2 the input still happens to be low, and during this time the small discharge of the output capacitance of Q_2 will be recharged. Using the clock to recharge is called *refreshing* the output. During clock pulse t_3 the inverter input happens to be high. The output goes low and stays low through successive refreshings until the input goes low again. The MOSFET's long time constant has become an asset. The advantage of dynamic gating is lower power consumption as power is only dissipated when Q_1 is on. This is also the time when Q_2 refreshes the output. Dynamic gating is a sampling method of handling data since the clock must be on. Dynamic metal-oxide semiconductor (MOS) gating is not limited to inverters; refreshing also works for NAND, NOR, and the other logic gates. In fact, dynamic gating is used with most MOSFET ICs.

Although the static and dynamic MOSFET gates have been *n*-channel, *p*-channel MOSFETs are equally feasible. In these the body is *n*-doped while the source and drain regions are both *p*-doped. The operation of the *p*-channel MOSFET is the same as for the *n*-channel except that the voltage polarities are reversed. Zero volts is still off but power and threshold voltages are negative; typical V_{DD} values range from -5 to -15 V. Level shifting circuits are required to make *p*-channel gates compatible with TTL gates, so that it is easier, and therefore less expensive, to manufacture *p*-channel gates. However, *n*-channel gates are approximately three times faster and are TTL voltage-compatible. For simplicity *n*- and *p*-channel MOSFETs have the symbol which is shown in Fig. 3.45c. Power supply polarity indicates whether *n*- or *p*-channel MOSFETs are used.

It is possible to combine both types into a single transistor, which is a *complementary MOS* (CMOS or COSMOS) device. Both gates and both

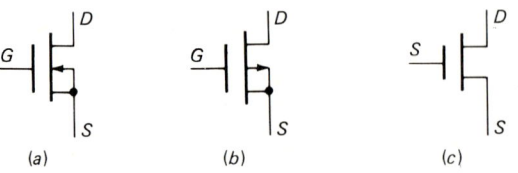

Figure 3.45 MOSFET types. (*a*) *n*-channel; (*b*) *p*-channel; (*c*) general-purpose symbol.

Figure 3.46
CMOS inverter. (*a*) Circuit; (*b*) simplified drawing.

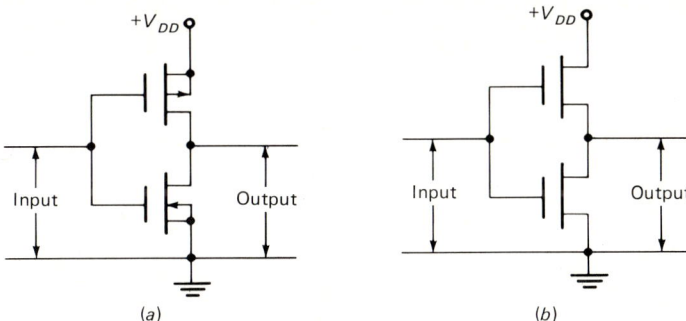

drains are connected together, as shown in Fig. 3.46. The source of the *n*-channel unit is connected to ground and the *p*-channel unit is connected to positive V_{DD}. Values of V_{DD} range from $+5$ to $+15$ V. When the input voltage is zero, the *n*-channel MOSFET is cut off. Therefore V_{DD} exists between the source and gate of the *p*-channel MOSFET. This causes the *p*-channel MOSFET to saturate and the output voltage to equal approximately V_{DD}. Thus with logic 0 input, the output is logic 1.

The circuit shown in Fig. 3.46 is a CMOS inverter. In a CMOS the off transistor has an extremely high resistance, and drain current is in the nanoampere range and CMOS gate power dissipation in the nanowatt range. This makes CMOS gates extremely attractive in battery-operated devices such as digital wristwatches and pocket calculators.

3.8 GATE COMPARISON

IC packaging is standard. The most common configuration is the *dual in-line package* (DIP). DIPs are rectangular with parallel rows of leads on the longer sides. Figure 3.47 shows a 14-pin DIP with numbered leads; other standard lead numbers are 16, 24, and 40. Typically, type number, date of manufacture, and the manufacturer's logo are printed on top. DIPs are available in either plastic or ceramic packages. Plastic satisfies commercial requirements, but

Figure 3.47 DIP.

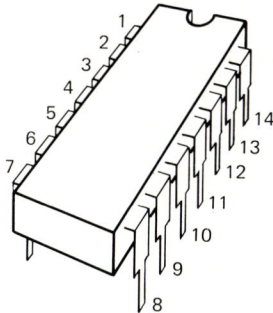

TABLE 3.6 Plastic DIP Characteristics

Leads	L, in	W, in	Wt, oz
14	0.75	0.25	0.03
16	0.75	0.25	0.03
24	1.25	0.55	0.17
40	2.0	0.60	0.25

the more rigorous military applications require hermetically sealed ceramic packages. Table 3.6 shows plastic DIP characteristics. Spacing between leads is 0.1 in. Ceramic DIPs are more expensive and slightly larger. The DIP configuration simplifies insertion and soldering on printed circuit boards.

One DIP may contain a single gate or upward of 10,000 gates. Packaging this much electronic instrumentation into small packages is a miracle of modern technology. ICs make small computers possible.

DIP packages are classified according to gate number.

- SSI (small scale integration) refers to DIPs containing between 1 and 12 gates.
- MSI (medium scale integration) is for 12 to 100 gates.
- LSI (large scale integration) is for more than 100 gates.

TTL and CMOS gates are mainly SSI and MSI. Most LSI packages are MOSFET. Two-input gates are SSI; for example, a 2-input NAND gate has 3 logic leads, 2 input and 1 output. Four independent NAND gates require 12 logic leads. These gates are all contained on a single silicon chip, which connects common power supply leads to all four gates. With 12 logic leads and 2 power supply leads, four 2-input NAND gates exactly fit into a 14-lead DIP. This SSI DIP is called a *quadruple* or quad 2-input NAND gate and the layout is shown in Fig. 3.48. Layout is the same regardless of manufacturer.

A quad 2-input NAND chip, including the area required for attaching the leads, occupied 0.05 in × 0.06 in, which is only 1.5 percent of the entire area

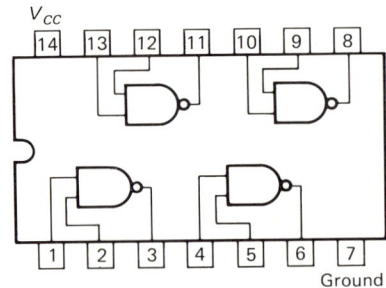

Figure 3.48 Quad 2-input NAND package.

of a 14-lead DIP. Increasing gate count does not change the area required for attaching leads very much. When chip area is doubled, the number of gates is tripled. This makes 14-pin MSI and LSI possible. DIPs with more leads will also be discussed.

Packages are standard but gate families vary considerably; TTL, MOSFET, and CMOS are based on different principles and have different characteristics. If a single gate type possessed all the advantages, it would not be necessary to discuss the others. If TTL were the fastest, smallest, used the least power, etc., all computers would use TTL exclusively.

This is not the case. Depending on application, characteristics such as speed and power consumption are traded. For example, switching speed is insignificant in a digital wristwatch, but low power drain increases battery life. On the other hand, speed is the most important characteristic of large computers, and increased power dissipation is willingly exchanged for fast switching. Trade-offs must be made. In some computers it is even advantageous to perform different functions with different gate types. Gate parameters affect each other, but it is possible to distinguish four categories: loading, power dissipation, propagation delay, and noise immunity.

Loading Impedance determines how many gates can be driven from a single gate. The maximum number of gates which can be driven from one gate is called *fan-out*. Since conditions are different for logic 0 and logic 1, fan-out also is different for logic 0 and logic 1. Figure 3.49a shows the driving gate at logic 0. Gates $B, C, D, \ldots, N$ are driven to ground through gate A. The sum of these currents flows through the saturation resistance of gate A. In this case the driving gate is a current *sink* for the driven gates. The value of the sink current is the output low current (I_{OL}). Excessive sink current causes overheating and increases the output voltage. In Fig. 3.49b, the driving gate is at logic 1. The output resistance is different, and gate A is a current *source*, so that source current flows from gate A to the driven gates. I_{OH} is the output high current.

Fan-in refers to the input levels and will also be different for logic 0 and logic 1. The input high current (I_{IH}) flows into a driven gate at a logic high. The input low current (I_{IL}) flows from a driven gate at logic low. Gates are

Figure 3.49
Gate loading. (*a*) Logic 0; (*b*) logic 1.

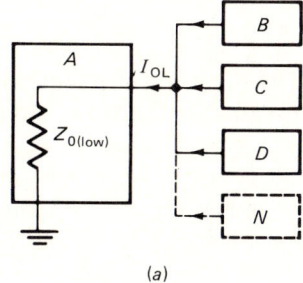
(a)

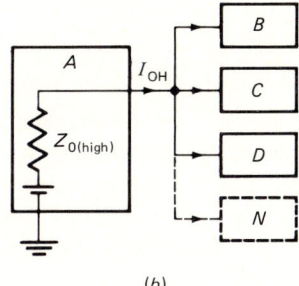
(b)

TABLE 3.7
TTL Load Currents

Current	Value
I_{IL}	1.6 mA
I_{IH}	40 µA
I_{OL}	16 mA
I_{OH}	800 µA

rated in terms of unit load (UL), which is the number of gates that can be driven:

$$UL_{high} = \frac{I_{OH}}{I_{IH}} \tag{3.8}$$

and

$$UL_{low} = \frac{I_{OL}}{I_{IL}} \tag{3.9}$$

These current levels for standard TTL gates with totem-pole output are shown in Table 3.7. Knowing the input and output currents, we can determine the UL ratings.

Example 3.4 Find the (a) high and (b) low fan-out of a standard TTL gate.

Solution (a) $UL_{high} = \dfrac{I_{OH}}{I_{IH}} = \dfrac{800 \; \mu A}{40 \; \mu A} = 20 \; UL$

(b) $UL_{low} = \dfrac{I_{OL}}{I_{IL}} = \dfrac{16 \; mA}{1.6 \; mA} = 10 \; UL$

This example shows that twice as many gates can be driven at logic high as at logic low. However, logic high and logic low are equally likely. Therefore the smaller number is the limiting value. Table 3.8 compares standard TTL UL rating with the Schottky equivalents.

TABLE 3.8
UL Ratings

UL	Standard	Schottky
High	20	25
Low	10	12.5

Since MOSFET and CMOS gates have higher input impedance, their UL values are determined more by propagation delay and noise immunity than by resistive loading.

Power dissipation Heat is dissipated when gates function. Gate type and packaging density affect temperature rise. Methods used to reduce heating are, in order of increasing complexity: conventional heat conduction, forced air cooling using fans, and air conditioning. The cost of keeping temperature rise within acceptable limits must be considered, so that the power dissipation of each gate type is a factor in making a final selection.

Since current is different when a gate is high and when it is low, power dissipation is also different. Gates are compared on an average power (P_{av}) basis. Average power assumes logic high and low are equally probable:

$$P_{av} = \frac{P_H + P_L}{2} \tag{3.10}$$

where P_H is the power dissipated when the gate is at logic high and P_L is the power dissipated when the gate is at logic low.

Example 3.5 A quad 2-input NAND gate operates from a 5-V power supply. The current drawn by each gate is 1 mA when the gate is high and 3 mA when the gate is low. Find: (a) the average power dissipated by each gate; (b) the average power dissipated by the entire DIP.

Solution (a) Since $P = VI$

$$P_H = V_{CC}I_H = 5 \times 1 \times 10^{-3} = 5 \text{ mW}$$

$$P_{V_L} = V_{CC}I_L = 5 \times 3 \times 10^{-3} = 15 \text{ mW}$$

and since

$$P_{av} = \frac{P_H + P_L}{2} = \frac{(5 + 15) \text{ mW}}{2}$$

$$P_{av} = 10 \text{ mW}$$

(b) Since there are 4 gates per chip

$$P_{chip} = 4 \times 10 \text{ mW} = 40 \text{ mW}$$

Table 3.9 compares gates in terms of power supply voltage and per gate power dissipation. These values are measured at 25°C. Temperature variation causes power dissipation to fluctuate. Variation in power supply voltage also affects power dissipation. Information of this type is listed in specification sheets.

TABLE 3.9 Gate Dissipation

Type	Voltage	P_{av}, mW
TTL	+5	10.0
Schottky TTL	+5	22.5
n-MOSFET	+5	1.0
p-MOSFET	−12	1.7
CMOS	+5	0.05

Figure 3.50
Pulse distortion. (*a*) Circuit; (*b*) waveforms.

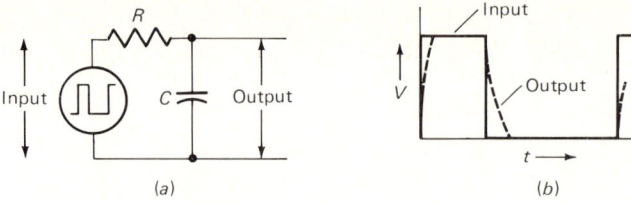

Propagation delay A pulse generator connected to an *RC* circuit is shown in Fig. 3.50 (*R* is the transistor output impedance, and *C* is a combination of transistor capacitance and stray wiring capacitance). If the *RC* time constant is much shorter than the *on* time of the pulse, the output voltage is a reasonable reproduction of the input. The situation is more complicated when transistors are involved. Since these capacitances must be charged and discharged for switching, transistor charge storage affects the input and output waveforms.

Charge storage has a different effect on turn-on and turn-off. Therefore, on and off delays are different. Also, if a common emitter is used, input and output waveforms are out of phase. Figure 3.51 shows an inverter input and the output waveforms. At an input between 0 and 5 V the output begins to go low. The time between input rising and output dropping is t_{PHL}, the time for the output pulse to go from high to low. A similar situation occurs when the input returns to low; there is a delay until the output returns to high. This time is t_{PLH}, the time for the output pulse to go from low to high. The *propagation delay* t_{pd} is the average of t_{PHL} and t_{PLH}:

$$t_{pd} = \frac{t_{PHL} + t_{PLH}}{2} \tag{3.11}$$

Figure 3.51
Inverter time delays.

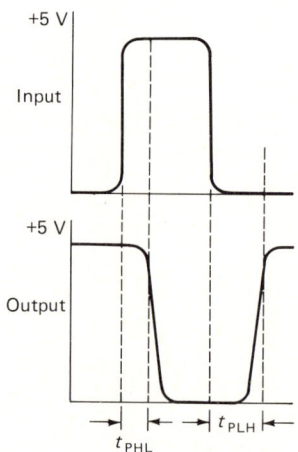

Example 3.6 Worst-case time delays for the standard TTL NAND gate are $t_{PHL} = 15$ ns and $t_{PLH} = 22$ ns.
Find the propagation delay.

Solution
$$t_{pd} = \frac{t_{PHL} + t_{PLH}}{2} = \frac{(15 + 22) \text{ ns}}{2} = 18.5 \text{ ns}$$

The Schottky equivalent is about twice as fast. As described, Schottky gates are faster because no saturation occurs. However, because of this a Schottky gate dissipates more power. The *speed-power product* (spp) is a method used in comparing gates:

$$\text{spp} = t_{pd} \cdot P_{av} \qquad (3.12)$$

Time is measured in seconds and power in watts. Since watts are joules per second, the dimension of speed-power product is:

$$\text{spp} = \text{seconds} \times \frac{\text{joules}}{\text{seconds}} = \text{joules}$$

Example 3.7 Find the speed-power product of a standard TTL NAND gate.

Solution From Example 3.6, $t_{pd} = 18.5$ ns
From Example 3.5, $P_{av} = 10$ mW

$$\text{spp} = t_{pd} \times P_{av} = 18.5 \times 10^{-9} \times 10 \times 10^{-3} = 185 \times 10^{-12} \text{ J}$$

The spp for the Schottky equivalent is about the same as for the standard TTL NAND gate. This means that speed improves at the same rate that power dissipation increases.

The value of t_{pd} is for a single gate. When several gates are connected in sequence, total propagation delay is the sum of the individual gate delays. Another problem involves one gate driving several gates, which increases load capacitance, and this in turn increases propagation delay. Also, since additional current is required to drive several loads, power dissipation increases. Loading, power dissipation, and propagation delay are interrelated. Table 3.10 gives a comparison.

TABLE 3.10 Gate Parameters

Type	P_{av}, mW	t_{pd}, ns	Fan-out, UL
TTL	10.0	100	10
Schottky TTL	22.5	4.75	12.5
n-MOSFET	1.0	100	30
p-MOSFET	1.7	300	10
CMOS	0.05	25	10

Figure 3.52 Gate voltage levels.

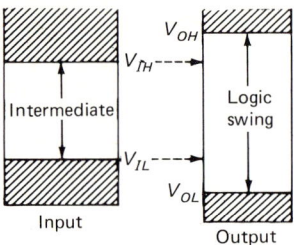

Noise immunity The ability of a gate to resist triggering on false signals is called noise immunity. False signals are a result of stray electric and magnetic fields which induce voltage on signal lines. Careful DIP layout and shielding reduce false triggering.

Noise immunity is determined by high and low logic voltages. If a gate logic low level is 0.2 V or less and the input stage is an *npn* base-emitter junction, the output switches when the input is about 0.6 V. This gate withstands an input noise pulse of less than 0.4 V.

Figure 3.52 shows noise immunity as a graph. Voltages are given in general terms and apply to all gates. On the input graph the input low voltage (V_{IL}) is the maximum input voltage which maintains a logic low. The minimum input voltage which can maintain a logic high is V_{IH}. In a good design the voltage range between V_{IH} and V_{IL} is avoided, since an intermediate voltage represents either logic state, and there is no way to interpret the output.

At the output V_{OL} is the maximum output low voltage and V_{OH} is the minimum output high voltage. The voltage range between V_{OH} and V_{OL} is the *logic swing*. For reliable operation the output levels must be nested within the corresponding input levels:

- V_{IL} must be greater than V_{OL}
- V_{OH} must be greater than V_{IH}

The dashed lines in Fig. 3.52 show these requirements. Voltages between these values are the safety margin. These values of noise voltage can be tolerated without false triggering. Frequently, the safety margin is called *noise margin* (NM). NM_L, defined by

$$NM_L = V_{IL} - V_{OL} \tag{3.13}$$

is the low noise margin and

$$NM_H = V_{OH} - V_{IH} \tag{3.14}$$

is the high noise margin.

TABLE 3.11 Gate Comparison

Type	P_{av}, mW	t_{pd}, ns	Fan-out	NM_H, V	NM_L, V
TTL	10.0	10	10	1.9	1.0
Schottky TTL	22.5	4.75	12.5	1.9	0.9
n-MOSFET	1.0	100	30	4.0	3.0
p-MOSFET	1.7	300	10	4.0	3.0
CMOS	0.05	25	10	4.5	4.5

Example 3.8 For the standard TTL NAND, typical voltages are $V_{IH} = 1.4$ V, $V_{IL} = 1.2$ V, $V_{OH} = 3.1$ V, and $V_{OL} = 0.2$ V. Find: (a) the low noise margin; (b) the high noise margin.

Solution (a) $NM_L = V_{IL} - V_{OL}$

$= 1.2 - 0.2 = 1.0$ V

(b) $NM_H = V_{OH} - V_{IH}$

$= 3.1 - 1.4 = 1.9$ V

Increasing fan-out lowers V_{OH} and raises V_{OL}; therefore excessive fan-out decreases noise margin. As described, power dissipation, propagation delay, and fan-out are related and although it is convenient to separate these quantities for discussion, all should be considered simultaneously. Selection of gate type requires trade-offs. Table 3.11 compares these quantities by showing typical values. Variations exist, and it is possible to improve one parameter at the expense of another.

SUMMARY

1. The switch is the basic computer circuit. Transistors switch rapidly and are useful electronic switches.

2. Gates are switches with one or more inputs and only one output. The AND-gate output is high only when all inputs are high.

3. An OR gate has a high output if any input is high. Gate input/output relationships are described by logic equations, logic symbols, truth tables, and timing diagrams.

4. The NOT gate, or inverter, has an output which is opposite to the input. AND, OR, and NOT gates are the basic types.

5. Three-state gates can have logic high or logic low outputs. Also, the output of a three-state gate can be an open circuit.

134 Computer Circuit Concepts

6. TTL gates use multi-emitter inputs and totem pole outputs. These gates have a low output impedance for both logic 1 and logic 0.

7. MOSFET gates are smaller and dissipate less power. Because of their long time constant, MOSFETs can be used for static or dynamic switching.

8. The important characteristics in selecting a gate type are fan-out, power dissipation, propagation delay, and noise margin. No gate type is superior in all characteristics and trade-offs are required.

PROBLEMS

3.1 A transistor with a 0.5-µA leakage current and a 15-mA saturation current has a 45-mV saturation voltage. This transistor is connected to a 5-V supply. Find: (a) the transistor *off* resistance; (b) the transistor *on* resistance; (c) the load resistor.

Answer: (a) 10 MΩ; (b) 3 Ω; (c) 333 Ω.

3.2 A switch is connected in series to a 100-V supply and a 500 Ω resistor. When the switch is open, the voltage across the load is 0.05 µV. When the switch is closed, the voltage across the load is 99.994 V. Find: (a) the *off* resistance and (b) the *on* resistance of the switch.

3.3 Using data given in Table 3.1 find: (a) the ratio of *off* resistance between a relay and a transistor; (b) the ratio of *on* resistance between a relay and a transistor; (c) the ratio of switching speed between a relay and a transistor.

Answer: (a) $10^6:1$; (b) $6.67 \times 10^{-3}:1$; (c) $>30 \times 10^6:1$.

3.4 The speed of light is approximately 186,000 mi/s. How many feet does light travel in 1 ns?

3.5 Show a truth table for a 2-input AND gate in which open (O) and closed (C) represent switch conditions, and on and off represent the lamp conditions.

Answer:

A	B	Lamp
O	O	Off
O	C	Off
C	O	Off
C	C	On

3.6 In terms of 1 and 0 prepare a truth table showing the condition for a 2-input OR and a 2-input AND gate.

3.7 Show a timing diagram for a 2-input AND gate.

Answer:

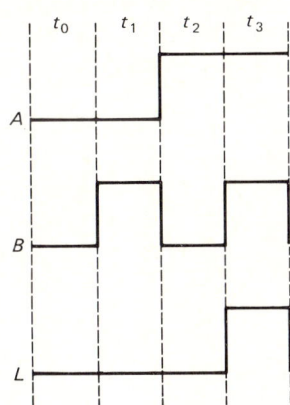

3.8 Show the conditions for 3-input OR and AND gates in terms of 1 and 0.

3.9 A chain of diode OR gates receives signals which are 5 V or 0 V. If the diodes have forward-biased voltage drops of 0.7 V, find the voltage output of (*a*) the first gate, (*b*) the second gate, and (*c*) the third gate.

Answer: (*a*) 4.3 V (*b*) 3.6 V (*c*) 2.9 V.

3.10 Show how a pair of 2-input AND gates can be connected to form a 3-input AND gate.

3.11 Describe the operation of the AND gate shown in Fig. 3.14 by considering all the possible input voltage conditions, assuming 5 V is high and 0 V is low.

Answer:

A	B	L
0	0	0.6
0	5	0.6
5	0	0.6
5	5	5.0

3.12 Show a schematic for a binary-to-decimal decoder.

3.13 Show two methods for connecting a 2-input NAND gate as an inverter.

Answer:

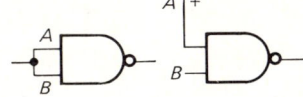

Either tie both inputs together or connect one input to the positive power supply terminal.

3.14 Show the schematic for a 1-of-8 decoder.

3.15 Show the truth table for a 3-input NAND gate.

Answer:

A	B	C	AND	NAND
0	0	0	0	1
0	0	1	0	1
0	1	0	0	1
0	1	1	0	1
1	0	0	0	1
1	0	1	0	1
1	1	0	0	1
1	1	1	1	0

3.16 Show two methods for connecting a 2-input NOR gate as an inverter.

3.17 Show a truth table for a 3-input NOR gate.

Answer:

A	B	C	OR	NOR
0	0	0	0	1
0	0	1	1	0
0	1	0	1	0
0	1	1	1	0
1	0	0	1	0
1	0	1	1	0
1	1	0	1	0
1	1	1	1	0

3.18 Show the symbol and truth table for a three-state gate with an inverted output which is enabled with a logic 1.

3.19 Show the symbol and truth table for a three-state gate with an inverted output which is enabled with a logic 0.

Answer:

Enable	Input	Output
0	0	1
0	1	0
1	0	high Z
1	1	high Z

3.20 Show the (*a*) schematic, (*b*) Boolean equation, and (*c*) truth table when two open-collector AND gates are connected together.

3.21 How does the circuit of Prob. 3.20 compare with a conventional 4-input NAND gate?

Answer: It is the same.

3.22 A 5-V TTL NAND gate has a 6 kΩ output impedance. Find the load voltage when the load impedance is (*a*) 4 kΩ; (*b*) 400 Ω.

3.23 A 5-V TTL NAND gate with a totem pole output stage has a 50-Ω output impedance. Find the load voltage when the load impedance is: (*a*) 4 kΩ; (*b*) 400Ω.

Answer: (*a*) 4.94 V; (*b*) 4.44 V.

3.24 Describe how a TTL NAND gate with totem pole output can be used as an inverter.

3.25 Five thousand MOSFETs can fit on a silicon chip which is 0.15 in × 0.15 in. Find (*a*) the area and (*b*) the dimensions of a typical MOSFET.

Answer: (*a*) 4.5×10^{-6} in²; (*b*) 0.00212 in × 0.00212 in.

3.26 Show a schematic for a MOSFET OR gate.

3.27 How many inverters can be packaged into a 14-lead DIP?

Answer: Six (2 leads per inverter × 6 = 12 leads + 2 leads for power = 14 leads).

3.28 How many 3-input NAND gates can be packaged into a 14-lead DIP?

3.29 Find I_{OH} and I_{OL} for a Schottky TTL gate.

Answer: Use Table 3.12.

$$I_{OH} = UL_H \times I_{IH} = 25 \times 40 \, \mu A = 1 \, mA$$
$$I_{OL} = UL_L \times I_{IL} = 12.5 \times 1.6 \, mA = 20 \, mA$$

3.30 A Schottky TTL NAND draws 2.7 mA when the output is high and 6.3 mA when the output is low. Find the average power dissipation per gate.

3.31 Time delays for a Schottky TTL NAND gate are t_{PHL} = 5 ns and t_{PLH} = 4.5 ns. Find the propagation delay.

Answer: 4.75 ns.

3.32 Find the speed-power product for a Schottky TTL NAND gate.

4
FLIP-FLOPS

4.0 INTRODUCTION

Computer circuits store data for various periods of time, and gate circuits can be adapted for storage. Gates must perform short- and long-term memory functions. Feedback converts gates to data storage devices.

This chapter begins with a discussion of regeneration as applied to gates. The basic storage circuit is the latch, which contains several gates. The latch discussion is followed by a description of flip-flops, which are more sophisticated latches. Improvements in flip-flops lead to the *J-K* flip-flop, which is an extremely versatile latch. Computers also require waveshaping, and latches perform this function.

4.1 REGENERATION

Pencil and paper calculations yield an answer. Besides displaying an answer, the sheet of paper serves other functions in solving a problem. A piece of paper contains original data, calculations, intermediate results, and finally an answer. In this sense one piece of paper is more versatile than a computer. Computers require separate circuits for each step in the problem-solving process. An input circuit, arithmetic circuit, storage circuit, and output circuit are needed to accomplish what can be done on a piece of paper.

A gate is the computer component which is as adaptable as a sheet of paper. Gates were originally introduced as high-speed Boolean components. Boolean operations are important, but a computer must perform all steps needed to solve a problem. Gates can perform all the necessary steps.

Consider storing the intermediate results which occur in every mathematical problem. Partial products in multiplication and partial quotients in division are two of many intermediate results in arithmetic operations. A sheet of paper can store intermediate results indefinitely, or else these results can be erased. In other words, paper can have either a long-term or a short-term memory. Since computers perform the same calculations, long- and short-term memory circuits are needed.

When gates implement Boolean expressions, outputs exist only as long as inputs are available. Boolean output is a consequence of input conditions and only remains while inputs exist. If inputs change, outputs change accordingly. Therefore, gates require modification to store information. The output condition of the most recent input condition cannot be saved, but a practical memory circuit must accept and store data even after the source is removed.

An ordinary light switch has a memory. After a light has been turned on, it remains on after the hand has been removed from the switch. Similarly, when the switch is turned off, the light remains off. A light switch remembers the last input as long as necessary. Light switches possess another useful feature. There are only two stable conditions, since a light is never partially on or partially off. In this sense a light switch is a binary component; *on* can represent binary 1 and *off* can represent binary 0.

A computer memory circuit has the same characteristics as a lamp but operates faster. A memory circuit must:

1. Accept input data signals
2. Store data until needed
3. Present data whenever necessary

The basic computer storage element is the *latch*, a circuit capable of storing a single bit. If the last signal presented to a latch is 1, the latch stores 1, and if it is 0, then the latch stores 0. Both states are stable, and therefore a latch is a *bistable* device.

The 2-input OR gate shown in Fig. 4.1 has bistable properties. One input lead is for data, and the other input is obtained directly from the output. Output is fed back to the input, and input and output are in phase. This is positive feedback, sometimes called *regeneration*. Feedback is not absolutely necessary to obtain memory, the lamp with switch being a memory circuit without regeneration; however, latches always use regeneration. With proper design, regeneration assures rapid transistion between high and low.

Suppose a low is placed on the data input of the circuit shown in Fig. 4.1. If the low is maintained for sufficient time, it reaches the output. As discussed, propagation delays for TTL OR gates are in the nanosecond range. Therefore if a data low is maintained for, say, 25 ns, a low exists at the OR output. This output low is fed back to the input along the regeneration connection. Once the low is fed back, the low on the data line can be removed but the output will continue to store a low. Output status can be "read" any time after the input data are removed and the output will still be low. Thus the latch stores a low.

Later on it may be necessary to store a high in this same latch. Placing a

Figure 4.1
OR gate latch.

Figure 4.2
AND latch.

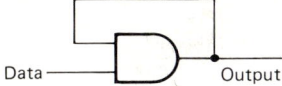

high on the data line permits such a change. After a high has been maintained for the required time, a high appears at the output and is simultaneously fed back to the input. At this time the latch stores a high. The high on the data line can then be removed.

So far so good, but a problem arises. Suppose the latch must be changed back to low, and a low is accordingly entered on the data line. After 25 ns nothing happens; entering a low on the data line cannot change the latch any more. This latch is an OR gate, and the fundamental OR property is a high output whenever at least one input is high. Once a high is fed back, it doesn't matter what signal is on the data line—as long as a high exists at the feedback input terminal, the output will remain high. It is difficult to regain control of this latch. Temporarily breaking the feedback line or removing power are possible solutions, but neither method is practical for computer circuits.

We might consider using the AND latch shown in Fig. 4.2. This latch also uses regeneration from output to input. Suppose we enter a high on the data line. Maintaining a high for sufficient time results in a high at the output and this high is fed back. At this time the data can be removed and the output will continue storing a high. Later on when a low is required, a low is entered on the data line. After this low reaches the output, a low is fed back. The latch now stores a low for as long as necessary. Again control is lost. The fundamental AND property is an output of high only when both inputs are high. As long as one AND-gate input remains low, the output is always low.

Since regaining control is difficult, one OR or AND is not a practical latch. However two important latch principles have been demonstrated:

1. Gates can store signals after the input is removed.
2. Regeneration can be used to store signals.

Previously the circuit shown in Fig. 4.3a was presented as the electronic version of the second Boolean theorem. Since $\bar{\bar{A}}$ is the same as A, the input and output of this circuit are always in phase. Regeneration is achieved by connecting a lead from the $\bar{\bar{A}}$ output back to the A input, as shown in Fig. 4.3b.

When a high is entered on the data line, the output of the first inverter is low. Since the output of the first inverter is the input to the second inverter, a low is input to the second inverter. The output of the second inverter is then

Figure 4.3
Two-inverter latch. (a) $\bar{\bar{A}}$ output; (b) adding regeneration.

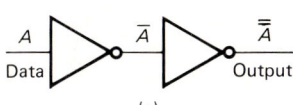
(a)

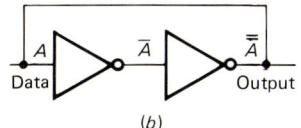
(b)

Figure 4.4
Cross-coupled latch.

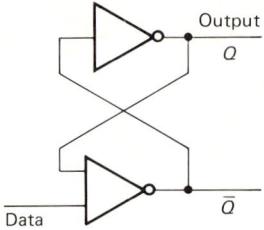

high. This high is fed back to the input. At this time, the input high can be removed, and this two-inverter latch continues to store a high. Similarly, when a low is entered, a low will output and be fed back to the input. When the low is fed back, this circuit outputs and stores a low. A 2-inverter latch uses twice as many gates as a single-gate latch. Although an inverter is less expensive than a 2-input gate, a 2-inverter latch costs more than a 2-input gate.

The 2-inverter latch may be more expensive but it has a useful feature. The output of the first inverter is the complement of the latch output. In other words, both the output and the invert of the output are always available. Some circuits require a signal but others require the complement, and in some certain situations both the signal and the complement are required. With a 2-inverter latch, no additional circuits are required to obtain the signal or the complement.

The 2-inverter latch has another interesting property—it is a symmetrical circuit, that is, the output of each inverter is the input to the other. Figure 4.4 emphasizes symmetry by redrawing the circuit to show cross-coupling between opposite input and output leads. The output is Q and the invert is $\bar{Q}$. These symbols are used by chip manufacturers to indicate complementary outputs.

The convenience of using gates instead of circuit schematics is obvious, as one symbol replaces an entire circuit. However, it is necessary to "look inside" the symbol and see what is happening. As has been described, the basic inverter circuit is a common emitter stage, and therefore the basic 2-inverter latch consists of two common emitter stages with positive feedback, as shown in Fig. 4.5.

Suppose the input to Q_1 is high. This high drives Q_1 into saturation. When Q_1 is saturated, the voltage at the collector of Q_1 is essentially zero, and this output is $\bar{Q}$. Since the collector is connected to the base of transistor Q_2 through a resistor, the base voltage at Q_2 is even lower. Therefore Q_2 is cut off, and the Q output is approximately the power supply voltage. In other words, when the input data are high, Q is high and $\bar{Q}$ is low. The input can be removed and both outputs "remember." Similarly, when the input is low, Q_1 cuts off and $\bar{Q}$ is high. This in turn raises the base voltage on Q_2, and Q_2 saturates. Thus, for an input low, Q is low and $\bar{Q}$ is high. Both bistable outputs are correct.

Figure 4.5
Two-transistor latch. (*a*) Circuit; (*b*) redrawn.

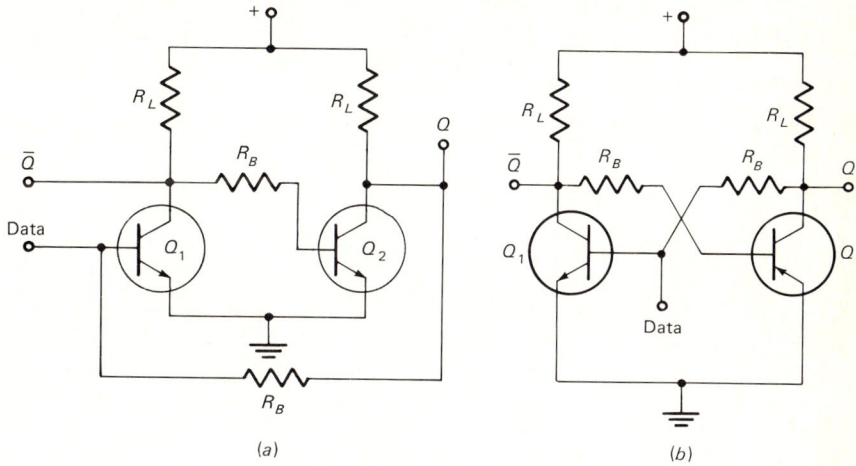

Like single-gate latches, the 2-inverter latch has practical problems. In fact, this circuit latches without any input when power is applied. Applying power to a circuit is a transient situation. Before power is applied, the circuit is inactive. When power is applied, the circuit goes from inactive to active. Even if perfectly matched components existed, there would still be several extra free electrons in one of the 2 inverters. With power applied, this unbalance acts as an input signal; one inverter is driven into saturation and the other into cutoff. Thus a 2-inverter circuit latches before a signal is applied. Moreover, it is impossible to predict which inverter will saturate and which will cut off; Q is just as likely to be high as low when power is applied. A group of 2-inverter latches constitutes a random number generator but not a practical latch.

4.2 NOR LATCH

A single-gate latch is partially successful. Circuit modification is required to prevent losing control. A 2-inverter latch, on the other hand, is useful because both the output and its complement are available. However, outputs latch as soon as power is applied. A practical latch combines the useful features of single-gate and 2-inverter latches, storing a high or low bit and providing complementary outputs.

Problem 3.16, which requires using a NOR gate as an inverter, contains the key to practical latching. Connecting both inputs together as shown in Fig. 4.6 is one method. Since the inputs are connected to each other, there is effectively only one input, and the output is the invert of this input.

Figure 4.6
NOR-gate inverter.

Figure 4.7
Alternate two-inverter latch.

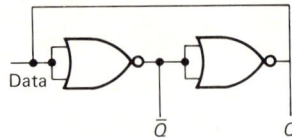

Although a more expensive 2-input NOR gate has been used to replace a less expensive single-input inverter, this alternate method of obtaining inversion is useful. The standard SSI NOR IC contains four NOR gates. Whenever a particular application requires fewer than four NORs, the unused gates can be used as inverters; therefore inverters are available without actual inverter ICs. When two NORs of a 4-gate IC are available, the 2-inverter latch shown in Fig. 4.7 can be constructed. This latch performs in exactly the same manner as the 2-inverter latch described in the previous section. The same advantages and the same problems exist when NOR gates replace true inverters. However, the NOR gate version is transformed into a practical device by separating the inputs. Disconnecting the input leads preserves inversion and simultaneously provides independent control leads.

Figure 4.8 shows a 2-NOR-gate latch with the inputs separated. The data line is labeled S for *set*, and the other open lead is labeled R for *reset*. This NOR latch is redrawn in the conventional manner and shown in Fig. 4.9a. R and S are the inputs while Q and $\bar{Q}$ are the complementary outputs. The block diagram is shown in Fig. 4.9b. This RS latch is practical.

Both outputs are connected to complementary input gates, which is required for regeneration; $\bar{Q}$ is fed back to the R gate input and Q is fed back to the S gate input. Inputs to the R gate are $\bar{Q}$ and R, and inputs to the S gate are Q and S.

Boolean equations for an RS latch are written by inspecting output and

Figure 4.8
Separating the input lines.

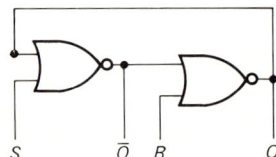

Figure 4.9
RS latch using NOR gates. (*a*) Logic diagram; (*b*) block diagram.

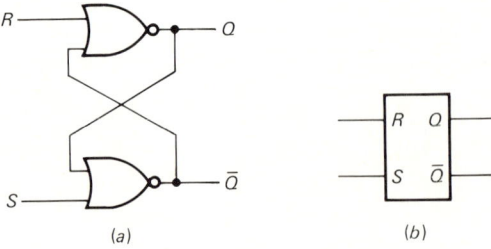

TABLE 4.1
2-Input NOR Gate

A	B	OR	NOR
0	0	0	1
0	1	1	0
1	0	1	0
1	1	1	0

input signals shown in Fig. 4.9a. For any NOR gate the output gate is the extended invert of the inputs. Therefore the equations for outputs are

$$Q = \overline{R + \bar{Q}} \tag{4.1a}$$

$$\bar{Q} = \overline{S + Q} \tag{4.1b}$$

Recall that an OR gate outputs a 1 if at least 1 input is 1 and that a NOR gate is the invert of an OR gate. For convenience, NOR-gate characteristics are presented again in Table 4.1.

Although the *RS* latch is a practical device, random complementary outputs are still generated when power is applied. Applying power may result in $Q = 0$, $\bar{Q} = 1$ when R and S are both 0. Under these circumstances crosscoupling returns 0 to the *S*-gate input and 1 to the *R*-gate input. These conditions are shown in Fig. 4.10a. A 0 and a 1 at the *R* gate result in $Q = 0$ and are consistent with the NOR-gate truth table. Similarly, 0s at the *S*-gate inputs result in $\bar{Q} = 1$. This is also consistent with the truth table, and conditions shown in Fig. 4.10a are stable. Figure 4.10b shows the reverse output conditions when R and S are at 0. These outputs are equally likely. The inputs to the R gate are both 0. This should and does result in a 1 at Q. At the S gate the inputs are 0 and 1, which should and does result in 0 at $\bar{Q}$. Both sets of output conditions are stable when R and S are 0. In particular, $S = 0$, $R = 0$ are the conditions for storing data.

Other control signal combinations can change the output conditions. In Fig. 4.11 $R = 0$ and $S = 1$; the output of the R gate is $Q = 1$; and the output of the S gate is $\bar{Q} = 0$. Feeding back $Q = 1$ means both *S*-gate inputs are 1. This should and does result in $\bar{Q} = 0$. Similarly, feeding back $\bar{Q} = 0$ means both inputs to the R gate are 0. A condition which should and does result in $Q = 1$.

Figure 4.10
Latch with $R = 0$ and $S = 0$.
(a) $Q = 0$, $\bar{Q} = 1$;
(b) $Q = 1$, $\bar{Q} = 0$.

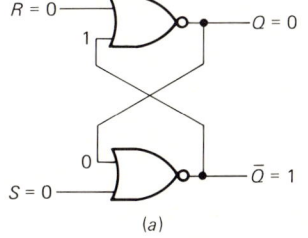
(a)

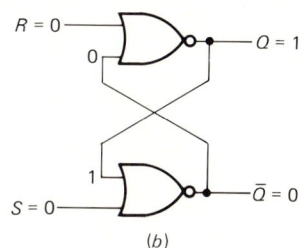
(b)

Figure 4.11
$R = 0$, $S = 1$ latch inputs.

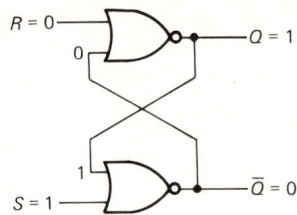

If the opposite outputs existed prior to applying $R = 0$ and $S = 1$, NOR-gate action would switch both outputs to the states in Fig. 4.11. In other words, if the outputs had initially been $Q = 0$, $\bar{Q} = 1$, NOR gates force the outputs to reverse when $R = 0$ and $S = 1$. Thus, $R = 0$ and $S = 1$ are the conditions which set 1 at Q and 0 at $\bar{Q}$.

The opposite set of control signals, $R = 1$ and $S = 0$, are shown in Fig. 4.12. The outputs are $Q = 0$ and $\bar{Q} = 1$ and are consistent. $\bar{Q} = 1$ feeds 1 back to the R-gate input. With both inputs at 1, the R gate should and does output $Q = 0$. Therefore $Q = 0$ feeds 0 back to the S-gate input. With both S gate inputs at 0, the S-gate should and does output $\bar{Q} = 1$. As before, these outputs either existed previously or else have been forced by applying $R = 1$ and $S = 0$. $R = 1$ and $S = 0$ set a 0 at Q and a 1 at $\bar{Q}$. When the control signals are opposite, Q has the same value as S and $\bar{Q}$ has the same value as R.

Two more points should be considered. The control signals must be maintained long enough to activate the NOR gates. The other point is more subtle. If truly identical gates existed, predicting latch outputs would be impossible. For matched gates, propagation times along with all other parameters would be equal. This means signals would reach the outputs at the same time and also be fed back at the same time. Under such circumstances it is equally likely that the outputs would be opposite to what has been described. Outputs would not be predictable—at one time they would be as desired and the next time they might just as well be the reverse. Unpredictable outputs cannot be the basis of circuit design.

Perfectly matched components do not exist. Every real NOR gate is slightly different from every other. Therefore slight differences in propagation time exist. Whichever NOR gate is faster drives the slower gate into the required conditions.

Figure 4.12
$R = 1$, $S = 0$ latch inputs.

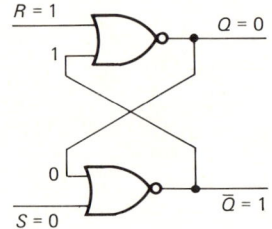

Figure 4.13 $R = 1$, $S = 1$ latch inputs.

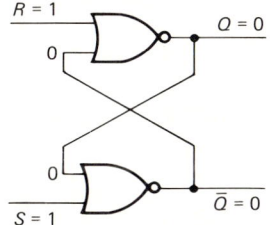

Three useful control signal combinations exist. However, with two control signals, four combinations are possible. The remaining combination, $R = 1$ and $S = 1$, is shown in Fig. 4.13. The inputs at each gate are 0 and 1. Under these circumstances both outputs are 0, and this is a disaster, as the first and most basic Boolean hypothesis is violated, namely, that an output and its complement must always be opposite. Since with R and S both equal to 1, Q and $\bar{Q}$ are both 0, this control signal combination must not be allowed to occur. This condition must be prevented when the computer is designed.

Table 4.2 summarizes the results obtained from a NOR-gate RS latch. The multiple outputs Q and $\bar{Q}$ are shown; the $+$ and $-$ subscripts of Table 4.2 refer to time. Q_+ and $\bar{Q}_+$ are the outputs after control signals activate the latch, and Q_- and $\bar{Q}_-$ are the outputs prior to circuit activation. Three of the four control signal combinations perform useful functions. When R and S are different, Q_+ is the same as S. This permits setting the latch output at either 0 or 1. After the latch is set, the desired output can be stored by applying 0 to both R and S. As long as R and S are maintained at 0, the latch continues to "remember" the correct bit. Whenever a change is necessary, appropriate control line signals are placed at R and S. After allowing sufficient time for the circuit to respond, both control signals are changed back to 0 and the new bit is stored.

Simultaneous 1s on the control lines must be avoided. Except for this condition, the two-NOR-gate RS latch solves the problems associated with single-gate or 2-inverter latches.

4.3 NAND LATCH

A good deal of computer design involves studying alternate versions of a circuit. Circuits which perform identical functions are compared on the basis

TABLE 4.2 RS Latch Conditions

R	S	Q_+	$\bar{Q}_+$	Comment
0	0	Q_-	$\bar{Q}_-$	Previous outputs are stored
0	1	1	0	Q_+ is the same as set (S)
1	0	0	1	Q_+ is the same as set (S)
1	1	0	0	Must be prevented

Figure 4.14 De Morganized *RS* latch.

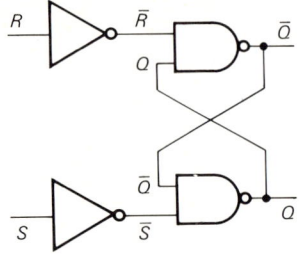

of cost, speed, and reliability. De Morgan's theorem is a particularly useful tool for obtaining equivalent forms; for example, it can yield alternative versions of the *RS* latch.

Inverting both sides of Eq. (4.1a) yields

$$\bar{Q} = \overline{R + \bar{Q}}$$

This equation can be De Morganized to obtain

$$\bar{Q} = \overline{\bar{R} \cdot \bar{\bar{Q}}}$$

which further reduces to

$$\bar{Q} = \overline{\bar{R} \cdot Q} \tag{4.2a}$$

Similarly, inverting both sides of Eq. (4.1b) yields

$$\bar{\bar{Q}} = \overline{S + Q}$$

which can be De Morganized to obtain

$$Q = \overline{\bar{S} \cdot \bar{Q}} \tag{4.2b}$$

Equations (4.2a) and (4.2b) contain extended inverts of ANDed quantities. Thus, an *RS* latch can be constructed from NAND gates. Equation (4.2a), the equation for $\bar{Q}$, can be implemented if $\bar{R}$ and Q are NAND inputs. Similarly, Eq. (4.2b), the equation for Q, is implemented when $\bar{S}$ and $\bar{Q}$ are NAND inputs. If R and S are latch inputs, inverters are required to obtain $\bar{R}$ and $\bar{S}$. Figure 4.14 shows the De Morgan equivalent circuit.

The De Morgan equivalent latch obtains its outputs from the opposite gates. In a NOR-gate latch Q is the output of the R gate and $\bar{Q}$ is the output of the S gate. Figure 4.14 shows reverse conditions for a De Morgan NAND latch, Q here being the output of the S gate and $\bar{Q}$ the output of the R gate.

Describing NAND gate *RS* latch operation requires a NAND-gate truth table, which is repeated in Table 4.3. As before, an AND gate outputs a 1 only when both inputs are 1, and NAND is the invert of AND. Since a NAND *RS* latch is the De Morgan equivalent of a NOR latch, it should function in the same manner. Figure 4.15 shows both possible sets of random output conditions when the control signals are 0.

TABLE 4.3
2-Input NAND Gate

A	B	AND	NAND
0	0	0	1
0	1	0	1
1	0	0	1
1	1	1	0

When R and S are 0, $\bar{R}$ and $\bar{S}$ are 1. In Fig. 4.15a both inputs to the R NAND gate are 1. As can be seen from the NAND truth table, if both inputs are 1, the output should be 0. This is shown in the logic diagram; truth table and logic diagram are in agreement. At the S NAND gate the inputs are 1 and 0. These signals should and do output a 0. As with the NOR latch, if both control signals to the De Morgan latch are 0, the outputs $Q = 1$ and $\bar{Q} = 0$ represent a stable state. Similarly, following the signals through the circuit shown in Fig. 4.15b shows that $Q = 0$, $\bar{Q} = 1$ is also a stable state. Thus as in the NOR latch, 0s at both inputs are the conditions for storing data.

In Fig. 4.16 the control signals are $R = 0$ and $S = 1$. In this case, tracing signals through the circuit shows that both inputs to the R NAND gate are 1. This should and does result in outputs of $Q = 1$ and $\bar{Q} = 0$. These same inputs produced the same outputs for the NOR RS latch. Also, as with the NOR latch, this De Morgan equivalent has the Q output corresponding to the S input and the $\bar{Q}$ output to the R input. Similarly if the outputs shown in Fig. 4.16 did not exist prior to applying $R = 0$, $S = 1$, NAND-gate action would cause these outputs to exist.

Figure 4.15
Initial De Morgan conditions. (a) $Q = 1$; (b) $Q = 0$.

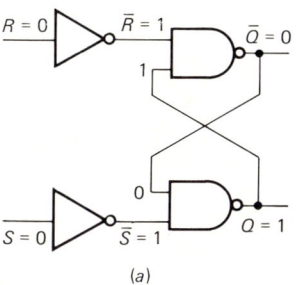
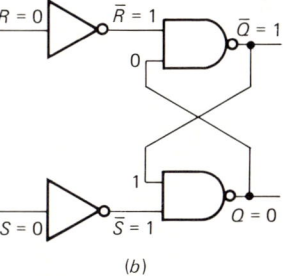

Figure 4.16
De Morganized latch with $R = 0$ and $S = 1$.

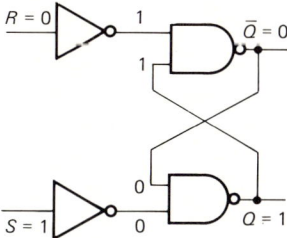

TABLE 4.4
De Morganized Latch Characteristics

R	S	Q_+	$\bar{Q}_+$	Comment
0	0	Q_-	$\bar{Q}_+$	Previous outputs are stored
0	1	1	0	Q_+ is the same as set (S)
1	0	0	1	Q_+ is the same as set (S)
1	1	1	1	Opposite of NOR latch but must also be prevented

When $R = 1$ and $S = 0$, the results are $Q = 0$ and $\bar{Q} = 1$. As before, applying opposite signals to R and S of the De Morganized circuit results in a latch which is set to the desired outputs. The three useful conditions of the RS NOR latch are the same as those that occur in the De Morgan NAND RS latch.

Applying $R = 1$ and $S = 1$ to the De Morgan latch also causes problems. With the NOR latch both outputs are 0. With the De Morganized latch both outputs are 1. Whether both outputs are 1 or 0, identical outputs at Q and $\bar{Q}$ are equally dangerous.

Table 4.4 summarizes the De Morgan latch conditions. The only difference between this table and the NOR latch table is the output signals which result when both inputs are 1. The gate type can be deduced by applying invalid inputs and observing whether both outputs are 1 or 0. However, the block diagram is the same whether NORs or NANDs are used. Whether or not Q is opposite from S is a matter of drafting convenience and does not relate to gate type. Figure 4.9b is the standard symbol for any type of RS latch.

Other RS latches are possible. NOR gates were originally selected because separate inputs and inversion are necessary characteristics. These characteristics also exist in a NAND gate, and a different NAND RS latch is shown in Fig. 4.17a. In this case the inputs to $\bar{Q}$ are R and Q, while the inputs to Q are S and $\bar{Q}$. Thus the equations for this NAND RS latch are

$$\bar{Q} = \overline{R \cdot Q} \tag{4.3a}$$

and

$$Q = \overline{S \cdot \bar{Q}} \tag{4.3b}$$

Figure 4.17 NAND RS latches. (a) Q_+ = reset; (b) Q_+ = set.

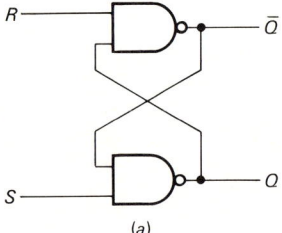
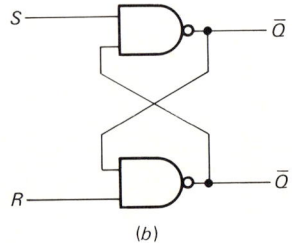

**TABLE 4.5
NAND RS Latch**

R	S	Q_+	$\bar{Q}_+$	Comment
0	0	1	1	Must be avoided
0	1	0	1	Q_+ is the same as reset
1	0	1	0	Q_+ is the same as reset
1	1	Q_-	$\bar{Q}_-$	Previous outputs are stored

These equations are different from those for the NOR latch and the De Morgan equivalent. However, the Q output is taken from the S gate and the $\bar{Q}$ output is taken from the R gate. Equations (4.3a) and (4.3b) are different from Eqs. (4.2a) and (4.2b) because the inverters are removed. The NAND RS latch uses the same number of gates as the NOR RS latch. Since NAND and NOR gates have different truth tables, NAND latch characteristics are different from those of the NOR latch and the De Morgan equivalent. In fact, NAND latch characteristics are obtained by considering the inputs as outputs of the inverters shown in Fig. 4.14. NAND RS latch characteristics for Fig. 4.17a are presented in Table 4.5. This latch has the same useful properties: 1s and 0s can be entered and stored, and also one set of input conditions results in the same output at both Q and $\bar{Q}$.

As shown in Table 4.5, this latch is not compatible with previous latches, as all outputs are reversed. However with the redesign shown in Fig. 4.17b, this NAND latch is compatible with the other latches. Interchanging R and S leads makes Q the same as S. Also, there is nothing wrong about storing with 1s as inputs instead of 0s. Similarly, preventing simultaneous 0s at the inputs is no better or worse than preventing simultaneous 1s. The NAND latches shown in Fig. 4.17 are as valid and cost-effective as NOR latches. In fact, a latch discussion can begin with NANDs and then progress to NORs. The important point is to investigate alternate forms for possible advantages.

Many applications require placing several latches in the same condition at the same time. This is accomplished by connecting all R inputs or all S inputs together, as shown in Fig. 4.18. The same signal is applied to all latches. Either a latch is in the desired state or else gate action forces the appropriate

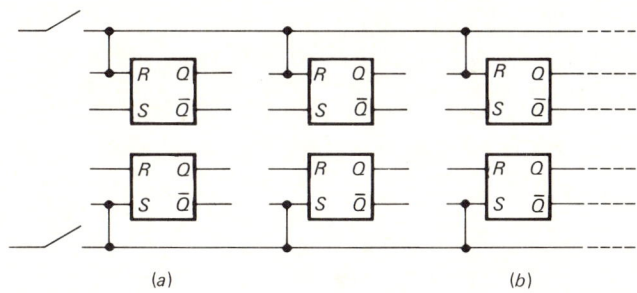

Figure 4.18 Synchronizing latch outputs. (*a*) Common R line; (*b*) common S line.

Figure 4.19
Electromechanical switch waveform.

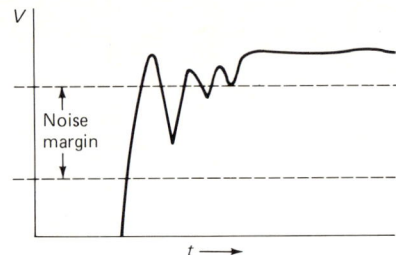

state. This circuit "zeroes" a chain of latches prior to use. Afterwards, independent bits can be entered and stored in each latch.

RS latches can eliminate transients when electromechanical switches are used as computer inputs. Switch closure is not instantaneous. As shown in Fig. 4.19, a switch contact vibrates for several milliseconds before stabilizing. Contact vibration is not particularly serious if the switch operates a lamp or turns on an amplifier. In such circumstances the transient is not noticeable. However, if the switch is part of a computer keyboard, voltage fluctuations generate additional pulses, which represent extra 1s and 0s. Therefore activating a mechanical switch does not input correct data. It is extremely difficult to eliminate mechanical switch bounce. Also, switch characteristics change with age and use. However, if a switch is attached to an *RS* latch as shown in Fig. 4.20, voltage transients are restricted to the switch and eliminated from the latch output. A spring-loaded momentary switch is shown attached to the *R* terminal of a latch. When the switch is activated, it remains at the *S* terminal only as long as the operator maintains pressure on the switch. *R* and *S* inputs are both connected to a 5-V supply through resistors, which keep the gate input current within safe limits. Since the switch is grounded, *R* is normally at 0 V and *S* is normally at 5 V. This is equivalent to $R = 0$ and $S = 1$; thus, the outputs are $Q = 1$ and $\bar{Q} = 0$. These are the initial conditions of the timing diagram shown in Fig. 4.21.

When an operator pushes the keypad button, the switch begins to move from *R* toward *S*. Once contact at *R* is broken, the signal at *R* is a logic 1. However, the voltage at *S* does not go to 0 until the switch reaches *S*. This time interval is t_{RS}. After the switch reaches *S*, the voltage at *S* is 0 but will

Figure 4.20
Switch debouncer.

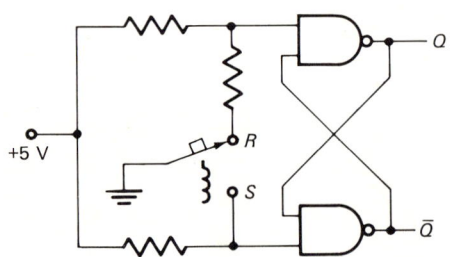

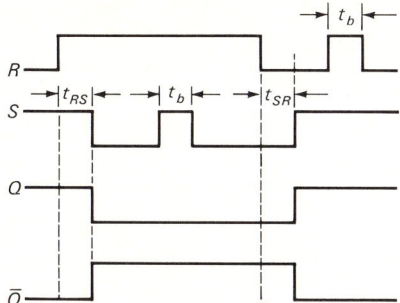

Figure 4.21 Debouncer timing diagram.

bounce for some time. The bounce interval is shown as t_b. As the switch touches S, this input goes to 0. $\bar{Q}$ switches to 1 and Q to 0 in the several nanoseconds required to operate the NAND gates. Since $\bar{Q}$ is fed back to the R gate, both inputs to the R gate are now 1 and this drives Q to 0, where it will remain as long as both R and $\bar{Q}$ are 1. Thus the latch output does not "see" t_b, and contact vibration cannot affect Q. When the switch is released, conditions are reversed. The contact now moves from S to R, and this interval is t_{SR} on the timing diagram. As the switch reaches R, the R input goes to 0; at this instant Q becomes 1, and this fed-back 1 drives $\bar{Q}$ to 0. Again contact vibration cannot affect Q because $\bar{Q}$ is 0 and S is 1, and thus Q is maintained at 1. This circuit is known as a *switch debouncer* or a *contact conditioner*. It does not prevent switch chatter but isolates chatter from the circuit output.

4.4 FLIP-FLOPS

Many computer circuits function in a specific order: the second step must occur after the first, the third step must occur after the second, and so on. In such cases the output is determined by two factors, input signal, and the previous signals. Circuits which operate in a specific order are called *sequential circuits.*

A single-lane road filled with automobiles has the same characteristics as a sequential circuit. No car can move until the car in front moves. Progress is determined by the time at which each car arrives, and no car can go faster than any other. By contrast, a multilane highway with only a few cars has the characteristics of a nonsequential circuit. Any car can be anywhere along the road at any time. Cars pass each other or not as determined by the speed which each driver selects independently. Nonsequential events occur in random order.

Previously described *RS* latches are nonsequential but can be converted to sequential circuits. Extra components are needed but no new circuit types are required. The components needed to construct a computer have been presented. Computer requirements are satisfied by strategically connecting some or all of the three basic circuits, namely, AND, OR, and NOT gates.

Figure 4.22
Computer clock voltage.

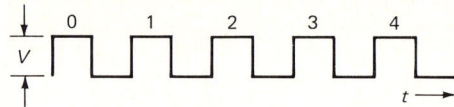

Latches, for example, are converted into sequential circuits by using gates to synchronize input signals to a master *clock*. Figure 4.22 shows the clock waveform. A clock generates a train of pulses and the waveform is binary, since the clock is either on or off. Circuits to be synchronized are connected to the clock. Clock period is accurate because the source is an oscillator whose frequency is determined by a quartz crystal.

A *synchronous RS latch* is shown in Fig. 4.23, in which AND gates are connected to the basic RS, R and S are the standard reset and set leads, and R_i and S_i are the reset and set inputs to the latch portion of the circuit. The clock pulse (C_p), is connected to both AND gates. A latch which is clocked is known as a *flip-flop*. Figure 4.23 is an RS flip-flop.

Flip-flop action is determined by control signals and clock pulse condition. As always, an AND gate output can only be a 1 when both inputs are 1. In order for a logic 1 to arrive at R_i, R and C_p must be 1 at the same time. Similarly, for S_i to receive a 1, S and C_p must be 1 at the same time. The Boolean equations for R_i and S_i are

$$R_i = R \cdot C_p \tag{4.4a}$$

and $\quad S_i = S \cdot C_p \tag{4.4b}$

The NOR gate outputs are

$$Q = \overline{R_i + \overline{Q}} \tag{4.5a}$$

and $\quad \overline{Q} = \overline{S_i + Q} \tag{4.5b}$

If Eqs. (4.4a) and (4.4b) are substituted into Eqs. (4.5a) and (4.5b), the results are

$$Q = \overline{R \cdot C_p + \overline{Q}} \tag{4.6a}$$

and $\quad \overline{Q} = \overline{S \cdot C_p + Q} \tag{4.6b}$

Figure 4.23
Clocked *RS* latch. (*a*) Logic diagram; (*b*) block diagram.

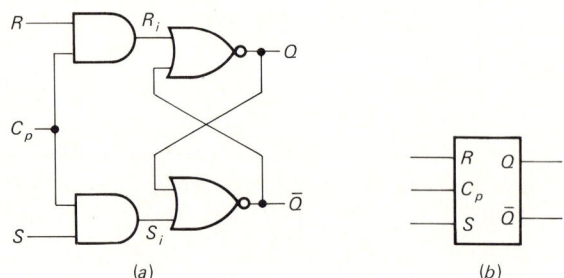

Figure 4.24
Initial flip-flop conditions.

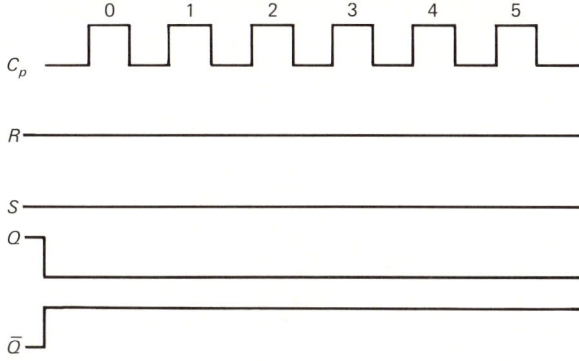

These equations completely describe an *RS* flip-flop. Although Eqs. (4.6a) and (4.6b) are manageable, it is the sequential aspect of *RS* flip-flops which is of interest, and therefore timing diagrams are more useful. In this case a display of time relationships between clock pulses, control signals, and outputs is easier to understand than Boolean equations.

The original latch discussion began with R and S at 0. This is also a convenient place to begin describing *RS* flip-flop operation. Either combination of Q and $\bar{Q}$ can be the starting point; Fig. 4.24 happens to show $Q = 1$ and $\bar{Q} = 0$.

Under these circumstances the original outputs will be stored indefinitely because R_i and S_i cannot change unless C_p and either R or S are 1 at the same time. Since both R and S are always 0, R_i and S_i will never change.

On the other hand, Fig. 4.25 shows a situation in which both R and S change states. However neither latch output changes because R and S are never logic 1 when C_p is logic 1. Therefore neither AND gate can change state and Q and $\bar{Q}$ continue to store initial conditions. Outputs cannot change unless a control signal and the clock pulse are logic 1 at the same time.

Figure 4.25
Clocked data storage.

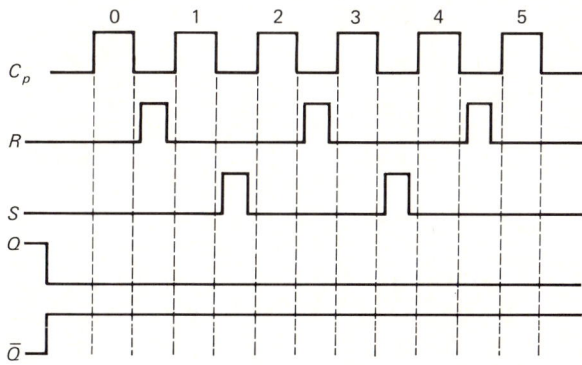

Figure 4.26 Flip-flop operation.

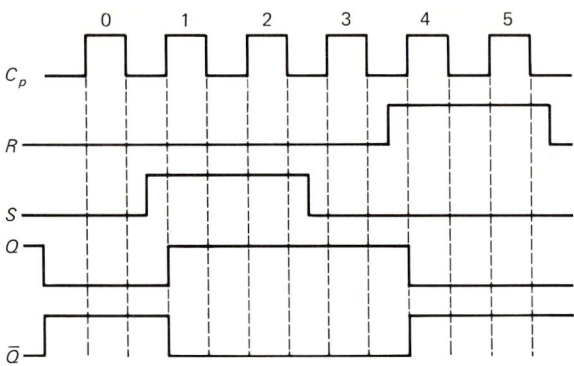

Circuits are useful when outputs change in response to inputs. Figure 4.26 illustrates an arbitrary set of conditions which result in change. During C_{p0}, or clock pulse number 0, R and S remain at 0. These are the conditions for storing data, and outputs retain the original values. During the interval between C_{p0} and C_{p1}, S goes to 1. Since the clock pulse is 0, the outputs are still at the original values. The first change in output occurs at the beginning of C_{p1}. This is the first time when S and C_p are simultaneously at 1. The simultaneous 1s cause a 1 at the output of the S AND gate, which is also S_i. Since S_i is now 1 and $\bar{Q}$ is still 1, $\bar{Q}$ is driven to 0. Driving $\bar{Q}$ to 0 also means that both inputs to the R_i NOR gate are 0. These inputs drive Q to 1. The outputs are now in the set condition, $Q = 1$ and $\bar{Q} = 0$.

In this case, S remains at 1 even after C_{p1} goes off. However, during the interval between C_{p1} and C_{p2} the clock is off. Neither the R nor the S AND gate can be activated. The outputs which resulted during C_{p1} are stored during the interval between C_{p1} and C_{p2}. During C_{p2} the input conditions are the same as during C_{p1}, and the outputs remain constant. The flip-flop remains set during C_{p2}.

At the end of C_{p2}, S remains on but C_p goes to 0. This disables the S AND gate and both R and S are now 0. These are the conditions for storing data. The same outputs continue to be stored. During the interval between C_{p3} and C_{p4}, R goes to 1. However, C_p is 0 at this time and nothing changes. At the beginning of C_{p4} both R and C_{p4} are 1. The result is a 1 at R_i. Now R_i is 1 and $\bar{Q}$, the other input to the R_i NOR gate, is 0. This causes Q to switch to 0. Switching Q to 0 also feeds 0 back to one input of the S_i NOR gate. Since the other input to this gate is also 0, $\bar{Q}$ switches to 1. This is the last change in the outputs because no other changes are shown for R or S. Thus the output will continue to store the reset conditions; $Q = 0$ and $\bar{Q} = 1$ remain. At a later time output changes can be implemented by changing control signals. $R = 1$ and $S = 1$ must still be prevented at the same time. When the clock is at logic 1, the truth table for a clocked flip-flop is the same as Table 4.2.

It is also possible to apply De Morgan's theorem to a flip-flop. Inverting both sides of Eq. (4.6a) results in

$$\bar{Q} = \overline{R \cdot C_p + \bar{Q}}$$

and De Morganizing this expression yields

$$\bar{Q} = (\bar{R} + \bar{C}_p) \cdot Q \tag{4.7a}$$

Similarly, inverting both sides of Eq. (4.6b) results in

$$\bar{\bar{Q}} = \overline{S \cdot C_p + Q}$$

which De Morganizes to

$$Q = (\bar{S} + \bar{C}_p) \cdot \bar{Q} \tag{4.7b}$$

Inverters are required to obtain $\bar{R}$, $\bar{S}$, and $\bar{C}_p$. The logic diagram for this latch is shown in Fig. 4.27.

This De Morgan RS flip-flop is more complicated than the original version shown in Fig. 4.23, for which only four gates of two different types were required. The De Morgan version uses seven gates of three different types.

However, additional use of De Morgan's theorem results in significant improvement. Equation (4.7a) contains $(\bar{R} + \bar{C}_p)$ and Eq. (4.7b) $(\bar{S} + \bar{C}_p)$. Both expressions can be rewritten. De Morganizing part of an equation is valid and has been discussed.

$$\bar{R} + \bar{C}_p = \overline{R \cdot C_p} \tag{4.8a}$$

and $\quad \bar{S} + \bar{C}_p = \overline{S \cdot C_p} \tag{4.8b}$

Substituting these forms into Eqs. (4.7a) and (4.7b) yields

$$\bar{Q} = \overline{(R \cdot C_p)} \cdot Q \tag{4.9a}$$

and $\quad Q = \overline{(S \cdot C_p)} \cdot \bar{Q} \tag{4.9b}$

Equations (4.9a) and (4.9b) are more compact than Eqs. (4.7a) and (4.7b). Compactness is an advantage in itself, but the real advantage of Eqs. (4.9) is shown in Fig. 4.28.

Figure 4.27
De Morganized flip-flop.

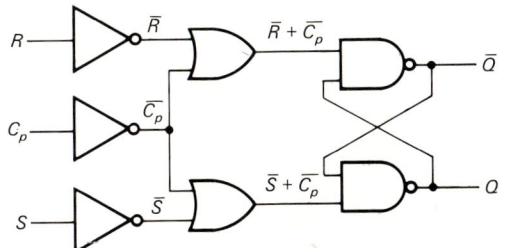

Figure 4.28
ALL-NAND flip-flop.

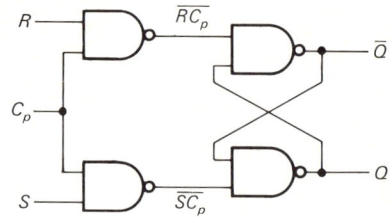

The original version used 2 NOR gates and 2 AND gates. The De Morgan equivalent shown in Fig. 4.28 also contains 4 gates but all are NAND. This all-NAND circuit is a practical application of De Morgan's theorem. Merely applying De Morgan's theorem in "cookbook" fashion resulted in a more complex circuit, requiring almost twice as many gates. However, strategically applying De Morgan's theorem to a portion of the more involved equations yields a simpler flip-flop with only one type of gate.

4.5 *D*-TYPE CIRCUITS

Thus far none of the circuits avoids the problem of simultaneous 1s at R and S. Depending on the particular circuit, simultaneous 1s result in outputs which are identical or else unpredictable. A circuit which prevents identical inputs at R and S is shown in Fig. 4.29a, where D is the only input and is connected directly to the S input of the latch or flip-flop and through an inverter to the R input. Therefore, S is the same as D, and R is always the invert of D. Figure 4.29b shows the truth table for the D-input circuit. There is a single input and two separate outputs. The D input eliminates any possibility of 1s at both R and S at the same time.

A D-input circuit can be connected in front of any latch or flip-flop. Figure 4.30 shows a D input connected to the clocked AND/OR latch of Fig. 4.23. In keeping with standard D-latch terminology, C_p has been replaced by E, which stands for *enable*. However, the meaning and function of E and C_p are identical; E is the clocked input to the D latch. The D input eliminates simultaneous 1s, but it also eliminates simultaneous 0s, the condition for storing data. Nevertheless, the D latch can store data because outputs cannot change when E is 0.

Figure 4.29
Data input circuit. (*a*) Schematic; (*b*) truth table.

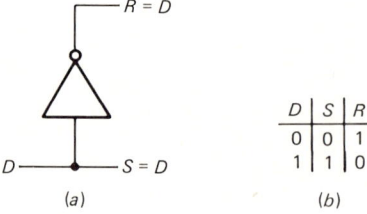

Figure 4.30
D latch. (*a*) Logic diagram; (*b*) symbol.

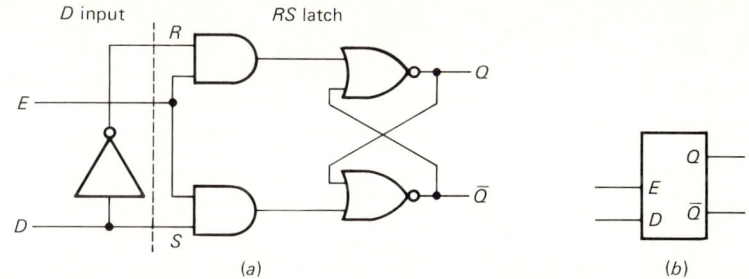

When clocked *RS* latches are used, Q follows S when C_p is 1. The same condition exists for a *D* latch. In fact, Boolean equations for a *D* latch are the same as for a clocked *RS* latch provided the necessary substitutions are made. S becomes D, R becomes $\bar{D}$, and C_p becomes E. With these substitutions Eq. (4.6) becomes

$$Q = \overline{\bar{D} \cdot E + \bar{Q}} \tag{4.10a}$$

and $\quad \bar{Q} = \overline{D \cdot E + Q} \tag{4.10b}$

With 2 inputs, *D* and *E*, there are four possible combinations. Table 4.6 shows the results. Although the *D* input guarantees opposite outputs, Q and $\bar{Q}$ are usually included. Table 4.6 shows that data is stored when *E* is 0. When *E* is 1, Q is the same as *D*.

Figure 4.31 shows a possible set of waveform conditions. The input happens to be an alternating sequence of 0s and 1s. As in previous discussions, initial conditions are taken as $Q = 0$ and $\bar{Q} = 1$. In other words, reset condition is the starting point.

A *D* latch cannot respond to data while *E* is 0. Therefore pulses D_0 through D_3 do not change the latch outputs. During D_4, *E* goes to 1. But D_4 is 0, and since Q follows D, outputs remain at the initial conditions. D_5 is 1 while *E* is 1. Therefore Q switches to 1 and $\bar{Q}$ switches to logic 0. The next data pulse, D_6, is 0. Since *E* is still 1, Q switches to 0 while $\bar{Q}$ switches to 1. *E* is still 1 when D_7, which is 1, switches on; therefore Q switches to 1 and $\bar{Q}$ back to 0. While D_7 is still 1, *E* switches off. In this case, Q and $\bar{Q}$ store the values which existed during D_7.

TABLE 4.6
D-Latch Truth Table

E	D	Q_+	$\bar{Q}_+$	Comment
0	0	Q_-	$\bar{Q}_-$	Stores previous condition
0	1	Q_-	$\bar{Q}_-$	Stores previous condition
1	0	0	1	Q follows D
1	1	1	0	Q follows D

Figure 4.31
D-latch waveforms.

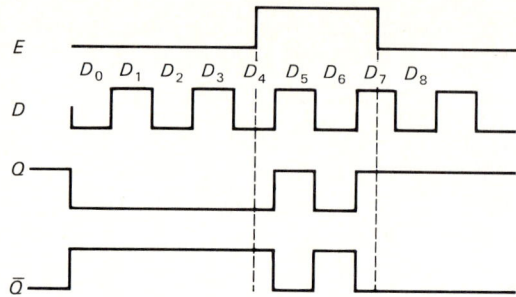

The description of the waveforms shown in Fig. 4.31 illustrates *D*-latch operation. When the enable pulse is on, data pass through the *D* latch. Then enable goes off, and the latch stores data which existed when enable went off. Enable can be considered as a "window" which transfers data when the window is open and stores data when the window is closed.

D latches are used as temporary storage devices. For example, partial sums or partial products, discussed in Chap. 2, can be saved in *D* latches until required for the next operation. ICs with 4- and 8-bit *D* latches are standard. Computers process standard-size collections of bits, which are called computer *words*. An 8-bit word is known as a *byte*. With some attempt at humor, a 4-bit word is called a *nibble*. Figure 4.32 is a pin-out of the 7475 IC, a 4-bit TTL *D* latch. MOS versions are also available. Latches in which $\bar{Q}$ follows *D* are just as possible, and both types exist. The latter condition is obtained by reversing input and inverter connections. Figure 4.33 shows both configurations. In either case *Q* follows *S* and $\bar{Q}$ follows *R*.

The *D* flip-flop is an important variation of the *D* latch. A *D* latch is converted to a *D* flip-flop by including an *RC* wave-shaping circuit at the clocked input, as shown in Fig. 4.34. If the time constant of the *RC* circuit is

Figure 4.32
4-Bit TTL *D* latch.

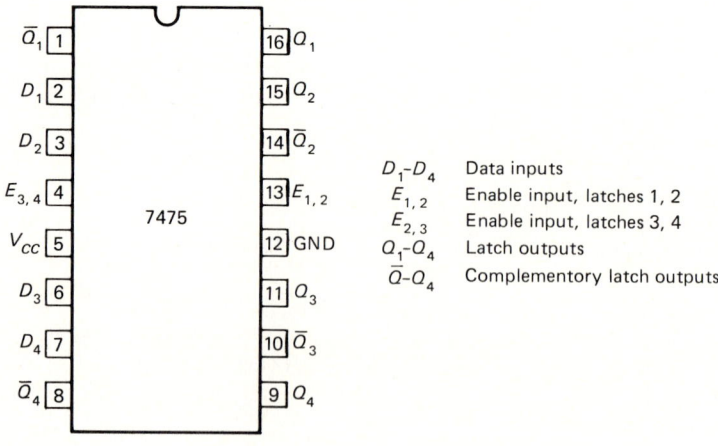

Figure 4.33
D-latch input. (*a*) *S* follows *D*; (*b*) *R* follows *D*.

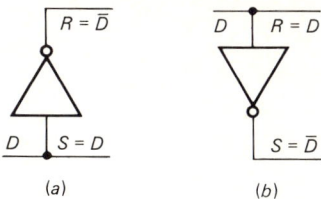

Figure 4.34
Differentiating circuit.

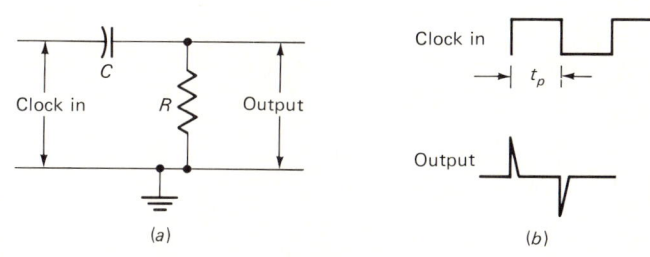

much smaller than t_p, the outputs are voltage spikes. A positive voltage spike occurs when the clock voltage switches from off to on, and a negative voltage spike occurs when the clock voltage switches from on to off.

The only difference between the symbols for a *D* latch and a *D* flip-flop, shown in Fig. 4.35, is clock identification. Latches use the letter *E* for enable, and flip-flops use C_p for clock pulse. Boolean equations for a *D* flip-flop are the same as for a *D* latch. While this is true, it obscures the difference between a *D* latch and a *D* flip-flop. Either AND gate requires simultaneous 1s to output a 1. Because of the differentiated clock input, simultaneous 1s can only exist during the short time that the positive voltage spike is above the V_{IH}, the minimum voltage required to activate the AND gate. Such action is called *edge triggering*—in this case, positive edge triggering. The *D* flip-flop shown in Fig. 4.35 can only be activated while the clock pulse switches from 0 to 1. It cannot be activated while the clock is at logic 1 or logic 0 or during the time that the clock switches from 1 to 0. On the other hand, a *D* latch can be activated during the entire time *E* is at logic 1.

Edge triggering changes *D*-circuit operation in such a way that the input is only "sampled" during the very short interval of the C_p leading edge. Figure

Figure 4.35
D flip-flop. (*a*) Logic diagram; (*b*) symbol.

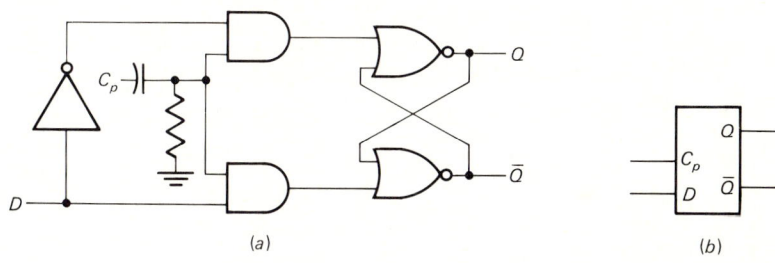

Figure 4.36
D flip-flop waveforms.

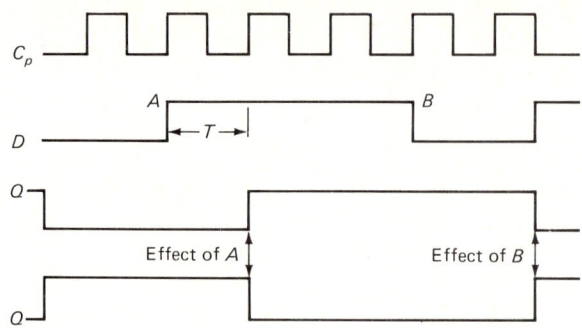

4.36 shows the effect of voltage sampling on a typical waveform. The D flip-flop can only be triggered when a positive-going edge of C_p coincides with a logic 1 at the D input. In Fig. 4.36 this occurs at A. Since the input is 0 just before this instant, the differentiated clock pulse and time A do not coincide. Therefore, previous outputs are retained. But the next time C_p has a positive-going edge, D is also positive, and Q goes positive at this instant. The effect of A shows up at the output one complete clock waveform later. During the next positive-going clock pulse, D remains positive and Q remains at logic 1.

At time B the input is 0, but the differentiated spike is not high until D is 0. Q remains at logic 1. However, the next time C_p has a positive-going edge, D is still 0, but now Q switches to 0. Once more it has taken a complete clock period for the output to react to a change at the input. In other words, Q is the same as D but delayed by one clock cycle. For this reason the D flip-flop is called a *delay flip-flop*, while the D latch is called a *data latch*.

Triggering is the only difference between a D latch and a D flip-flop. A logic 1 activates a latch while a positive-going edge activates a flip-flop. As a practical matter the value of capacitance necessary to create a differentiating circuit is too large to be built into an IC chip. When using ICs, gates generate the equivalent of a differentiated voltage spike. This is accomplished by exploiting a situation which was originally described as a problem.

When gates implement Boolean equations, required signals may not occur at the same time. As shown in Fig. 4.37, this situation occurs when signals pass through unequal numbers of gates. A is a series of clock pulses and goes directly from C_p to the output AND gate. B, the other signal path, passes through an extra AND gate. The output of B has the same shape as A but is

Figure 4.37
Generating voltage spikes.

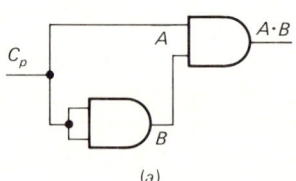
(a)

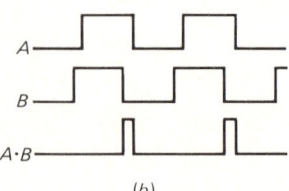
(b)

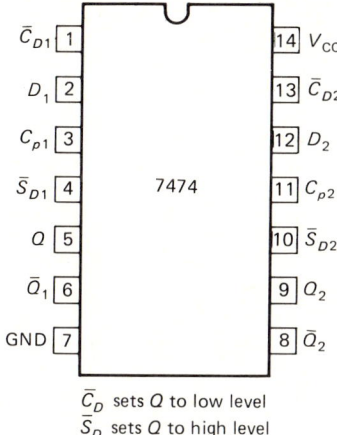

Figure 4.38 Dual TTL *D* flip-flop.

$\bar{C}_D$ sets Q to low level
$\bar{S}_D$ sets Q to high level

delayed by propagation delay. An AND gate can only output a 1 when both inputs are 1. As shown in Fig. 4.37*b*, simultaneous 1s only exist for a short time. Therefore the output $A \cdot B$ is a series of narrow voltage spikes and is equivalent to positive differentiated voltage spikes. Gates only generate positive voltage spikes while *RC* differentiation causes both positive and negative spikes. However, negative spikes do not affect gates and are not required. In effect, gates replace capacitors used in conventional differentiating circuits. *D* flip-flops are used extensively. Figure 4.38 shows the pinout of the 7474, which is TTL. MOS versions also exist.

4.6 *T*-TYPE CIRCUITS

The *RS* NOR circuit in Fig. 4.9 is the basis of all latches and flip-flops. Modifications such as clocking, *D* input, and edge triggering satisfy specific requirements. For example, the clocked *RS* latch in Fig. 4.23 was obtained by adding a clocking input to the basic *RS* circuit. This latch stores and transmits bits on a synchronous basis. Similarly, Fig. 4.28 is a De Morgan clocked *RS* latch. Operation is the same but only one gate type is required.

Another modification is obtained by connecting the *S* input directly to the $\bar{Q}$ output and the *R* input to the *Q* output. In this case, *T* corresponds to C_p or *E* of the previous circuits. *T* is an input which receives clocked pulses and is called a *T* flip-flop. *T* stands for *toggle* and Fig. 4.39 shows an all-NAND-gate *T* flip-flop. *T* is the only input. *R* and *S* are part of an internal set of positive feedback connections.

It is not possible to predict which output of a *T* flip-flop will be high when power is first applied. *S* at 0 and *R* at 1 are merely the same conditions used to describe other latches and flip-flops. Without any loss of generality, *T* can be taken as 0. The condition for each gate terminal when *Q* is low and $\bar{Q}$ is high is shown in Fig. 4.40*a*. Because of the feedback paths, *R* must be 0 when *Q* is 0 and *S* must be 1 when $\bar{Q}$ is 1. Signals at each gate terminal are consistent with

Figure 4.39
T flip-flop. (a) Logic diagram; (b) block diagram.

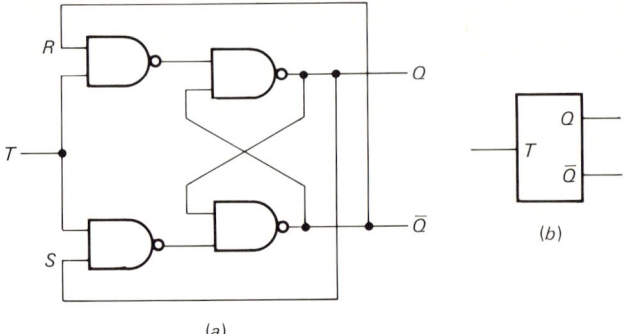

the NAND-gate truth table shown in Fig. 4.40b. Consider what happens when T switches to logic 1. The R gate is not affected because 0 and 1 have the same effect on a NAND gate as 0 and 0. At the same time a 1 and 1 at the S gate will alter the output of the S NAND gate. The 1 at T has been "steered" to the input NAND gate, which can be switched. When the S-gate output changes, all the other gates are affected, and as a result the outputs reverse. After stabilizing, Q is 1 and $\bar{Q}$ is 0. These conditions continue as long as T remains 1. In fact, as shown in Fig. 4.41, the outputs remain at Q = 1 and $\bar{Q}$ = 0 even after T switches back to 0.

Outputs remain unchanged because 0 at the common lead to both input gates is the condition for storing data. A 0 at T has the same effect as a 0 at C_p of a flip-flop or E of a latch. Figure 4.41b shows the waveforms, beginning with the initial low at T and proceeding through a high at T and back to a low at T. Conditions are different the next time T goes high. At this time a 1 at T has no effect on the S gate but causes the R gate output to reverse. Now Q is 0, and $\bar{Q}$ is 1. These outputs remain even after T returns to 0. On the next high at T, the S gate output switches but the R gate cannot. In a T flip-flop outputs alternate every time T goes high because the input automatically is steered to the gate which can be switched. A T flip-flop toggles each time T goes from 0 to 1. Waveforms are as shown in Fig. 4.42.

One interesting effect of toggling is an output which lasts exactly twice as

Figure 4.40
Initial T flip-flop conditions. (a) Logic levels; (b) truth table.

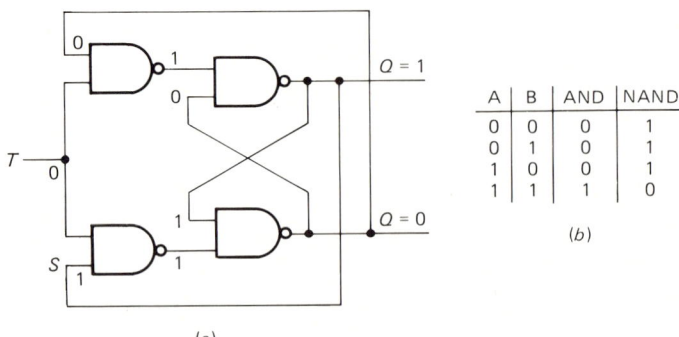

Figure 4.41 Continued T flip-flop conditions. (a) Signals; (b) waveforms.

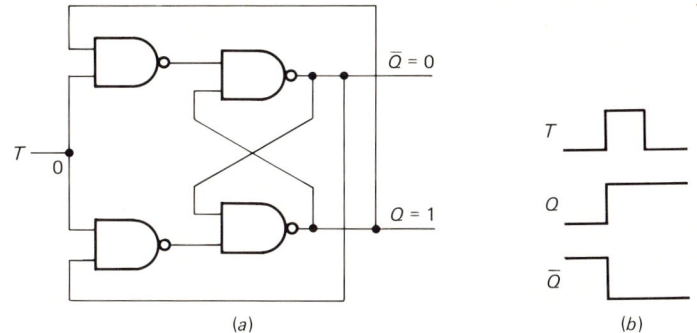

long as the input. In other words, the output frequency (prf_{out}) of a T flip-flop is half the input frequency (prf_{in}):

$$\text{prf}_{\text{out}} = 0.5\, \text{prf}_{\text{in}} \tag{4.11}$$

Another feature is shown in Fig. 4.42b. The output is a square wave whether or not the input is a square wave because the output switches each time that T goes positive.

Example 4.1 Assuming that the T input shown in Fig. 4.42b is high for 40 µs and low for 60 µs, find (a) prf_{in} and (b) prf_{out}.

Solution (a) The asymmetrical input period T_{in} is the sum of $t_{\text{on}} + t_{\text{off}}$.

$$T_{\text{in}} = 40\ \mu s + 60\ \mu s = 100\ \mu s$$

and since $\text{prf} = 1/T$

$$\text{prf}_{\text{in}} = 1/100\ \mu s$$
$$= 10\ \text{kpps}$$

(b) Using Eq. (4.11), the symmetrical output frequency is

$$\text{prf}_{\text{out}} = 0.5\, \text{prf}_{\text{in}}$$
$$= 0.5 \times 10\ \text{kpps}$$
$$\text{prf}_{\text{out}} = 5\ \text{kpps}$$

Figure 4.42 T flip-flop waveforms. (a) Symmetrical input; (b) asymmetrical input.

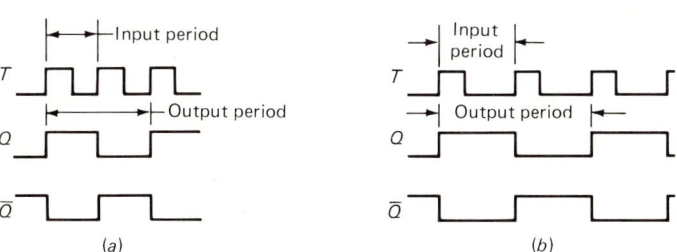

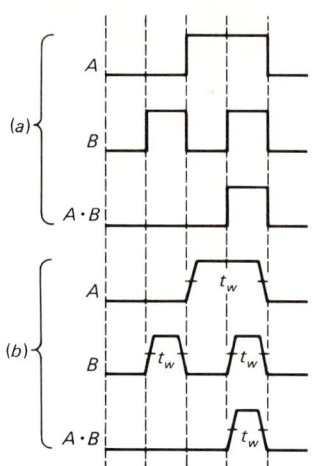

Figure 4.43
AND-gate waveform. (*a*) Ideal; (*b*) trapezoidal.

T flip-flops are also available which trigger when T goes from high to low. Specific application determines which toggling mode should be selected. This T flip-flop is correct in principle, but modifications are needed to prevent timing problems. Timing is more critical at higher frequencies.

Thus far, timing diagrams have been based on rectangular pulses, which change from one value to another in zero time. This condition cannot occur in actual circuits. For one thing all circuits contain capacitance, either as discrete components or else in the form of shunt capacitance between components and between semiconductor junctions. Capacitances must be charged or discharged before a circuit switches from one voltage level to another. It also takes time for current carriers to move across semiconductor junctions. Thus, computer circuits do not process square waves. While actual waveforms are complicated, trapezoids are practical approximations. Ramp voltages indicate transitions between low and high levels. Figure 4.43 compares the ideal "brickwall" input and output pulses with the actual trapezoidal waveform for a 2-input AND gate. Pulse width t_w is the time during which the pulse is above 1.5 V.

Even for the relatively simple 2-input AND gate, problems occur when real waveforms are considered. Variations in input rise and fall times affect output pulse width. Also, variations in triggering levels of individual gates cause a wide range of output pulse widths. When both inputs have low trigger levels, the output is a wide pulse; when both inputs have high trigger levels, the output is a narrow pulse. Timing problems become more complicated when stages are cascaded. All gates have propagation delays and each gate type has a specific range. Table 4.7 compares propagation delays (in nanoseconds) for some Schottky gates.

Propagation delays are:

$$t_p = t_{p(G1)} + t_{p(G2)} + \ldots \tag{4.12}$$

TABLE 4.7
Typical Propagation Delays

IC No.	Type	t_{PLH} Min	t_{PLH} Max	t_{PHL} Min	t_{PHL} Max
74S00	Quad 2-input NAND	2.0	4.5	2.0	5.0
74S02	Quad 2-input NOR	2.0	5.5	2.0	5.5
74S04	Hex inverter	1.0	3.5	1.0	4.0
74S08	Quad 2-input AND	2.5	7.0	2.5	7.5
74S32	Quad 2-input OR	2.0	7.0	2.0	7.0

where the subscripts in parentheses refer to the individual gates. Table 4.7 allows a quantitative analysis of propagation delay.

Example 4.2 Compare propagation delays of the equivalent circuits illustrated in Fig. 1.21b and c.

Solution Each signal in Fig. 1.21b passes through one AND gate and one OR gate. Using Table 4.7

$$t_{PLH(min)} = 2.0 + 2.5 = 4.5 \text{ ns}$$

and $t_{PHL(max)} = 7.0 + 7.0 = 14.0 \text{ ns}$

Values for t_{PHL} are slightly longer.

These are the same values for signals *A* and *C* in Fig. 1.21c. However, signal *B* only passes through one AND gate, and this results in

$$t_{PLH(min)} = 2.5 \text{ ns} \quad \text{and} \quad t_{PHL(max)} = 7.0 \text{ ns}$$

and again t_{PHL} is slightly longer.

This example demonstrates that a narrow signal pulse can exit the circuit without interacting with the remainder of the circuit. It also shows the wide range of propagation delays which can occur in a relatively simple circuit. Also, as described in Chap. 3, propagation delay is not an independent parameter. Noise immunity, fan-in, fan-out, and power dissipation affect each other and also propagation delay.

When latches and flip-flops are considered, timing problems are even more complicated because these circuits are regenerative. Signals pass from the input through the circuit to the output and are also fed back to the input. In such cases time delays can result in oscillation or complete shutdown.

Settling times must be considered when regenerative circuits are investigated. *Setup time* t_s is the time just before a clock or enable pulse during which data must be kept at the input to guarantee recognition. *Hold time* t_h is the time immediately after the clock or enable pulse during which data must be kept at the input lead to ensure recognition. Figure 4.44 shows low-setup

Figure 4.44
Settling time waveforms.

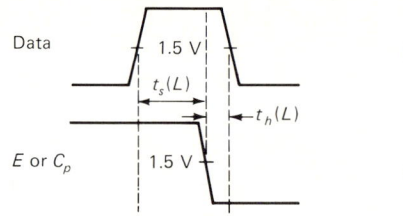
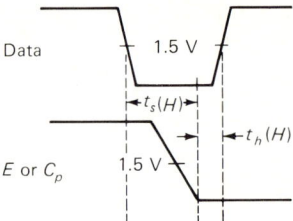

$t_{s(L)}$ and high-hold $t_{h(H)}$ timing diagrams for latches and flip-flops which accept data when clock pulses are logic high.

Typically the output of one flip-flop is transferred to another, and flip-flop chains can be rather long. Propagation delays and settling time must be considered because attempting data transfer while flip-flops are between states causes unpredictable results. One incorrect data transfer anywhere along the chain invalidates whatever data exist along the entire chain.

A partial solution to propagation and settling time problems has already been presented. Synchronizing latches and flip-flops with clock pulses ensures that data can only be processed when the clock pulses are high. A long clock pulse eliminates signal propagation and settling time delays. However, increasing clock pulse length is a mixed blessing. Longer clock pulses result in slower circuits.

Racing is another regenerative circuit timing problem. Latches and flip-flops have at least one pair of feedback paths and *T* flip-flops have two. The output changes only after all the delays have occurred. For the circuit shown in Fig. 4.39 the signal path contains two NAND gates. Once *Q* goes high, this high is fed back to the *R* input lead. This change at *R* can in turn result in a low at *Q* and the output will continue to alternate between low and high as long as *T* remains high. The flip-flop will then have become a square-wave generator and will no longer function as a *T* flip-flop. Racing can be eliminated by making the cumulative time delay t_d greater than t_p, the time that the toggle pulse is high. This makes the feedback signal high while the toggle pulse is still high, and outputs will be stable. However with real ICs, t_d is quite short compared with t_p; IC speed then becomes a problem rather than a solution.

A method of delaying the feedback pulse is required. Delay-line techniques using discrete components are not practical on an IC chip. Solutions are presented in the next section.

4.7 J-K FLIP-FLOPS

Timing problems are resolved by cascading two flip-flops which are clocked from opposite-phase signals. Complementary clocking is achieved by driving one flip-flop directly from C_p while the other is driven through an inverter. The input flip-flop is called a *master*, and the output flip-flop is called a *slave*.

Figure 4.45
RS master/slave flip-flop.

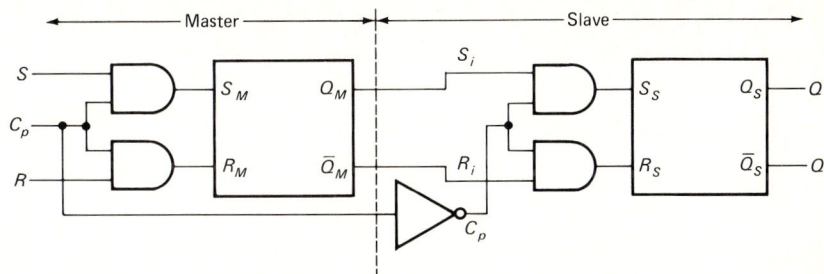

Figure 4.45 is a functional diagram of an *RS* master/slave flip-flop. Gates show the clocked inputs while latches are shown in block diagram form. The inputs to the slave flip-flop are the outputs of the master.

The inverter is the key to master/slave operation. $\bar{C}_p$ is high when C_p is low and vice versa. Therefore, master and slave can never be enabled at the same time. When $\bar{C}_p$ is high, data are transferred from the master outputs Q_m and $\bar{Q}_m$ to the slave inputs R_i and S_i. At this time C_p is low and new data cannot be entered. When C_p is high, conditions are reversed. The master flip-flop is enabled and data at R and S enter. But $\bar{C}_p$ is now low and no data are transferred from master to slave. In other words, when an *RS* master/slave flip-flop accepts input data, the input is on and the output is off. Similarly, when the output is on the input is off.

Figure 4.46 shows the time difference of events in master/slave operation. At t_1, C_p reaches V_1. At this time $\bar{C}_p$ is low enough to disable the input AND gates of the slave. Then at t_2, C_p is high enough to enable the input AND gates of the master. R and S accept data between t_2 and t_3. At t_3 the clock pulse is low enough to disable the master. Whatever data were input at t_3 are locked in. When C_p drops to V_4, the complementary voltage $\bar{C}_p$ is high enough to enable the slave inputs. Because of the time difference between these four voltages, a master/slave cannot simultaneously input and output data. The master can accept data between V_2 and V_3. Then when the voltage is below V_1 and V_4, the slave outputs data.

With some modification this circuit can be converted to the extremely versatile *J-K* master/slave flip-flop. As shown in Fig. 4.47, master input AND gates have three leads instead of two. Also, there is feedback from $\bar{Q}$ to S and from Q to R. These feedback lines are the same as in the *T* flip-flop. *J* and *K* can be used as clock control input leads. A high at *J* sets the outputs, while a high at *K* resets. As with conventional latches, simultaneous 1s at the inputs

Figure 4.46
Clock pulse waveform.

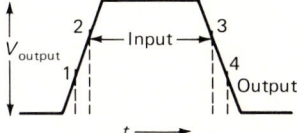

Figure 4.47
J-K master/slave flip-flop.

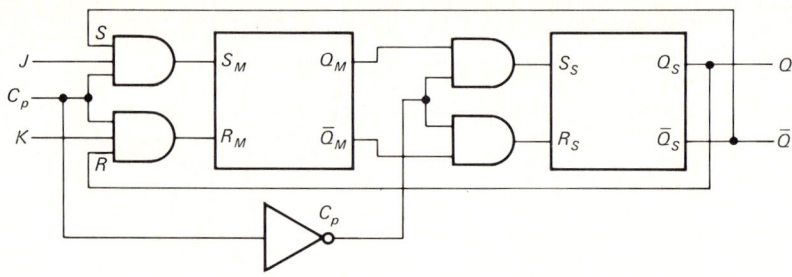

must be avoided. Data can be accepted at J and K when C_p is high. Then, when C_p is low, these inputs are transferred to the outputs. Master/slave action prevents simultaneous inputs at J and K from appearing at the outputs. In this way a J-K can either be initialized or else can accept data. If the J and K inputs are both low, the outputs cannot change regardless of C_p. Therefore, simultaneous 0s at J and K are the conditions for storing data.

Because of the feedback connections, the J AND gate is at the same state as $\bar{Q}$. Also, the K AND gate is at the same state as Q. Since one of these outputs is low, one of the input AND gates is also low. That particular gate is disabled. Therefore, if J and K leads are both high, a high at C_p is automatically steered to the input, which causes the outputs to switch. Used in the manner, the J-K circuit performs as a T flip-flop but without timing problems. The master/slave connection guarantees a feedback signal which is always in the correct time sequence. As long as C_p is high for a sufficient time to activate the circuit, propagation delays and racing do not disturb a J-K master/slave flip-flop.

Admittedly, the J-K flip-flop is a complex circuit. In effect an entire RS flip-flop plus an inverter is used as a delay line which does not attenuate the feedback signals. This form of delay line occupies considerable area on an IC chip. However, it is the most efficient solution to the timing problems. J-K input conditions provide the synchronous applications summarized in Table 4.8.

A J-K flip-flop can be made even more versatile. On addition of an external inverter a J-K circuit converts to a D flip-flop. Moreover, the application can be altered by switching the signals to J and K. For example, after inputs have been entered, these data can be stored by applying lows at both J and K.

TABLE 4.8
J-K Options

J	K	Q_-	$\bar{Q}_+$	Application
0	0	Q_-	$\bar{Q}_-$	Store previous data
0	1	0	1	Reset or accept data
1	0	1	0	Set or accept data
1	1	$\bar{Q}_-$	Q_1	Toggle

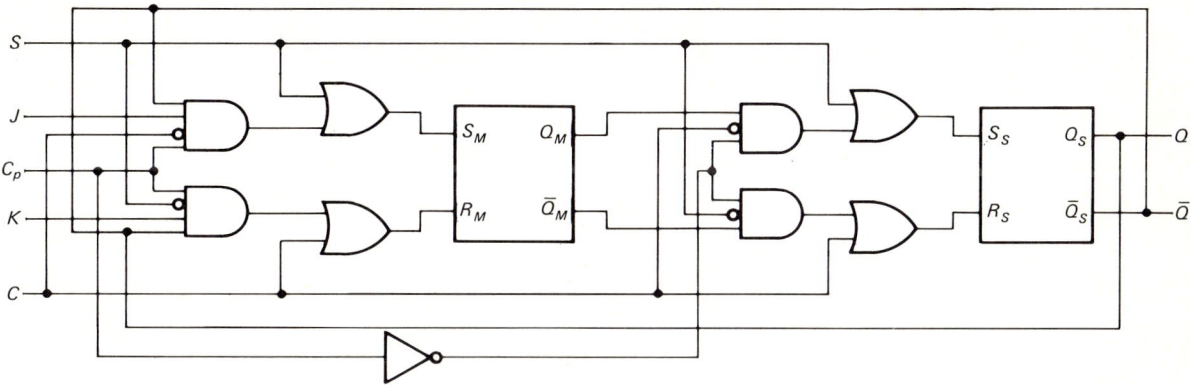

Figure 4.48 *J-K* master/slave flip-flop with separate *C* and *S* functions.

When data is entered, the data must exist long enough to coincide with a clock pulse edge. To ensure this overlap, data must remain at the input at least as long as an entire clock pulse period. For example, if the clock pulse is on for 40 ns and off for 60 ns, data must be maintained for at least 100 ns. Also if data changes while the clock pulse is high, results are unpredictable.

Descriptions of latch and flip-flop operation usually assume a circuit in the reset conditions. However, when power is first applied, reset occurs half the time and set occurs the other half. *J* and *K* inputs provide capability of initializing a flip-flop. After power is applied, either a high at *K* forces the outputs into reset or else the circuit is already reset. While *J* and *K* inputs can be used for initializing, this limits the *J-K* flexibility. Therefore ICs are manufactured which perform the reset and set functions independently of the *J* and *K* inputs. Typically, such leads are labeled *clear* (*C*) and *set* (*S*). Figure 4.48 shows what is involved in adding separate *C* and *S* functions to a *J-K* circuit.

Including the "simple" clear and set functions increases circuit complexity, as they only react with the active flip-flop. Therefore provision is required for initializing the master and the slave; specifically, OR gates which output to R_M, S_M, R_S, and S_S are needed. Two OR gates have the clear signal and the respective AND gates as inputs, and the other 2 OR gates have the set signal and remaining AND gates as inputs. Additional inputs are required at the master and slave input AND gates. These leads receive initializing signals through 4 inverters, which must also be added. An input to either *S* or *C*, maintained for at least one complete clock pulse, initializes the circuit.

Either activating level is available. Figure 4.49*a* shows a block diagram of a positive edge-triggered flip-flop. In this case the output switches state when C_p goes from low to high. Bubbles at *S* and *C* indicate that set or clear are obtained from low rather than high. A low to initialize *J-K* flip-flops is fairly typical. Figure 4.49*b* shows a negative edge-triggered flip-flop which also uses

Figure 4.49
J-K flip-flop symbols. (a) Positive edge trigger; (b) negative edge trigger.

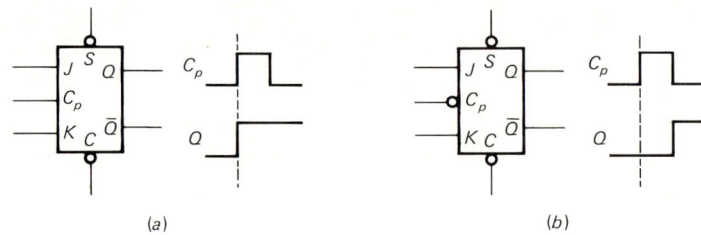

lows to set or clear. In either case seven signal leads are required. This means that two independent flip-flops require 14 signal leads and two more leads for power.

Such 16-pin dual *J-K* flip-flops exist in both TTL and MOS. The TTL 7476 and the MOS 4027 are dual negative edge-triggered *J-K*s. Removing set inputs while retaining clear inputs eliminates one lead from each flip-flop. This less flexible version fits into a 14-pin DIP. Figure 4.50 compares pin-outs of the 14-pin 7473 and the 16-pin 7476. The Schottky versions of the 7473 and 7476 have a maximum clock rate of 30 MHz and a maximum propagation delay of 20 ns.

Because of the variety of input signal combinations, the *J-K* master/slave is the most popular flip-flop. Choices of triggering and initializing increase circuit flexibility.

4.8 WAVESHAPING CIRCUITS

Circuits in this chapter are classified as *multivibrators*. A multivibrator is a regenerative circuit with two possible states. If either state can be maintained indefinitely, the circuit is a *bistable multivibrator*. Latches and flip-flops are bistable multivibrators. Two symmetrical feedback paths yield rapid tran-

Figure 4.50
Dual *J-K* flip-flops.

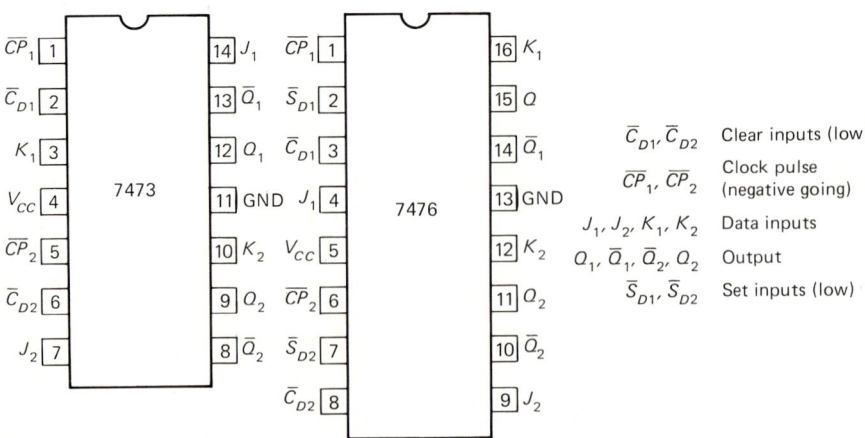

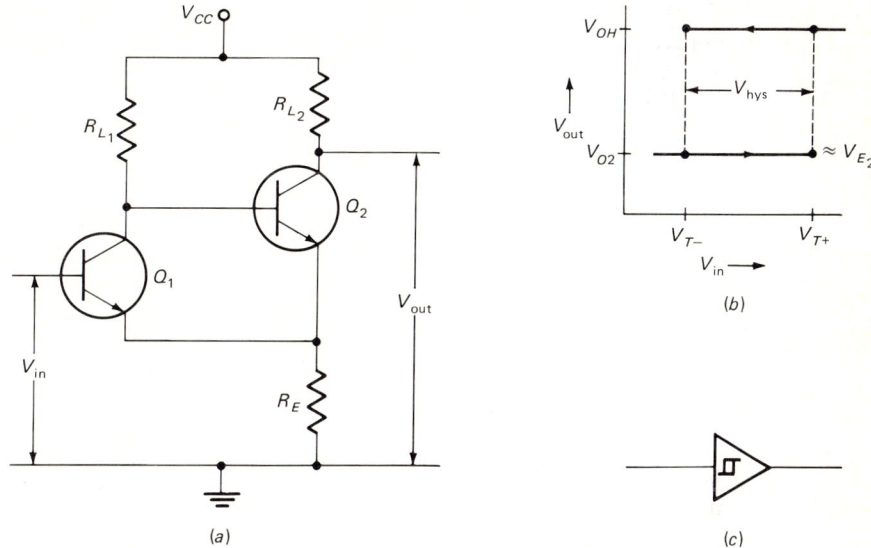

Figure 4.51
Schmitt trigger. (*a*) Schematic; (*b*) input–output characteristic; (*c*) symbol.

sitions and ensure the stability of both states. Other multivibrators contain one feedback path or else two feedback paths which are not symmetrical; these circuits are *monostable* multivibrators. Bistable multivibrators store and transfer pulses, while monostable multivibrators "clean up" and adjust waveforms, which other circuits then process.

The *Schmitt trigger*, shown in Fig. 4.51a, is a single-feedback monostable circuit. Here R_E is the regenerative component. Voltage across R_E in output transistor Q_2 is fed back to input transistor Q_1. If the input voltage to a Schmitt trigger is 0 or low, Q_1 is cut off. With Q_1 cut off the base of Q_2 is high enough to drive Q_2 into saturation. When Q_2 is saturated, the output low voltage is

$$V_{OL} = V_{Q_2(\text{sat})} + I_{C_2(\text{sat})}R_E$$
$$\simeq I_{C_2(\text{sat})}R_E = V_{E_2} \tag{4.13a}$$

As V_{in} increases, a voltage is reached at which Q_1 barely begins to conduct. Since Q_2 is still saturated, current flowing through R_E does not change appreciably. Thus, V_{out} remains the same. V_{in} continues to increase and Q_1 conducts harder. The base voltage at Q_2 drops sharply when Q_2 cuts off. The value of V_{in} which drives Q_2 to cut off is V_{T+}, the upper trigger level. With Q_2 cut off and Q_1 saturated, the output high voltage is

$$V_{OH} \simeq V_{CC} \tag{4.13b}$$

When V_{in} drops below V_{T+}, Q_1 remains saturated. As V_{in} continues to drop, Q_1 goes back into conduction while Q_2 is still cut off. With further decrease in V_{in}, Q_1 cuts off and Q_2 switches into saturation. This voltage is

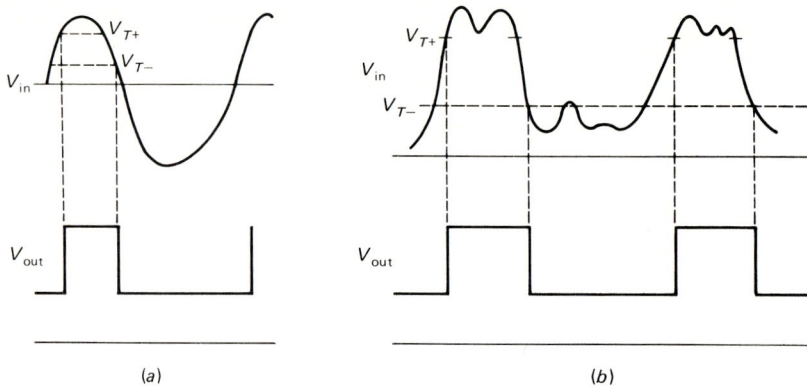

Figure 4.52 Schmitt trigger applications. (*a*) Waveshaping; (*b*) noise elimination.

V_{T-}, the lower trigger level. Once V_{T-} is reached, any further decrease in V_{in} cannot affect the circuit. The output voltage is given by Eq. (4.13a). Input/output voltage characteristics of a Schmitt trigger are shown in Fig. 4.51b, and the symbol is shown in Fig. 4.51c. *Hysteresis* is the difference between upper and lower trigger levels.

$$V_{hys} = V_{T+} - V_{T-} \tag{4.13c}$$

Schmitt triggers can decrease rise and fall times. Digital counters and timers require steep voltage rises for reliable triggering. Sine waves rise and fall gradually and cannot trigger a counter circuit. A Schmitt trigger "sharpens" a sine wave, as shown in Fig. 4.52a.

Another application is preventing noise from causing accidental triggering. As shown in Fig. 4.52b, noise can occur at high or low voltage levels. But a Schmitt trigger cannot go to V_{OL} until V_{T-} is reached and cannot return to V_{OH} until V_{T+} is reached. Therefore a Schmitt trigger will not change states unless both the upper and lower trigger levels are experienced. Ideally a single Schmitt trigger can be used to eliminate both high- and low-level noise. However, this is rarely the case because noise levels and trigger levels are not related. Schmitt triggers are also used for signal squaring and level discrimination.

Several IC Schmitt triggers exist. The 7414 is a 14-pin TTL hex Schmitt trigger inverter, and the 7413 is a dual 4-input Schmitt NAND gate. Signal characteristics are the same and are shown in Table 4.9. The 4093 is an MOS quad 2-input Schmitt NAND, and the 40014 is an MOS hex Schmitt inverter.

In bistable multivibrators, feedback paths are resistive. The Schmitt trigger, a monostable multivibrator, has a single feedback path, which is also resistive. The *one-shot*, another monostable multivibrator, is shown in Fig. 4.53a. A one-shot has unsymmetrical feedback paths. R_F is a resistive feedback from collector of Q_1 to base of Q_2, and C is capacitive feedback from collector of Q_2 to base of Q_1. One state can be maintained indefinitely but the other state is only partially stable. The duration of the partially stable state is

TABLE 4.9
TTL Schmitt Parameters

Characteristic	Max	Min
V_{T+} (V)	1.5	2.0
V_{T-} (V)	0.6	1.1
t_{PLH} (ns)	27	22
t_{PHL} (ns)	22	23

determined by the values of R and C, which are not related to the trigger pulse width used to activate the one-shot. In fact, the partially stable output pulse can be longer or shorter than the trigger pulse.

In the absence of an external trigger, R drives Q_1 into saturation. With Q_1 saturated, the collector of Q_1 is low and the base of Q_2 is even lower; thus Q_2 is cut off. The stable state is a low at Q and a high at $\bar{Q}$. These conditions exist if no signal or a high signal exists at input T. With Q_2 cut off, the voltage across C is $V_{CC} - V_{BE(sat)}$, or about 4.3 V for a 5-V supply.

If T goes low, Q_1 is driven into cutoff. This raises the voltage at the base of Q_1 to V_{CC} and drives Q_2 into saturation. With Q_2 saturated $\bar{Q}$ is low. The voltage across the capacitor changes to about $-5 + 0.7$, or -4.3 V. Since Q_1 is now cut off, the capacitor can only discharge through R. The instantaneous voltage V_C across the capacitor is the negative-going trigger pulse $V_{CC}e^{-t/RC}$ less the initial capacitor charge $V_{CC}(1 - e^{-t/RC})$

$$V_C = V_{CC}e^{-t/RC} - V_{CC}(1 - e^{-t/RC}) \tag{4.14a}$$

This capacitor voltage discharges toward 0, and Q_2 cuts off approximately when V_C reaches 0. As Q_2 cuts off, Q_1 is driven back into saturation, and the

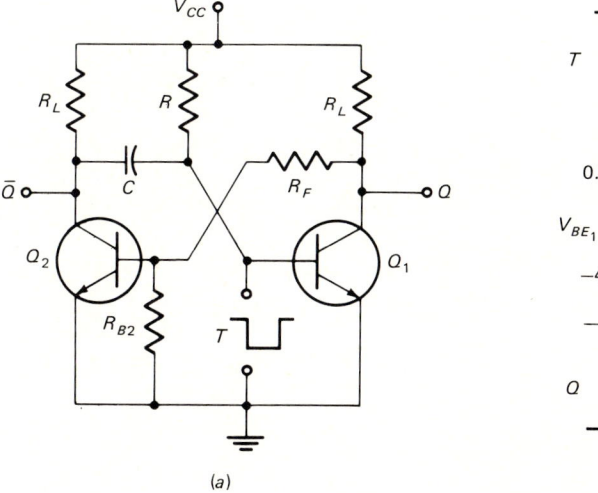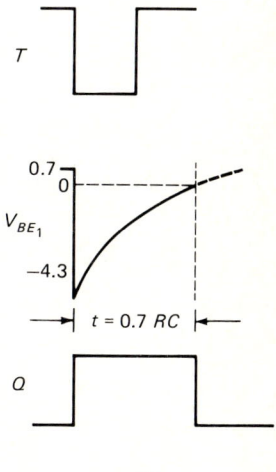

Figure 4.53
One-shot. (*a*) Schematic; (*b*) waveforms.

one-shot returns to the stable state. The time required for C to discharge is determined by setting Eq. (4.14a) equal to 0

$$0 = V_{cc}e^{-t/RC} - V_{cc}(1 - e^{-t/RC})$$

or
$$V_{cc} - V_{cc}e^{-t/RC} = V_{cc}e^{-t/RC}$$

$$V_{cc} = 2V_{cc}e^{-t/RC}$$

then $\frac{1}{2} = e^{-t/RC}$

Inverting both sides yields

$$2 = e^{t/RC}$$

Taking natural logarithms of both sides results in

$$0.639 \simeq t/RC$$

which is approximated as

$$\frac{t}{RC} = 0.7$$

Finally $\quad t = 0.7\,RC \qquad (4.14b)$

is the time for Q_1 remaining high. Figure 4.53b shows one-shot multivibrator waveforms. In this case the output lasts longer than the trigger pulse. However, the time during which Q_1 remains high depends only on R and C. These are design parameters and Eq. (4.14b) is independent of trigger pulse width.

Both transistors in a one-shot are inverters, and it is possible to use NAND gates as inverters. Figure 4.54 shows a NAND gate one-shot. NAND-gate numbers correspond to the transistor discussion. This gated one-shot has the same period, and it is possible to substitute NOR gates.

Figure 4.54
One-shot using NANDs.
(a) Schematic; (b) redrawn.

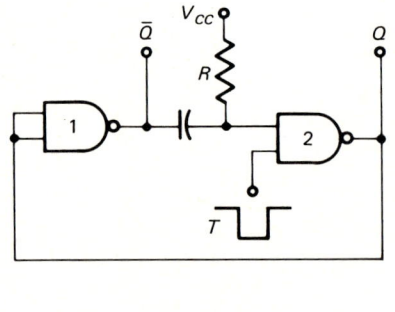
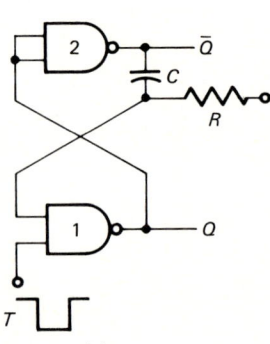

(a)   (b)

Example 4.3 A NAND-gate one-shot is triggered by a negative pulse, which remains low for 40 ns and repeats every 150 ns. Using an R of 5 kΩ, (*a*) find the value of C which will make Q high for 70 ns, and (*b*) sketch T, Q, and $\bar{Q}$.

Solution (*a*) Since $T = 0.7\,RC$

$$C = \frac{T}{0.7\,R} = \frac{70 \times 10^{-9}}{0.7 \times 5 \times 10^3}$$

$$C = 20 \text{ pF}$$

(*b*) The waveforms are shown in Fig. 4.55.

This example demonstrates a one-shot *pulse stretcher*, or delay. Pulse stretching ensures that pulses which must overlap do in fact coincide. Another application is based on the impossibility of retriggering until the circuit returns to the stable state. This can be exploited to eliminate all but the first pulse in a train of pulses, for example, when a mechanical switch is activated. With proper selection of R and C, Q will be high longer than the switch oscillates. It is also possible, within the limits of propagation delay and settling time, to obtain one-shot pulses which are narrower than the input pulses.

IC one-shots offer greater flexibility than can be obtained by constructing one-shots from individual gates. The TTL 74121 is a 14-pin one-shot which can be triggered by either negative or positive signals. Positive edge triggering is accomplished with an internal Schmitt trigger. IC capacitors are not practical, and the 74121 requires an external capacitor. The 4047, 4528, and 4538 are CMOS one-shots.

One-shot multivibrators shape input signals. The output of a one-shot is of constant duration even though the input width may vary. Besides pulse stretching, one-shots can also generate a fixed time delay. One-shots can be used as pulse generators provided the input is a square wave.

The *astable* multivibrator is an extension of the one-shot circuit. As shown in Fig. 4.56, both feedback paths are capacitors. Just as in a one-shot, a transistor is cut off while the other is saturated. Cutoff and saturation times

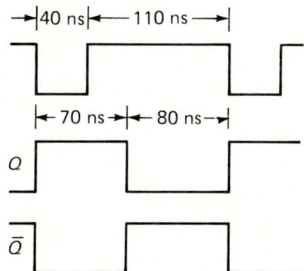

Figure 4.55 Waveforms of Example 4.3.

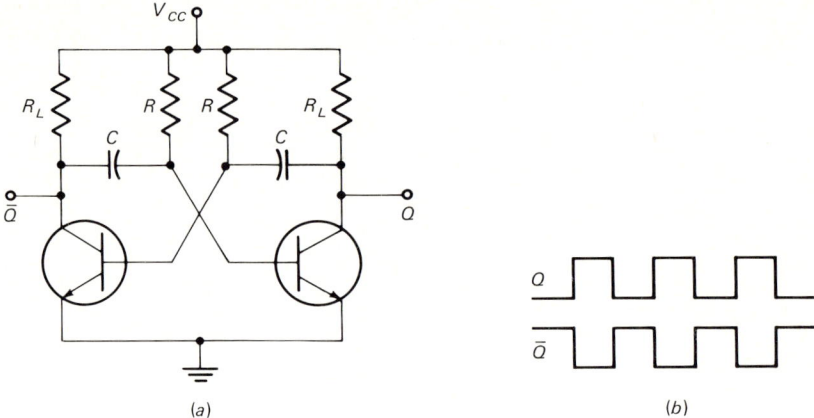

Figure 4.56
Astable multivibrator. (*a*) Schematic; (*b*) typical waveforms.

are determined by *R* and *C*. The output of an astable multivibrator is a self-sustained sequence of pulses, and symmetrical or asymmetrical outputs are possible.

SUMMARY

1. A latch is a circuit which stores a bit. Regeneration is used in latch design.

2. A two-NOR-gate latch has complementary outputs. It can accept and store either a 0 or a 1.

3. Cross-coupled NAND gates can also be used. An *RS* latch can isolate contact vibration of a mechanical switch from an electronic circuit.

4. Sequential circuits operate in a specified order. *RS* latches are converted into sequential circuits called flip-flops by including a clock lead at the input.

5. *D* latches pass data while Enable is on and store data when Enable is off. *D* flip-flops are edge-triggered and the output is the same as the input but delayed by one clock pulse.

6. *T* flip-flops toggle each time the input goes positive. The output frequency of a *T* flip-flop is half of the input frequency.

7. The master/slave flip-flop solves propagation delay and racing. Depending on control signals, a *J-K* circuit can perform *RS* and *T* flip-flop functions.

8. Asymmetrical feedback is used to adjust the waveform shape. Schmitt triggers and one-shots are the most popular monostable multivibrators.

PROBLEMS

4.1 Propagation delay and required pulse duration of an inverter amount to 15 ns. How long does it take a two-inverter latch to store a bit?

Answer: 30 ns.

4.2 Two high-speed Schottky NOR gates are used as a two-inverter latch. If each NOR gate has a 3-ns propagation delay, when does the latch function?

4.3 A high-speed Schottky OR gate has a propagation delay of 3 ns. What time is required for such a gate to latch?

Answer: 3 ns.

4.4 What happens if a single NOR gate is used as a latch?

4.5 What happens when a single NAND gate is used as a latch?

Answer: Control is lost when 0 is fed back.

4.6 Discuss whether or not two OR gates can be used as an *RS* latch.

4.7 Determine if the conditions shown in Fig. 4.51b are correct.

Answer: $\bar{R} = 1$ and $Q = 0$ should generate $\bar{Q} = 1$.
$\bar{S} = 1$ and $\bar{Q} = 1$ should generate $Q = 0$.

4.8 Describe the signals in Fig. 4.16 when $R = 1$ and $S = 0$.

4.9 For the NAND latch shown in Fig. 4.17 describe the conditions when $R = 0$ and when $S = 1$.

Answer: $R = 0$ and $Q = 0$ make $\bar{Q} = 1$.
$S = 1$ and $\bar{Q} = 1$ make $Q = 0$.

4.10 Assume the original outputs of an *RS* flip-flop are $Q = 0$ and $\bar{Q} = 1$. Sketch Q and $\bar{Q}$ to the same time scale.

4.11 C_p, R, and S are the same as in Problem 4.10 but the original outputs are $Q = 1$ and $\bar{Q} = 0$. Sketch Q and $\bar{Q}$.

Answer:

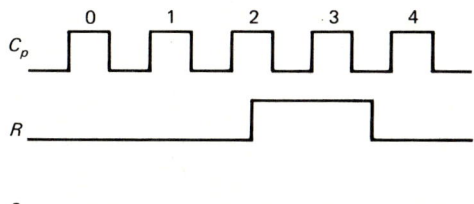

4.12 Conditions for an *RS* flip-flop are as shown. Sketch both outputs.

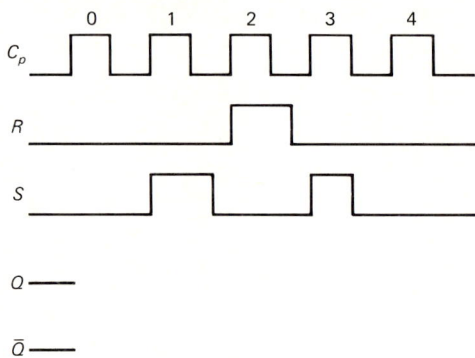

4.13 Assume that a *D* latch starts in the reset condition. Sketch the outputs for the conditions shown.

Answer:

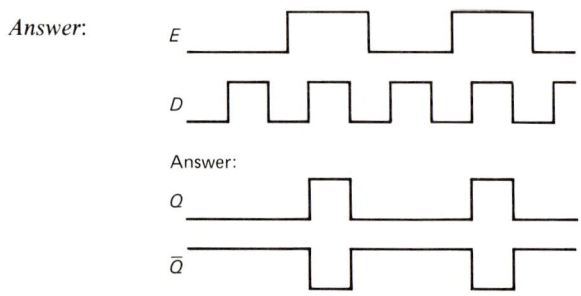

4.14 A set of conditions for a *D* flip-flop is as shown.

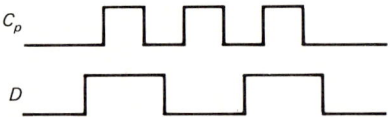

Assuming reset is the starting condition, sketch the outputs.

4.15 Another *D* flip-flop starting in the reset condition has C_p and *D* as shown. Sketch the outputs.

Answer:

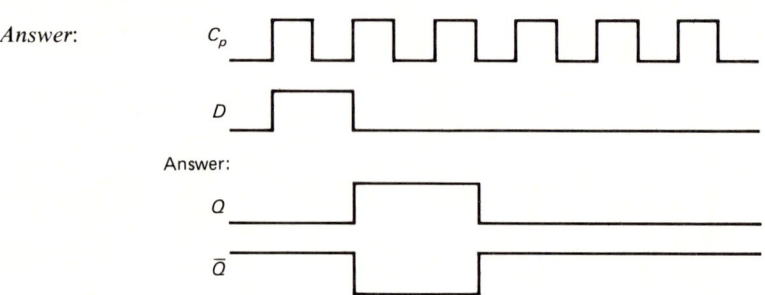

4.16 The input to a *T* flip-flop is as shown. Sketch the output waveform and determine its frequency.

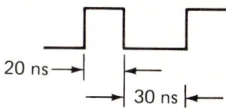

4.17 A logic circuit consists of the signal paths as shown. Assuming that the Schottky gates listed in Table 4.7 are used, what is the shortest t_p which can "sneak through" the inverter without interacting with signals *A* and *B*?

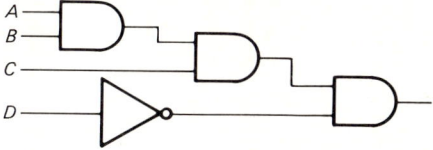

Answer: 4 ns.

4.18 It takes 30 ns for an input to a *T* flip-flop to reach the output. How long should the feedback delay be?

4.19 Using an external inverter show the connection for converting a *J-K* to a *D* flip-flop.

Answer:

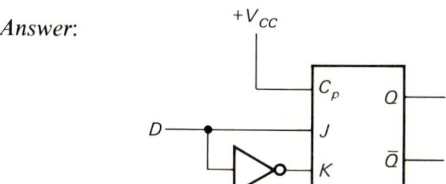

4.20 Show the connection needed to convert a *J-K* into a *T* flip-flop.

4.21 A 10 MHz asymmetrical waveform is applied to a *J-K* which is being used as a *T* flip-flop. What is the output?

Answer: 5 MHz square-wave.

4.22 Show the logic signals for the flip-flop of Fig. 4.41.

4.23 Use Table 4.9 to find the minimum hysteresis of a TTL Schmitt trigger.

Answer: 0.4 V.

4.24 A one-shot multivibrator in which $R = 10 \text{ k}\Omega$ and $C = 50$ pF is driven by a 20-kHz squarewave. Determine *T* and *Q*, and then sketch the waveforms.

4.25 What value of capacitance must be placed in parallel with *C* of the previous problem to obtain a 0.5-μs output?

Answer: 21.4 pF.

5
COUNTERS

5.0 INTRODUCTION

Circuits for counting events are used in computers, frequency meters, timers, and digital multimeters. Flip-flops are the basis of counting.

Asynchronous counters operate in numerical sequence. The output of each flip-flop activates the next flip-flop throughout the entire length of the counter chain. When synchronous counters are used, all flip-flops are triggered simultaneously. Control logic determines which flip-flops are activated on each particular count.

Modifications to synchronous counter circuits result in registers, which temporarily store data. A variety of registers are available. Data can be input sequentially or simultaneously, and the same possibilities exist at the output terminals. Registers can also rotate the data sequence. Besides temporarily storing data, registers can perform some calculations and interface between a computer and the outside world.

5.1 ASYNCHRONOUS COUNTERS

Many situations require counting. One important application is keeping track of the sequence of steps in a computer program. Frequency meters, digital watches, and production line totalizers are other applications of counting. Regardless of application, the bistable multivibrator is the basic counting circuit.

When the J and K inputs are connected to a logic high, a flip-flop changes state once for each input pulse. In this mode the J-K flip-flop is a trouble-free T flip-flop. The J-K configuration is one of the most popular circuits for constructing counters.

The input is connected to C_p of the first stage in a chain of flip-flops. Then the Q output of each stage is connected to C_p of the next stage. Thus, counts "ripple" up to the next higher stage, and this circuit is called a *ripple* counter. Ripple counters can count random or *asynchronous* events as well as periodic or synchronous events.

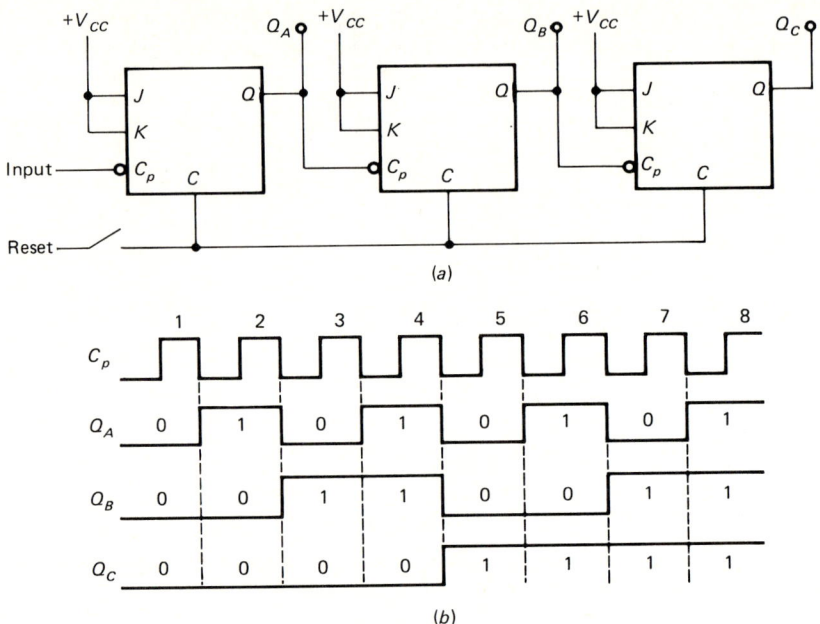

Figure 5.1
Three-bit ripple counter.
(*a*) Schematic;
(*b*) timing diagrams.

Figure 5.1*a* shows a three-stage ripple counter built with negative edge-triggered flip-flops. When power is first applied, it is unlikely that all three Q outputs will be low. Therefore clear leads are connected together. A momentary reset pulse clears the flip-flops which happen to have highs at Q; flip-flops with lows at Q are undisturbed. Resetting "zeroes" the counter. The first flip-flop output is Q_A, the second flip-flop output is Q_B, and the last is Q_C. It is standard practice to name outputs in alphabetical order with the least significant output as A and the others in ascending order.

If LEDs are connected to these flip-flop outputs, all three lamps will be off after the counter has been initialized. The first negative-going input pulse drives Q_A high. In terms of the *CBA* sequence, the outputs are now 001. The second flip-flop cannot go on at this time because negative edge-triggered flip-flops are used. When Q_A is high, C_p of the second flip-flop is also high. When the second negative-going input pulse occurs, Q_A flips back to low. This negative-going pulse triggers flip-flop B; Q_B is now high, and the *CBA* output is now 010.

On the third negative input pulse, Q_A goes high once more. Q_B is not affected because a high at C_p of the B flip-flop cannot trigger. Therefore, after the third negative input the *CBA* sequence is 011. As additional input pulses occur, the circuit continues to count in binary. After the count reaches 111, the next negative edge resets all three flip-flops. Counting begins again from 000. Figure 5.1*b* shows the timing diagram and *CBA* sequence. A 3-bit counter counts the first eight events, 0 through 7. Similarly, a 4-bit counter displays the first 16 events, 0 through 15. The maximum count C_{max} for a counter with N flip-flops is given by

$$C_{max} = 2^N - 1 \tag{5.1}$$

Example 5.1 A photocell triggers an assembly-line bottle filler. The maximum filling rate is 10 bottles per minute. How many counter stages are required to count a 1-h output?

Solution In 1 h the maximum number of bottles cannot exceed

$$10 \, \frac{\text{bottles}}{\text{min}} \cdot \frac{60 \, \text{min}}{\text{h}} = 600 \, \frac{\text{bottles}}{\text{h}}$$

Therefore
$$600 = 2^N - 1$$
$$601 = 2^N$$
$$N \log 2 = \log 601$$
$$0.301 \, N \simeq 2.779$$
$$N \simeq 9.23$$

Of course, building 0.23 flip-flops is impossible. Since more than 9 flip-flops are needed, a 10-bit counter is required.

A 3-bit ripple counter is practical, and the output is binary. However, if this counter is to be used to indicate floors in an elevator, decimal lights individually numbered 0, 1, 2, ..., 7 are more convenient. Converting binary numbers to decimal is called *decoding*. A 3-bit decoder can be built with 3-input AND gates.

The trick is to turn on only the correct decimal lamp represented by each binary sequence. That is, 000 should only turn on lamp 0, 001 should turn on only lamp 1, 010 should turn on only lamp 2, etc. As with other digital circuits, the truth table comes first. Then a circuit which satisfies the truth table is constructed.

Figure 5.2 shows truth table and decoder circuit. The truth table contains the decimal number, the CBA binary equivalent, and the binary equivalent in terms of the variables with inverted form. For example, the CBA equivalent of 5 is 101. With inverts this is $C \cdot \bar{B} \cdot A$. Therefore the C, A, and $\bar{B}$ are connected to the AND gate 5. The $\bar{B}$ output is high at this count since the B output is low. Thus all 3 inputs to gate 5 are high and the output goes high.

Each AND gate is connected to the appropriate inverting and noninverting outputs of the 3-bit counter. Only one AND gate is activated on each count. Some IC flip-flops do not present inverted states. In this case, inverters are required to obtain $\bar{Q}_C$, $\bar{Q}_B$, and $\bar{Q}_A$. The elevator decoder uses individual output lamps for each digit. This same decoding principle is also used to drive

Figure 5.2
Three-bit-to-decimal decoder. (*a*) Truth table; (*b*) schematic.

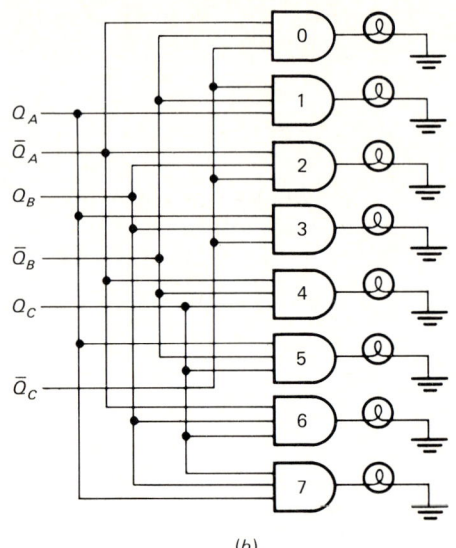

Decimal	Counter output	Q terminals
	C B A	
0	0 0 0	$\bar{C}\cdot\bar{B}\cdot\bar{A}$
1	0 0 1	$\bar{C}\cdot\bar{B}\cdot A$
2	0 1 0	$\bar{C}\cdot B\cdot\bar{A}$
3	0 1 1	$\bar{C}\cdot B\cdot A$
4	1 0 0	$C\cdot\bar{B}\cdot\bar{A}$
5	1 0 1	$C\cdot\bar{B}\cdot A$
6	1 1 0	$C\cdot B\cdot\bar{A}$
7	1 1 1	$C\cdot B\cdot A$

(*a*)   (*b*)

more sophisticated outputs, such as segment displays, printers, and CRT displays.

Binary counters can test logic gates. For example, there are four possible input combinations for a 2-input gate. Each combination must be tested to ensure proper operation. If the output of an astable multivibrator is connected to a 2-bit counter, the counter output cycles through 00, 01, 10, 11. Thus the Q_A and Q_B outputs step through all combinations. Similarly, outputs of a 3-bit binary counter can step through all eight combinations needed to test a 3-input gate.

Binary ripple counters are also used as precision frequency dividers. As described in the last chapter, the output frequency of a flip-flop is exactly half of the input frequency. In Fig. 5.1*b* the Q_A output frequency is $f_{in}/2$ and the Q_B output is half of the Q_A, or $f_{in}/4$. Similarly, the Q_C output is $f_{in}/8$. Frequency division by 2 continues as additional counter stages are added. In general the output frequency of the Nth stages is

$$f_N = \frac{f_{in}}{2^N} \tag{5.2}$$

Example 5.2 The output of a 10 kHz quartz crystal oscillator is connected to a 3-bit ripple counter. What frequencies are available?

Solution In addition to the 10 kHz source frequency, Eq. (5.2) can be used to find the frequencies available from the counter. At the Q_A output:

$$f_1 = \frac{10 \times 10^3}{2^1} = 5 \text{ kHz}$$

At the Q_B output:

$$f_2 = \frac{10 \times 10^3}{2^2} = 2.5 \text{ kHz}$$

and at the Q_C output:

$$f_3 = \frac{10 \times 10^3}{2^3} = 1.25 \text{ kHz}$$

Sometimes counters are specified in terms of the frequency division. A 2-bit counter is a divide-by-4, a 3-bit counter is a divide-by-8, etc. If the fundamental frequency is accurate, frequency division generates frequencies of any desired precision. Such circuits are used in precision oscillators known as frequency *synthesizers*. Frequency division with binary counters is used to generate electronic music. An *octave* is the range between a frequency and twice that frequency. The standard musical tone A is 440 Hz. One octave above this A at 880 Hz is the next A, etc. By starting with an appropriately high frequency, many octaves of A and of other notes can be obtained by frequency division. Sound produced from different musical instruments varies because each instrument has a different fundamental frequency and different harmonics. Electronically, different instruments are simulated by combining the correct fundamental frequency with the appropriate harmonics.

Addition is another application of counter circuits. Counting is a rudimentary form of addition. If 7 pulses are entered into a counter and followed by 6 more pulses, the counter displays a count of 13. While this is appropriate for adding small numbers, it is not fast enough for adding large numbers.

5.2 DOWN COUNTERS

Circuits which count upwards from zero are called *up counters*. In some applications it is more appropriate to count down from a preset number, and such circuits are called *down counters*. For example, scoreboards at football and basketball games indicate remaining time rather than elapsed time.

Up counters can be converted to down counters. Output bits are taken from the Q terminals in an up counter. Taking the outputs from the $\bar{Q}$ terminals is one way of making a down counter. When a 3-bit counter is initialized, the CBA count is 000. At the same time the $\bar{C}\bar{B}\bar{A}$ count is 111. After the first input pulse the CBA count is 001, the $\bar{C}\bar{B}\bar{A}$ count is 110, etc. Thus a sufficiently long counter of the type shown in Fig. 5.3a using $\bar{Q}$ outputs can be preset to 15 min. The counter would display the time remaining in a football game quarter. A down counter can be preset by triggering the appropriate preset or clear inputs of each flip-flop.

An alternate down-counting circuit in which each flip-flop is triggered from

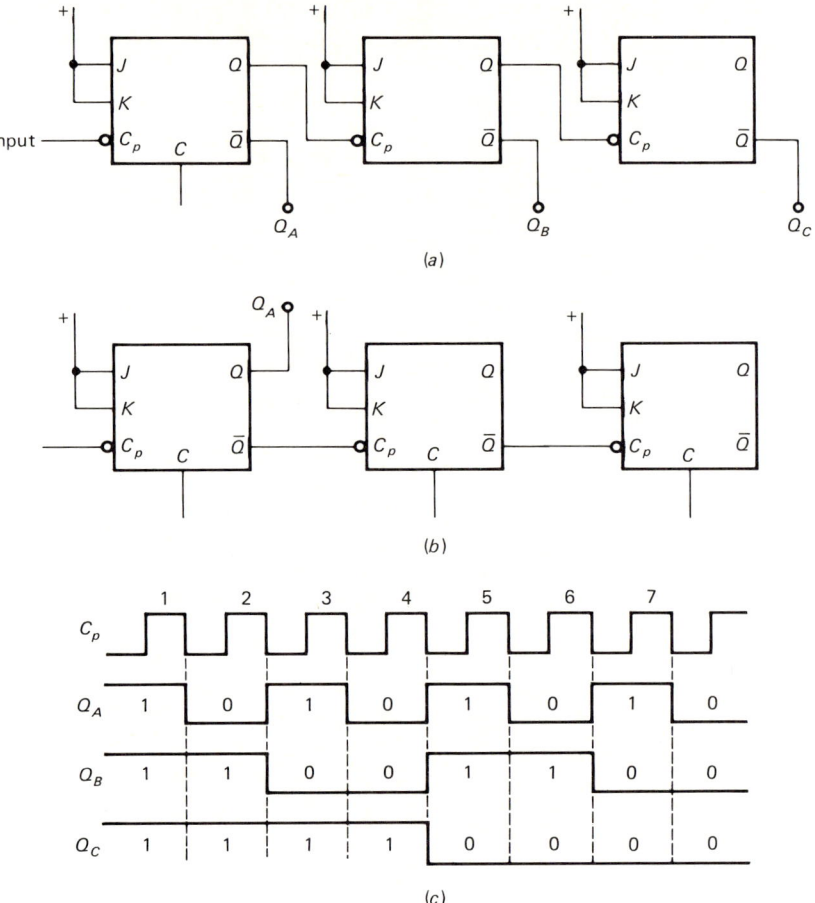

Figure 5.3 Three-bit down counting. (a) $\bar{Q}$ outputs; (b) $\bar{Q}$ triggering; (c) waveforms.

the previous $\bar{Q}$, is shown in Fig. 5.3b. We assume that the 3-bit down counter shown in Fig. 5.3b is initialized at a 111 count. The first negative-going pulse drives Q_A low, and the CBA count is now 110. Since Q_A is low after the first count, $\bar{Q}_A$ must be high. On the second negative-going input, flip-flop A reverses. Since Q_A goes low, this negative edge triggers the flip-flop B so that Q_B goes low. Thus, after the second input pulse the CBA count is 101, which is 2 less than 111. The CBA count continues to decrease by 1 for each incoming negative pulse until the counter reaches 000. When the next pulse triggers the counter, it resets to 111. Down counting begins again. Figure 5.3c shows a down counter timing diagram.

Because a down counter with $\bar{Q}$ triggering has the same output terminals as an up counter, a counter which counts either up or down can be built. The same flip-flops that count up can count down, and logic gates determine which operation occurs. Figure 5.4 shows a 3-bit up/down counter with a

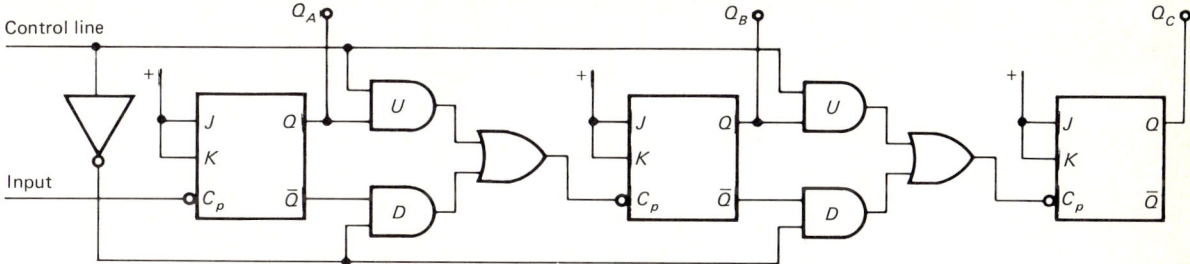

Figure 5.4 Ripple up/down counter.

single line to select up or down counting. A combination of AND with OR gates determines whether trigger pulses are taken from Q or $\bar{Q}$. AND gates for counting up are labeled U, and AND gates for counting down are labeled D.

One input lead of each U gate is connected to the Q output of each flip-flop, and the other input lead is connected to the control line. Similarly, one input lead of each D gate is connected to $\bar{Q}$ and the other D input is connected to the control line through an inverter.

When the signal on the control line is high, U gates are enabled and D gates are disabled. With the U gates enabled, the Q signal activates the OR gate connected to the C_p lead of the next flip-flop. The result is sequential up counting. When a low is on the control line, U gates are disabled and D gates are enabled through the inverter. Enabling D gates means that $\bar{Q}$s activate the D gates, which in turn trigger the OR gates connected to C_p leads. Since C_p is triggered by $\bar{Q}$, down counting takes place.

If a high on the control line is represented by C, then the Boolean equation for triggering all stages of an up/down counter except the first is

$$C \cdot Q + \bar{C} \cdot \bar{Q} = C_p \tag{5.3}$$

Table 5.1 is the truth table for this control logic equation.

Table 5.1 verifies that C_p will only be high when both control signal and flip-flop output are high at the same time. A Karnaugh map for Eq. (5.3) will show that 1 OR gate and 2 AND gates constitute a minimum gate circuit for controlling up/down counting.

Just as an up counter can perform slow addition, a down counter can perform slow subtraction. If the preset in a down counter is 13, then 7 counts

TABLE 5.1 Up/Down Counter Logic

C	Q	$C \cdot Q$	$\bar{C}$	$\bar{Q}$	$\bar{C} \cdot \bar{Q}$	$C \cdot Q + \bar{C} \cdot \bar{Q}$
0	0	0	1	1	1	1
0	1	0	1	0	0	0
1	0	0	0	1	0	0
1	1	1	0	0	0	1

Figure 5.5
Propagation delay in a 3-bit ripple counter.

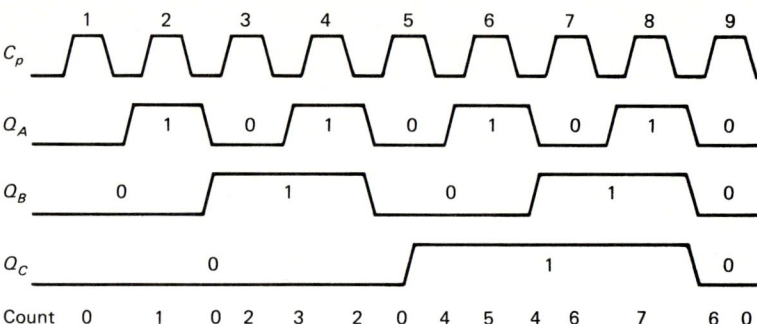

later the down counter will display 6. Therefore, depending on the control line, an up/down counter can perform either addition or subtraction. However, an up/down counter is even slower than an up or down counter because of the additional propagation delays caused by the AND and OR gates.

As discussed, propagation delays can cause a considerable difference between the real and ideal circuit performance. Since ripple counters consist of chains of flip-flops, these counters are particularly susceptible to cumulative propagation delays. Figure 5.5 shows a 3-bit ripple counter waveform with propagation delays included.

Consider a circuit initialized at 000. A negative-going input pulse activates flip-flop A, but Q_A does not go high until the time required for propagation delay. When Q_A does go high, the CBA count is 001. On the second negative-going pulse, Q_A should go low and Q_B should go high. Owing to the difference in propagation delays there is an intermediate count. Q_A only requires one propagation delay to change states, while Q_B must experience two propagation delays to switch. As a result Q_A goes low before Q_B goes high. This results in a count of 000 between 001 and 010. On the third input pulse the count goes from 010 directly to 011. However, differences in delay times of the flip-flops result in counts of 010 and 000 before the count actually reaches 100 after the fourth input pulse. More incorrect counts continue to occur throughout the counting sequence. For a 3-bit ripple counter there are two extra 0s, one extra 2, one extra 4, and one extra 6 before the counter resets to the correct 0. Extra counts are even more troublesome when counters of 4 or more bits are considered.

Unwanted signals in digital circuits are called *glitches*. As a result of differences between propagation delays, glitches always occur in ripple counters, but they are not always dangerous. "False" counts do not last as long as correct counts. Therefore glitches at relatively low counting rates such as occur in digital clocks can be eliminated from the numerical display. Either incorrect counts are automatically deleted or else the display is designed not to be activated until the count has existed longer than any of the incorrect counts. While false counts must occur, frequency division in ripple counters is still exact. Only one output is used for each division, and cumulative propagation delays do not degrade frequency division.

As counting rates go higher, glitches become more important. Ripple counters become useless when counting period approaches flip-flop propagation delay. Total propagation delay T_d is the time required for a counter to complete its response to an input pulse. T_d is longest when a ripple counter must go from the maximum count back to zero. For this condition the count pulse must ripple through the entire counter. As an example, in a 3-bit counter, T_d will be a maximum when the count resets to 000 from 111. The shortest input counting period T_{max} must be less than the sum of the individual flip-flop propagation delays. Assuming that the same flip-flop type is used throughout the counter

$$T_{max} < n \cdot t_{PHL} \tag{5.4a}$$

and the reciprocal of T_{max} is the highest possible counting rate

$$f_{max} < \frac{1}{T_{max}} \tag{5.4b}$$

A large selection of J-K flip-flop ICs with various options exist in both TTL and MOS. The 7473 is a 14-pin TTL package which contains two independent J-K master/slave flip-flops. The t_{PHL} is 40 ns.

Example 5.3 A 3-bit ripple counter is to be constructed using 7473s. What is the maximum counting rate?

Solution Since there are two flip-flops per 7473 chip, $1\frac{1}{2}$ packages are required. With $n = 3$ and since t_{PHL} is 40 ns

$$f_{max} < \frac{1}{3 \times 40 \times 10^{-9}} < 8.33 \text{ MHz}$$

If counters operate well below the maximum rate, ripple counters are useful. Self-contained ripple counters are available in TTL and MOS. The 7493 is a 14-pin TTL package which contains an internally connected 3-bit ripple counter and an independent 1-bit counter. When used individually, these counters divide by 8 and divide by 2. If the divide-by-2 circuit is externally connected to the divide-by-8 circuit, a divide-by-16 circuit is obtained. The t_{PHL} of the 3-bit counter section of the 7473 is 51 ns and the t_{PHL} for the single flip-flop is 18 ns.

5.3 SYNCHRONOUS COUNTERS

Cumulative time delays can be eliminated by triggering all flip-flops in a counter at the same time. Counters in which all flip-flops are triggered simultaneously are called *synchronous*. Since clock pulses in a synchronous

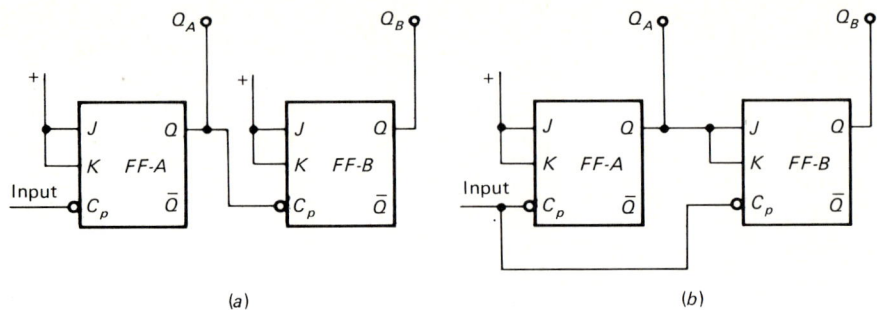

Figure 5.6
2-bit counters. (*a*) Asynchronous; (*b*) synchronous.

counter arrive at each flip-flop on each count, control logic is needed to determine which flip-flops change state as the count progresses. Independent adjustment of *J* and *K* logic levels on each count activates the appropriate flip-flops.

Figure 5.6 compares the schematics of a 2-bit ripple counter and a 2-bit synchronous counter. Reset lines are deleted for clarity. Ripple counter operation has been discussed: all *J*s and *K*s are connected to a logic high, and the output of each stage triggers the next higher stage. On the other hand, all the C_p's of a synchronous counter are connected together, but the *J* and *K* connections of each flip-flop are different.

For a 2-bit synchronous counter the *J* and *K* terminals of flip-flop *A*, J_A and K_A, are connected to a logic high, while J_B and K_B are connected to the Q_A output. When the synchronous counter is reset, Q_A and Q_B are low, and the *BA* output is 00. J_B and K_B are both low since Q_A is low. The first negative-going pulse reaches both C_p of *A* and C_p of *B* at the same time. Flip-flop *A* changes state because J_A and K_A are permanently high; however, flip-flop *B* cannot change since J_B and K_B are still low. Thus the *BA* output after the first pulse is 01. On the second input pulse both flip-flops change state because flip-flop *A* is always enabled and J_B and K_B are now enabled. The *BA* count after the second pulse is 10. On the third pulse J_B and K_B are both low since Q_A is low. Thus only flip-flop *A* can change state, and the *BA* count after the third pulse is 11. On the fourth pulse J_B and K_B are high, both flip-flops change state, and the count returns to 00.

Since both flip-flops are triggered at the same time, cumulative time delays do not occur. Maximum time delay occurs when flip-flop *B* must be activated, which requires turning on the internal J_B and K_B AND gates. Therefore the maximum operating time is the propagation delay of a single flip-flop plus the propagation delay of a single AND gate.

Design of a 2-bit synchronous counter does not require much analysis. However, it becomes increasingly difficult to determine *J* and *K* connections for successively higher flip-flops. A systematic procedure is needed; this is relatively easy since the design of all digital circuits follows the same pattern. Definition of the problem leads to a truth table. An attempt is then made to

TABLE 5.2
J-K Excitation Table

$Q_- \rightarrow Q_+$	J	K
0 → 0	0	d
0 → 1	1	d
1 → 0	d	1
1 → 1	d	0

minimize the Boolean equation(s), and finally the circuit is constructed for evaluation.

In this case the problem is to design an *n*-bit synchronous counter. The truth table for a synchronous counter is the same as for a ripple counter. The counter truth table must be correlated with the J-K truth table, and correlation must be performed for each flip-flop. The J-K truth table presented in the previous chapter could be used, as Table 4.8 does present all the relationships among J, K, Q_-, and Q_+. In this case, however, conditions are reversed; J and K are triggered from changes in Q_- to Q_+ of the previous stage, and it is more convenient to arrange the J-K data in a slightly different form called an *excitation* table. An excitation table shows the Q changes on the left and the necessary J and K conditions on the right. Table 5.2 is the J-K excitation table in terms of 1, 0, and d, which represents the don't care condition.

Consider that Q is low and must remain low after C_p is triggered. This Q condition is $Q_- \rightarrow Q_+$, or 0 → 0 on the excitation table, and is satisfied by making J low, while K can be either value. Similarly if Q is low and must switch to high after C_p is triggered, the change is 0 → 1, J must be high and K can be either value, etc. Remembering the J-K excitation table is straightforward. The Q before to Q after sequence is the normal binary count, and J is 0-1-d-d, while K occurs in reverse order.

Excitation table characteristics are transferred onto Karnaugh maps for each J and K. The sequence on the Karnaugh maps must be in the same order as the synchronous counter. For a 3-bit counter, the map is in the *CBA* sequence shown in Table 5.3. After each map is filled in, conventional looping is used, and the reduced Boolean equations represent the logic circuits for the

TABLE 5.3
CBA Karnaugh Maps

BA \ C	0	1
00	0	4
01	1	5
11	3	7
10	2	6

TABLE 5.4
J_A-K_A **Conditions**
a. **Counter Truth Tables**
b. **Karnaugh Maps**

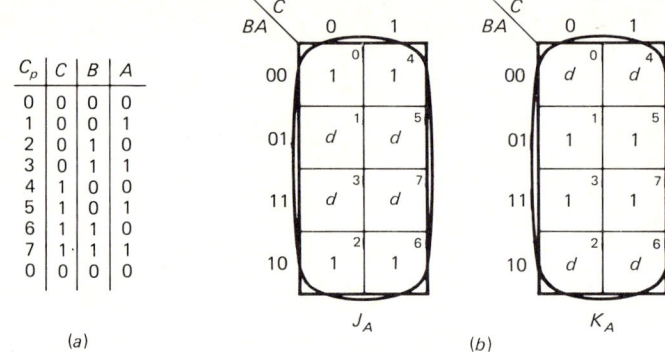

C_p	C	B	A
0	0	0	0
1	0	0	1
2	0	1	0
3	0	1	1
4	1	0	0
5	1	0	1
6	1	1	0
7	1	1	1
0	0	0	0

(a)

(b)

J and K terminals. Even a 3-bit synchronous counter requires six Karnaugh maps and demonstrates synchronous counter design.

Example 5.4a Determine the J_A and K_A equations for a 3-bit synchronous counter.

Solution Table 5.4a shows the truth table for a 3-bit counter, and Table 5.4b shows excitation table conditions transferred onto J_A and K_A Karnaugh maps.

For flip-flop A, use column A of the truth table. For example, as C_p switches from 0 to 1, Q_A switches from 0 to 1. Referring to the excitation table shows that when Q changes from 0 to 1, J must be 1 and K can be anything. Therefore place a 1 in square 0 of the J_A map and a d in square 0 of the K_A map. On the next count C_p switches from 1 to 2 and Q_A switches from 1 back to 0. From the excitation table a change from 1 to 0 requires J_A to be d and K_A to be 1. These conditions are placed in box 1 of both the J_A and K_A maps. As C_p switches from 2 to 3 and Q_A switches accordingly, place a 1 in the box 2 of the J_A map and a d in the box 2 of the K_A map. The remaining squares of both maps are completed in the same way. For square 7, Q_A changes from 1 to 0, which requires d in the J_A map and 1 in the K_A map. In this case the entire J_A and K_A maps are included in single loops. The reduced equations are

$$J_A = 1 \quad \text{and} \quad K_A = 1$$

These equations are the required J_A and K_A connections. The next step is determination of J and K connections for flip-flop B.

Example 5.4b Determine the J_B and K_B connections.

Solution Using column B of the counter truth table, follow the same procedure as in Example 5.4a. In the B column when C_p switches from 0 to 1, Q_B remains at 0. Referring to the excitation table shows J must be 0 and K is d. Place a 0 in

TABLE 5.5 J_B-K_B **Conditions**

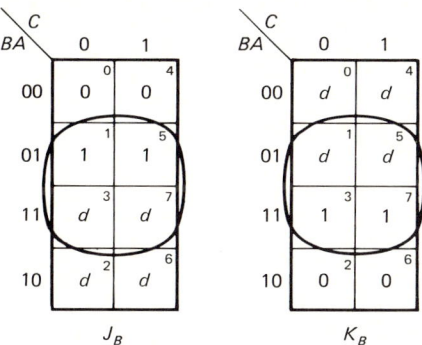

box 0 of the J_B map and a d in the box 0 of the K_B map. The remaining squares are completed in the same manner. For boxes 7, Q changes from 1 to 0, which requires J to be d and K to be 1. Table 5.5 shows the complete Karnaugh maps.

Karnaugh maps for both J_B and K_B are looped around the same squares. The reduced Boolean equations are

$$J_B = A \quad \text{and} \quad K_B = A$$

Equations resulting from Example 5.4b show the required connections for the J_B and K_B terminals. The equations for flip-flop C remain to be determined.

Example 5.4c Determine the J_C and K_C connections.

Solution For flip-flop C the C column of the truth table is combined with the J-K excitation table to obtain the Karnaugh maps shown in Table 5.6. These maps show a subtle but very important point. Each map contains an entire column of d's. In principle, reduction yields a single variable. However, a useful loop must contain "real" information. A string of don't cares does not

TABLE 5.6 J_C-K_C **Conditions**

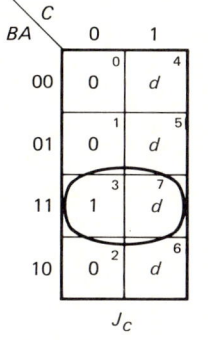
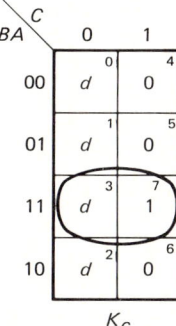

contain any data concerning actual changes of state. Valid loops must contain at least a single 1 or, in cases where the inverse equation is more efficient, at least a single 0. With this in mind, the most efficient loops are as indicated in Table 5.6 and the resulting Boolean equations are

$$J_C = A \cdot B \quad \text{and} \quad K_C = A \cdot B$$

Equations for each J and K terminal have been determined. The last step is the schematic to implement the equations.

Example 5.4d Construct the schematic for a 3-bit synchronous counter.

Solution For flip-flop A the design requires $J = 1$ and $K = 1$. These conditions mean J and K are always high and can be satisfied by connecting J_A and K_A to the positive terminal of the power supply. The equations for flip-flop B are $J_B = A$ and $K_B = A$. This means J_B and K_B must both be connected to Q_A. The equations for flip-flop C are $J_C = A \cdot B$ and $K_C = A \cdot B$. These conditions require ANDing Q_A with Q_B and connecting the resultant output to J_C and K_C. Figure 5.7 shows the completed schematic.

Counters are chains of flip-flops. In synchronous counters, activation is controlled by J-K logic since all C_p's are activated on each count. Maximum delay is the propagation delay of a single flip-flop plus that of the associated AND gate t_g. This applies whether the flip-flop is internal, as in flip-flop B, or external, as in flip-flop C. Typically, AND-gate propagation delay is about 20 ns.

$$T_{\max} = t_{\text{PHL}} + t_g \tag{5.5a}$$

This equation is independent of the number of stages since all flip-flops are activated simultaneously and all AND gates are activated at the end of flip-flop propagation delay. Maximum count frequency is the reciprocal of $T_{\max}$.

$$f_{\max} = \frac{1}{T_{\max}} \tag{5.5b}$$

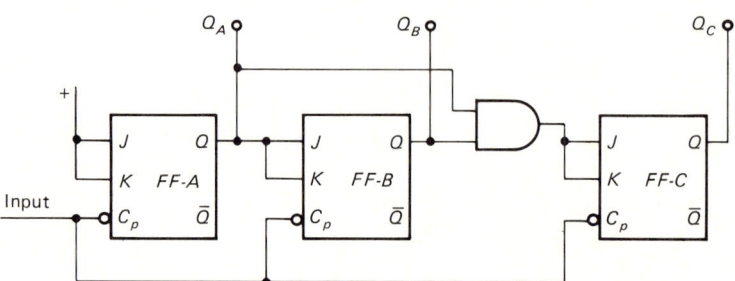

Figure 5.7 3-bit synchronous counter.

Example 5.5 Assuming flip-flops with 40 ns propagation delays are used in both cases, find the maximum counting rates of (a) a 4-bit ripple and (b) a 4-bit synchronous counter.

Solution (a) Maximum counting rate of a ripple counter is given by Eq. (5.4b) as follows:

$$f_{max} < \frac{1}{4 \times 40 \times 10^{-9}} < 6.25 \text{ MHz}$$

(b) Equations (5.5) are for a synchronous counter of any length

$$f_{max} = \frac{1}{40 \times 10^{-9} + 20 \times 10^{-9}} = 16.7 \text{ MHz}$$

In this case the synchronous counter is about $2\frac{1}{2}$ times as fast as the corresponding ripple counter.

The synchronous counter which has been described is an up counter. Conversion to down counting is performed by triggering J and K from $\bar{Q}$. With the inclusion of the required AND and OR gates a synchronous up/down counter can be constructed by using the same approach as for a ripple up/down counter.

Synchronous counters with more than 3 bits can be designed by using the excitation table method. A 4-bit counter, for example, requires the complete 3-bit circuit plus a fourth flip-flop whose J and K terminals must be triggered by $A \cdot B \cdot C$. This condition requires a 3-input AND gate. J-K ICs with multiple input AND gates are available: the 7472 is a TTL flip-flop which has a 3-input J AND gate and a 3-input K AND gate. This permits construction of synchronous counters without adding external AND gates. In a 4-bit synchronous counter the 3-input leads of flip-flop D would individually be connected to Q_A, Q_B, and Q_C. When flip-flop C is connected, two inputs of the three inputs are tied together to create a 2-input AND gate.

Just as with ripple counters, a large selection of self-contained synchronous counter chips are available in both TTL and MOS.

5.4 MODULO COUNTERS

Ripple and synchronous counters are *natural modulus* counters. A natural modulus counter counts to the binary limit determined by the number of flip-flops. The range of a 3-bit counter is from 000 to 111, which means the first eight decimal numbers, 0 through 7, are counted. A 3-bit counter has a natural modulus of 8. Similarly, a 4-bit counter has a natural modulus of 16. The natural modulus of binary counters is always a power of 2: 4, 8, 16, etc. Some applications require counting to nonbinary numbers. Decade counters, for example, count through the first 10 numbers, while digital clocks must be able to count to 12, 60, and in some cases 24. Counters which do not have a natural modulus are called *modulo*, or *mod*, counters.

198 Computer Circuit Concepts

Mod counter design is based on natural modulus counters. Natural modulus counters automatically count to the maximum number, but mod counters must be prevented from reaching the maximum number. In other words, a mod counter skips some of the numbers which a natural counter goes through.

Suppose a mod 6 counter is desired. Since the maximum count of a mod 6 counter is 5, a binary counter with a higher natural count than 5 is required. A 3-bit counter counts to 7 and is the starting point. Gating resets this counter before the natural count reaches 6. Skipping counts is the basis of mod counter design.

Example 5.6 Design a mod 10 ripple counter.

Solution A mod 10, or decade counter, counts from 0 through 9 and then resets. Thus the binary counter which will be appropriate for conversion into a decade counter must have a natural count greater than 9. A 3-bit counter is inadequate since the natural limit is 7. A 4-bit counter is satisfactory because the limit is 15.

In binary, a decade counter counts to 1001 and resets when the outputs are

D C B A
1   0   1   0

This counter must reset when the Q outputs of flip-flops D and B are high at the same time.

First a conventional 4-bit ripple counter is constructed, and then an AND gate is added. AND-gate inputs are Q_D and Q_B, and the reset line is the output. The first time that Q_D and Q_B are both high is when the count reaches 1010. As this count is reached, the AND gate resets the counter. Figure 5.8 shows the schematic and timing diagram. Observe that the count momentarily reaches 1010. However, the counter returns to 0000 as soon as the AND gate resets the flip-flops. It is not really necessary to reset all the flip-flops since some are at 0, but it is standard practice to connect all the resets together.

Mod counters use AND gates to skip the required number of counts. As described, several IC flip-flops have J and K AND gates with up to 3 inputs each. These ICs can be used instead of adding external gates. Logic is the same in either case. Many IC J-K flip-flops require a low rather than a high for reset. In such cases Clear is high and must be driven low for Reset. The correct reset is obtained by using NAND instead of AND.

Natural-modulus ripple counters divide the input frequency by powers of 2. Mod ripple counters can divide the input frequency by any desired number. Frequency division by 10 is used in decade frequency synthesizers, and cascading divide-by-10 counters yields synthesizers with several orders of magnitude. It is also possible to cascade different mod counters. For example, connecting a divide-by-10 to a divide-by-6 results in a divide-by-60 counter.

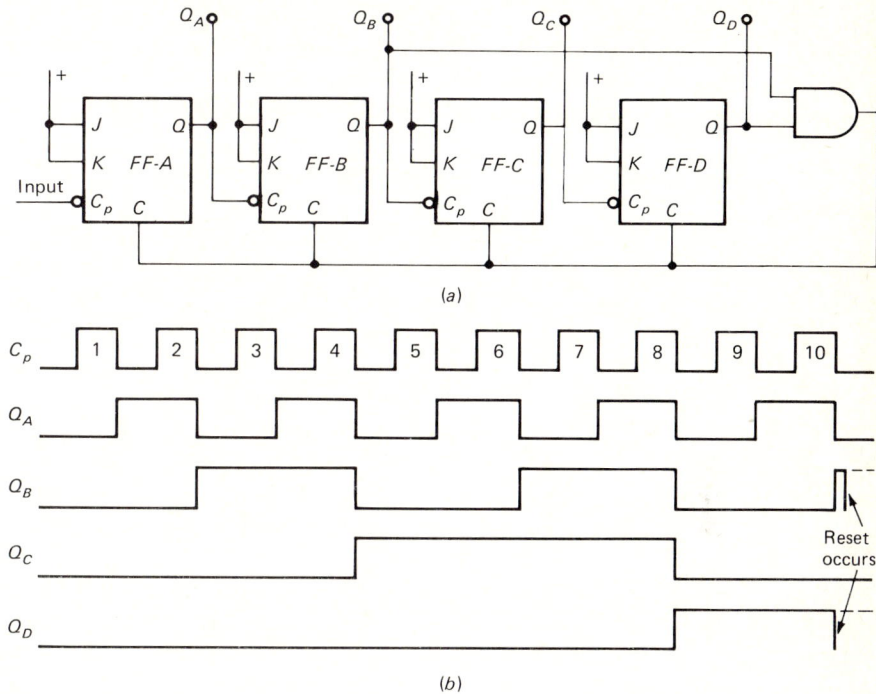

Figure 5.8
Mod 10 ripple counter.
(*a*) Schematic; (*b*) timing diagram.

This is the basis of digital clocks which operate from power mains, the 60-Hz line frequency being divided by 60 to yield one pulse per second. Additional mod circuits yield minutes and hours.

Mod ripple counters have the same timing problems as natural-modulus ripple counters. Cumulative delays can cause false counts and reset the counter prematurely. Also, variation in propagation delay can prevent some flip-flops from resetting during the AND-gate pulse. This condition is avoided by connecting the AND gate to a one-shot multivibrator. The stretched pulse is made longer than the longest time required for reset.

Many IC counters exist. The 7490 is a 14-pin chip with a divide-by-5 counter and an independent flip-flop. If the flip-flop output is externally connected to the input of the divide-by-5 counter, the result is a divide-by-10 counter; this chip is called a 2×5 counter. Ripple counter chips are also available in 2×6 and 2×8.

Synchronous counters are faster and have fewer problems than ripple counters. Skipping counts also applied to the design of synchronous mod counters.

Example 5.7 Design a mod 10 synchronous counter.

Solution Design is based on a 4-bit synchronous counter. Table 5.7 shows a decade

TABLE 5.7 Synchronous Decade Counter Data

C_p	D	C	B	A		Q_-	Q_+	J	K
0	0	0	0	0		0	0	0	d
1	0	0	0	1		0	1	1	d
2	0	0	1	0		1	0	d	1
3	0	0	1	1		1	1	d	0
4	0	1	0	0					
5	0	1	0	1					
6	0	1	1	0					
7	0	1	1	1					
8	1	0	0	0					
9	1	0	0	1					
0	0	0	0	0					

counter truth table and the J-K excitation table. Complete J_A and K_A Karnaugh maps are shown in Table 5.8. Although 4 bits are required, there are no numbers higher than 9. Remaining squares are left blank and represent don't cares. In both cases the maps are completely enclosed by single loops. Therefore

$$J_A = 1 \quad \text{and} \quad K_A = 1$$

Table 5.9 shows the J_B and K_B maps. As in natural modulus synchronous counting, maps are formed by combining column B of the counter truth table with the excitation table. In this case J and K maps of the same flip-flop are different.

$$J_B = A \cdot \bar{D} \quad \text{and} \quad K_B = A$$

J_C and K_C maps are derived from column C of the counter truth table. The results, shown in Table 5.10, are:

$$J_C = A \cdot B \quad \text{and} \quad K_C = A \cdot B$$

TABLE 5.8 J_A-K_A Maps

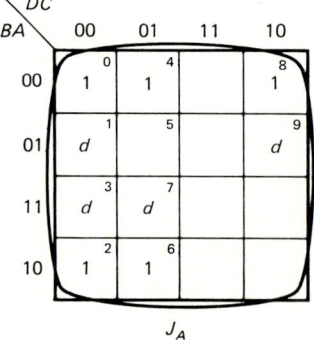
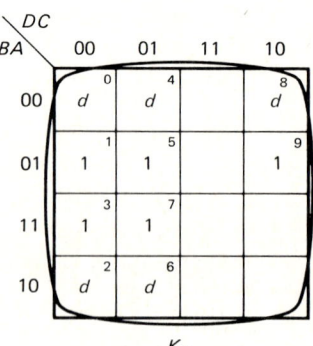

TABLE 5.9
J_B-K_B Maps

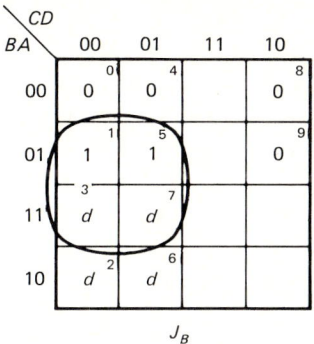
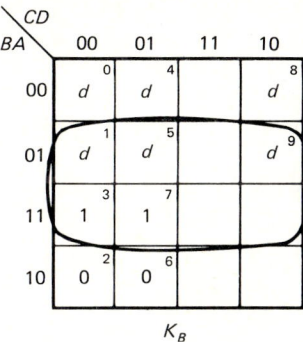

TABLE 5.10
J_C-K_C Maps

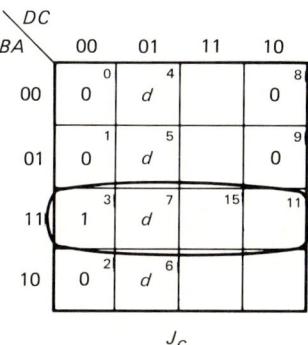
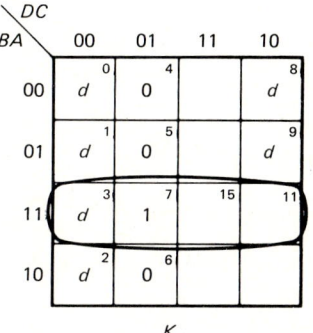

Column D is the highest, and Table 5.11 shows the J_D and K_D maps, which are different. The results are

$$J_D = A \cdot B \cdot C \quad \text{and} \quad K_D = A$$

A schematic based on these eight equations is the final step and is shown in Fig. 5.9. As in natural synchronous counters, all C_p's are tied together. The timing diagram for a mod 10 synchronous counter is the same as for a mod 10 ripple counter.

TABLE 5.11
J_D-K_D Maps

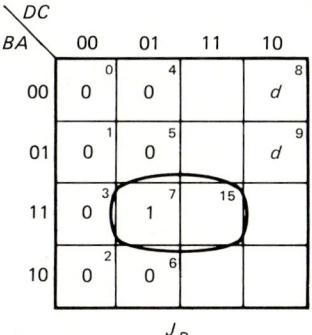
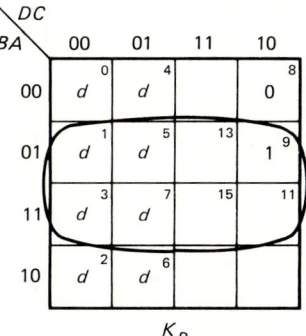

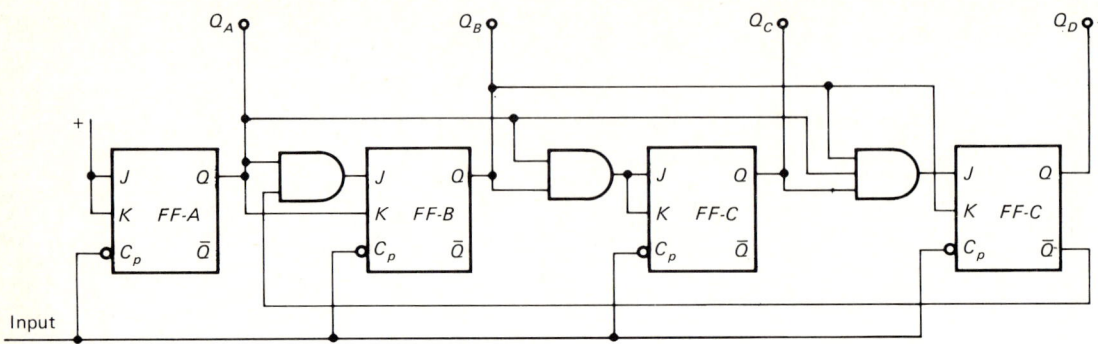

Figure 5.9 Mod 10 synchronous counter.

IC synchronous counters are available in either divide-by-10 or divide-by-16. Many of these chips are up/down counters. Also, many IC synchronous counters have independent preset terminals, which permits entering a binary number into the circuit before counting begins. The result is a programmable variable mod counter.

The number C_n must be less than C_{max} of the particular chip. If a preset number C_{pre} is entered, the counter begins with C_{pre} and counts to C_{max}. Therefore

$$C_{pre} = C_{max} - C_n + 1 \tag{5.6}$$

The 1 accounts for 0 as a valid number.

Example 5.8 The 74161 is a presettable divide-by-16 counter. What are the conditions on the present leads P_D, P_C, P_B, and P_A if the counter must count to 12?

Solution C_{max} for a divide-by-16 counter is 15_{10}, or 1111_2. Since 12_{10} is 1100_2, C_{pre} can be determined.

$$
\begin{array}{rcrrrr}
C_{max} = & & 1 & 1 & 1 & 1 \\
-C_n = & -1 & 1 & 0 & 0 \\
\hline
 & & 0 & 0 & 1 & 1 \\
+ & & 0 & 0 & 0 & 1 \\
\hline
C_{pre} = & & 0 & 1 & 0 & 0 \\
\end{array}
$$

Therefore the preset conditions are:

$$
\begin{array}{cccc}
P_D & P_C & P_B & P_A \\
0 & 1 & 0 & 0 \\
\end{array}
$$

Unfortunately IC nomenclature is not standard. Some manufacturers use numbers instead of letters as subscripts. These details as well as all the technical specifications and performance characteristics are given in manufacturers' catalogs, which are an absolute must when working with chips.

5.5 HYBRID COUNTERS

Ripple and synchronous counters are practical. Selection is a compromise, the "best" circuit not always being the one with superior performance characteristics. Optimization is a trade-off between performance and cost. In fact, counters exist which are compromises between ripple and synchronous. These circuits contain features of both counters and are called *hybrid counters*.

One hybrid counter combines ripple and synchronous counter stages in the same circuit. Practical versions contain a synchronous divide-by-3 counter, which is small enough to be designed by trial and error. However, standard synchronous counter technique can be applied to any mod.

Example 5.9 Design a synchronous mod 3 counter.

Solution A mod 3 counter steps to 2 and then resets. This requires a 2-bit counter. Figure 5.10 shows the counter truth table, Karnaugh maps, and resulting schematic.

Figure 5.10 Mod 3 synchronous counter. (*a*) Truth table; (*b*) Karnaugh maps; (*c*) schematic.

C_p	B	A
0	0	0
1	0	1
2	1	0
0	0	0

(a)

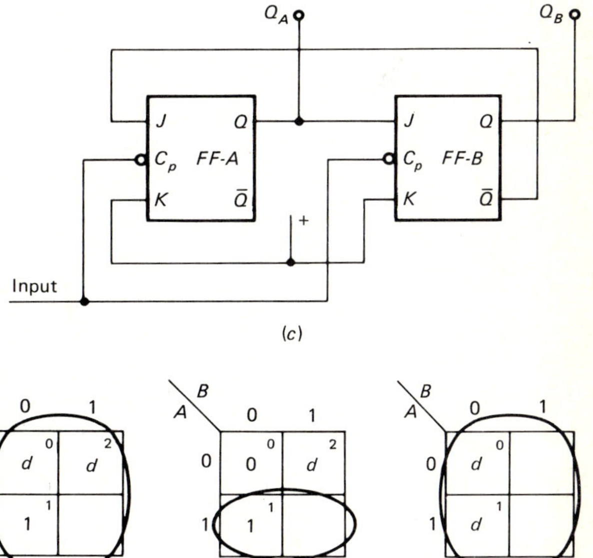

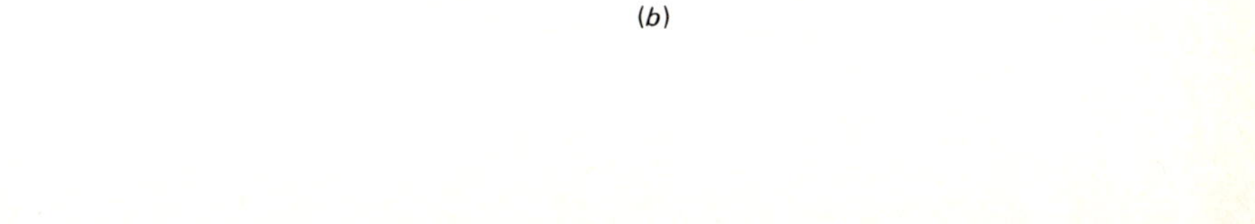

(b)

$J_A = B$ $K_A = 1$ $J_B = A$ $K_B = 1$

(c)

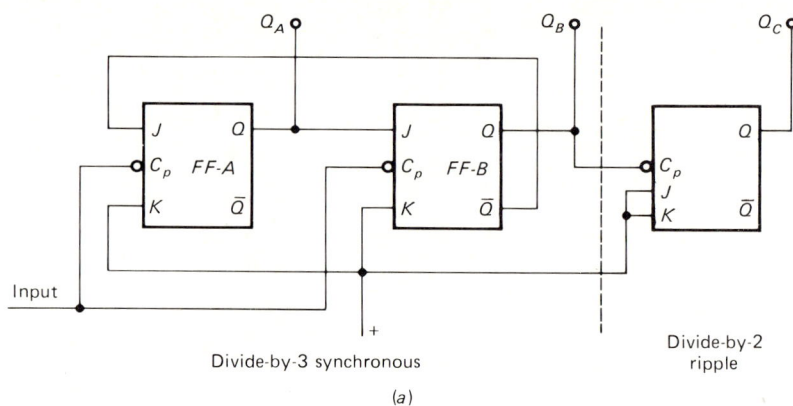

Figure 5.11 Mod 3 to mod 2 hybrid counter. (*a*) Schematic; (*b*) truth table.

C_p	C	B	A	Decimal
0	0	0	0	0
1	0	0	1	1
2	0	1	0	2
3	1	0	0	4
4	1	0	1	5
5	1	1	0	6
0	0	0	0	0

(*b*)

The cascading of counter stages to increase maximum count has been described. If the output of a mod 3 synchronous counter is connected to a mod 2 counter, a mod 6 counter results. A mod 2 counter is a single-stage C_p-triggered flip-flop, and C_p triggering from the previous stage is ripple counting. Thus if Q_B of a synchronous mod 3 counter is connected to C_p of a flip-flop, the result is the hybrid mod 6 counter in Fig. 5.11.

The counting sequence of this mod 6 counter is not what might be expected. The synchronous output is in standard binary sequence. On the third count, Q_B goes low, which triggers flip-flop C. This means the *CBA* count after 010 is 100. From the second to the third pulse the decimal equivalent goes from 2 to 4, not from 2 to 3. On the fourth pulse only flip-flop *A* can be activated, and the counter output switches to 101. The fifth count switches flip-flop *A* low and flip-flop *B* high, which results in an output of 110. On the sixth count all three flip-flops are reset and the sequence repeats.

As shown in Fig. 5.11*b*, the *CBA* output is not in standard sequence. There is no 3, and while the counter reaches 6, C_{max} for a mod 6 counter is 5. This hybrid counter is "incorrect." The output is binary in the sense of consisting of 1s and 0s, but counting is not in base 2. Out-of-sequence counting is not automatically wrong. This particular circuit is perfectly adequate for divide-by-6 operation. Also, a hybrid counter with an out-of-sequence output can drive a correct decimal display. As shown in Fig. 5.12, the decoding circuit is designed to produce correct decimal outputs by rearranging the decoding

Figure 5.12 Correcting the hybrid display. (a) Actual count; (b) simplified decoder schematic.

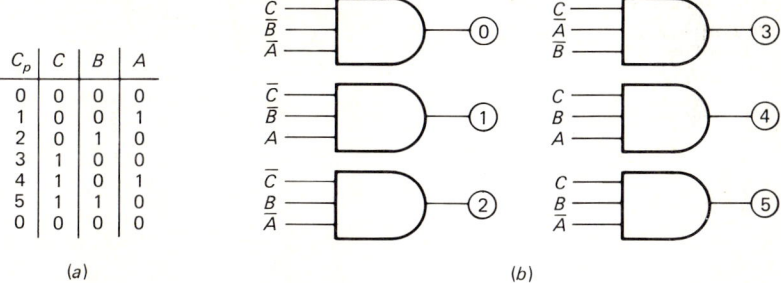

logic. When the CBA count is 100, the correct decimal count is 3. Therefore if the $C \cdot \bar{B} \cdot \bar{A}$ gate is connected to drive the decimal output, a 3 is displayed on the third count. The fifth count can be corrected in the same manner.

If true binary counting is required, the hybrid counter can be rewired. A divide-by-2 ripple counter is connected to the pulse source and this output drives a divide-by-3 synchronous counter, as shown in Fig. 5.13a. Since a divide-by-2 stage is now the input, Q_A switches states on each count, which is normal binary sequences. The synchronous divide-by-3 section is driven from Q_A each time Q_A goes low. Flip-flops B and C maintain the same state for two counts. The resulting count is the standard binary sequence shown in Fig. 5.13b.

Figure 5.13 Rearranged hybrid mod 6 counter. (a) Schematic; (b) truth table.

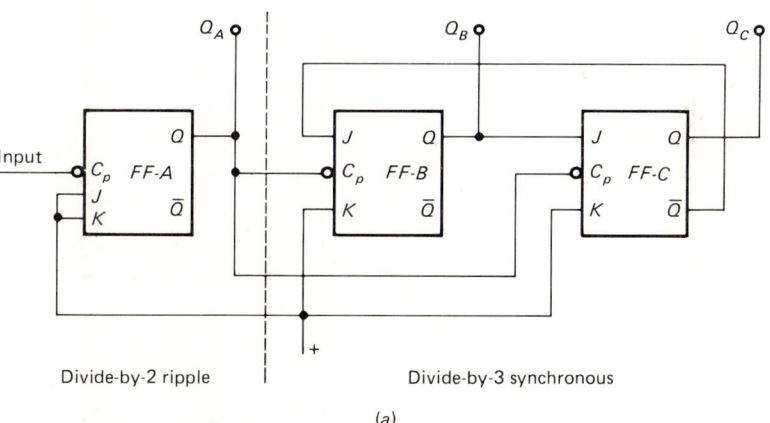

A mod 6 counter, whether hybrid or conventional, is a vital part of a digital clock. Hybrid mod 6 counters use 3 flip-flops. This is also true of conventional ripple or synchronous mod 6 counters. However, ripple and synchronous mod 6 counters require an AND gate to skip two counts, while a hybrid circuit accomplishes the same result without gates. The gateless hybrid counter uses not only fewer components but also fewer component types. Unfortunately a gateless hybrid counter cannot be constructed in any mod but must contain a synchronous mod 3 counter section. This limits gateless hybrid counting to mods which are multiples of 3. Gateless mod 9, mod 18, mod 27, etc., counters can be constructed by cascading synchronous mod 3 counters. Gateless hybrid counters can contain any number of input ripple counters. For example, if a 2-bit ripple counter drives the mod 3 synchronous section, the result is a mod 12 counter. Similarly, with a 3-bit ripple counter as input the result is a mod 24 counter.

The counter obtained by combining ripple and synchronous counter stages is not the only hybrid counter. In a *ring* counter each stage contains features of both synchronous and ripple counters. The C_p's of all stages are connected together as in a synchronous counter, but the output of each stage drives only the next stage, as in a ripple counter. If J-K flip-flops are used, the J and K terminals of each stage are connected together and driven from the Q output of the previous stage. Ring counters can also be constructed from the less expensive D flip-flops. In which case the C_p's are also connected together and the D terminal of each stage is driven from the previous Q output. Figure 5.14 shows both versions.

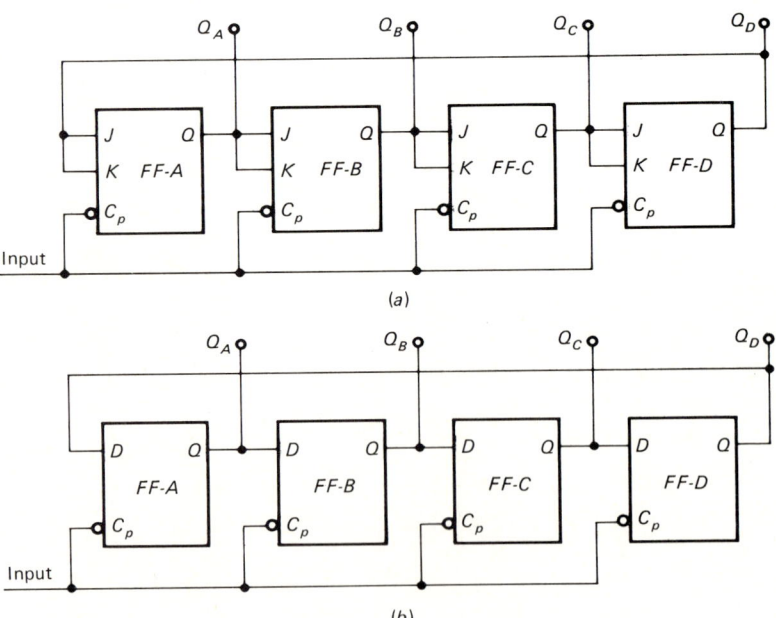

Figure 5.14
4-Bit ring counter.
(a) *J-K* flip-flops;
(b) *D* flip-flops.

In either case the output from the MSB is fed back to the LSB. This connection completes the ring. For clarity, Reset is omitted as in previous schematics. However, resetting a ring counter is somewhat different, as each flip-flop except the LSB starts at 0 but the LSB contains a 1. Reset is accomplished by connecting the P lead of flip-flop A to the C leads of all other flip-flops. A negative-going pulse on this reset line initializes the ring counter to the correct output states.

The $DCBA$ "zero" count of a 4-bit ring counter is 0001. When a count pulse is applied to a ring counter, Q_A goes low. Owing to the propagation delay, Q_A remains high long enough for C_p to activate flip-flop B. However, the other flip-flop outputs are all low, and only Q_B is high as a result of the first pulse. The $DCBA$ count after the first pulse is 0010. Since Q_B is now the high output, flip-flop C is driven high after the second pulse, giving a count of 0100. Similarly the count after the third count pulse switches to 1000. This high at Q_D enables flip-flop A so that the count after the fourth input pulse is 0001. At this time the counter is initialized and the same counting sequence repeats when additional counts occur. Figure 5.15 shows the truth table and timing diagram of a 4-bit ring counter.

In ring counters, a single 1 propagates down the ring while all remaining flip-flops are 0. A ring counter is comparable to the tooth of a gear, which steps one position for each input. In fact this is the principle of gear-driven counters used in turnstiles and automobile odometers. Gear-driven counters were also used in the electromechanical calculators which preceded electronic computers.

This particular ring counter counts the first four numbers, 0 through 3, and requires 4 flip-flops. Ring counters need as many flip-flops as the number to be counted; in other words, gateless ring counters can be constructed in any mod by connecting the appropriate number of flip-flops. A mod 10 ring counter requires 10 flip-flops while a mod 10 ripple or synchronous counter only requires 4. Despite the lower flip-flop count, a complete ripple or synchronous counter may actually require more components than a ring

Figure 5.15
Ring counter characteristics. (*a*) Truth table; (*b*) timing diagram.

C_p	D	C	B	A
0	0	0	0	1
1	0	0	1	0
2	0	1	0	0
3	1	0	0	0
4	0	0	0	1

(a)

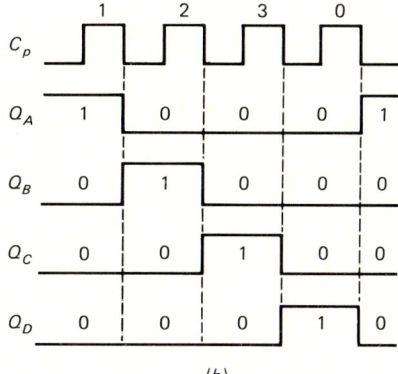

(b)

counter, as they require decoding circuits to display the decimal equivalent. However, the position of the single 1 in a ring counter represents the number and can directly drive a decimal display. Thus a ring counter requires more flip-flops but is gateless for any mod and is self-decoding. The ring counter is a practical example of a binary sequence which is not base 2.

As with other counters, ring counters can be cascaded to increase the count. The self-decoding feature of cascaded ring counters was used in display circuits of early frequency meters and electronic counters. Decade ring counters were used and the output of each flip-flop illuminated the appropriate display digit. The output of the most significant flip-flop of each decade activated the next decade. For example, a 5-decade ring counter goes from 0 to 99,999 and requires 50 flip-flops but no decoding circuits. On the other hand a 5-decade ripple or synchronous counter only requires 20 flip-flops but each decade requires at least 10 AND gates and possibly 10 inverters for decoding purposes. For the same number of decades a self-decoding ring counter requires 2.5 times as many flip-flops. With the advent of inexpensive 7-segment display chips, the self-decoding advantage of ring counters is no longer economically advantageous.

However, ring counters have important computer applications. The unique position of the 1 controls sequencing circuits. The output of the LSB activates the first step, the output of the next higher bit activates the equipment during the second stage, and so on. Ring counters are also used to pulse *stepper* motors, which are special motors designed to rotate in small increments as each pulse occurs.

Since only one flip-flop is on at any one time, the propagation delay for a ring counter is the same as for a synchronous counter. Nevertheless, a ring counter is much faster than a synchronous counter. In an n-bit synchronous counter some or all flip-flops may be on at any particular count, whereas in a ring counter only one of the n bits is on regardless of the count. Thus each flip-flop of an n-bit ring counter is on for only $1/n$th of the time, and thus an n-bit ring counter is n times as fast as the corresponding synchronous counter:

$$f_{\max(\text{ring})} = n \cdot f_{\max(\text{synch})} \tag{5.7}$$

Example 5.10 Compare maximum counting rates of 4-bit synchronous and ring counters. Assume both circuits use 20-MHz flip-flops.

Solution The maximum counting rate of a synchronous counter of any length is the same as that for one flip-flop. In this case it is 20 MHz. For a 4-bit ring counter Eq. (5.7) is applicable

$$f_{\max(\text{ring})} = 4 \times 20 \text{ MHz} = 80 \text{ MHz}$$

Ring counters can also be constructed to count down. Down counting is accomplished by using the $\bar{Q}$ output to enable the next higher flip-flop throughout the ring. It is also possible to construct an up/down ring counter by using the same combination of AND gate–to–OR gate logic already described.

The *shift*, or Johnson, counter is an important variation of the ring counter. A shift counter has two feedback paths. Q of the MSB is connected to K of the LSB, and $\bar{Q}$ of the MSB is connected to J of LSB. Because of the double feedback paths, shift counters are also called *twisted* ring counters. A 4-bit schematic is shown in Fig. 5.16a.

The reset line is omitted for clarity, but counting begins with a negative-going pulse, which as in conventional counters is applied to all C leads. Thus all Q outputs are initially low, and all $\bar{Q}$ outputs are initially high. All Js are

Figure 5.16 Mod 8 twisted ring counter. (*a*) Schematic; (*b*) truth table; (*c*) timing diagram.

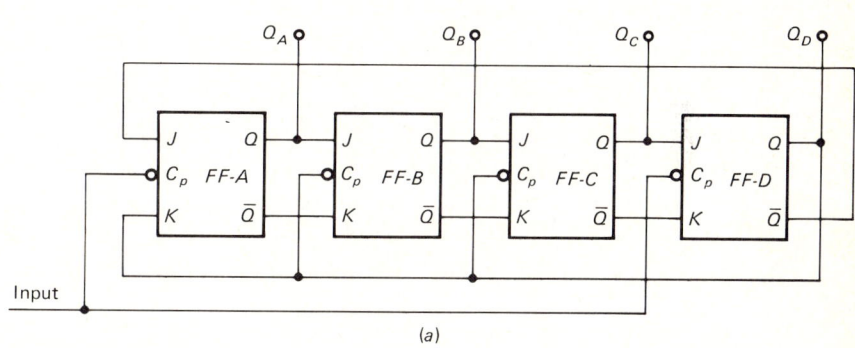

(a)

C_p	D	C	B	A
0	0	0	0	0
1	0	0	0	1
2	0	0	1	1
3	0	1	1	1
4	1	1	1	1
5	1	1	1	0
6	1	1	0	0
7	1	0	0	0
0	0	0	0	0

(b)

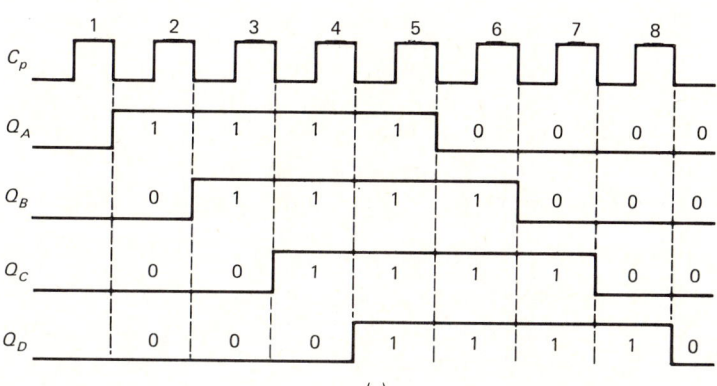

(c)

low except J_A, and all Ks are high except K_A. Since J_A is high and K_A is low, the first clock pulse drives Q_A high, but no other flip-flops can change. Thus the *DCBA* count after the first pulse is 0001. J_A and K_A retain their original states. As a result of the first pulse J_B is high and K_B is low. The second pulse drives Q_B high, and the count switches to 0011. The second clock pulse simultaneously enables flip-flop *C*, and the count after the third clock pulse is 0111. Similarly, the fourth clock pulse drives Q_D high, at which time the count is 1111. When Q_D is high, J_A is low and K_A is high. Therefore the fifth clock pulse drives Q_A low and the count is now 1110. Additional clock pulses will continue to propagate lows along the counter. The count goes to 1100 and next to 1000. Finally, on the eighth clock pulse the counter resets and counting begins again.

The twisted ring counter is another example of a binary code which is not in the conventional base 2 system. Although other ring counters are self-decoding, the twisted ring counter requires decoding. During operation of a twisted ring counter, the flip-flops are all set during the first half of the counting sequence and reset during the second half. Thus a gateless twisted ring counter can be built in any even mod, and several IC versions exist. The 4022 is a 4-bit mod 8 twisted ring MOS IC, and the 4017 is a 5-bit mod 10 package.

While the natural modulus of any twisted ring counter is even, it is possible to construct an odd mod gateless circuit. An odd-count twisted ring counter has a modulus which is 1 less than the even modulus count; for example, an odd-count 4-bit twisted ring counter is mod 7. To obtain any odd mod, J_A is connected to the $\bar{Q}$ output of the MSB as in an even mod twisted ring counter. However, the K_A signal is obtained from Q output of the flip-flop next to the MSB. This eliminates the all-1s count, and the count is 1 less than the natural count for an even number of stages.

5.6 SERIAL INPUT REGISTERS

Flip-flops are also used to construct *shift registers*, which are circuits for temporary data storage. Computer data is organized in words containing a specific number of bits. Typical registers contain 4, 8, 16, or more flip-flops, and several methods of entering and removing data exist. Since all flip-flops in a register are clocked simultaneously, registers are extensions of synchronous counter circuits.

One method of entering data into a register is a single bit at a time. Such registers are called *serial input* types. Serial-input registers only require one data line and are sometimes called *single-rail* registers. In serial input, the least significant data bit enters during the first clock pulse. Successive clock pulses input additional bits and simultaneously downshift bits which have already been entered. This process continues, bits following one another until the entire word has been loaded into the register. Entry of a word in serial form requires as many clock pulses as there are bits in the word. While all flip-flops

Figure 5.17
Serial-input shift register.

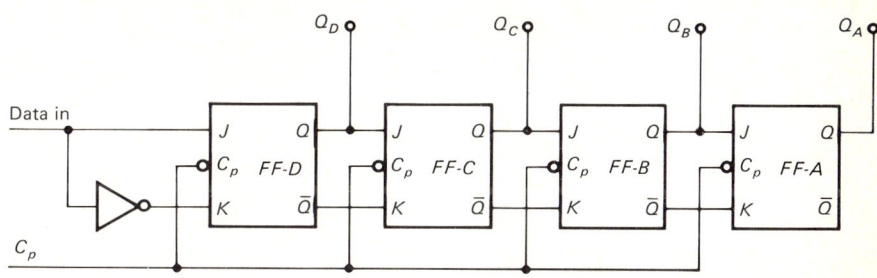

in a serial register are clocked simultaneously, the loading process has the sequential characteristics and operating time of a ripple counter. Figure 5.17 shows a 4-bit serial-input register. The reset line is omitted for clarity.

Since clock pulses arrive at all flip-flops simultaneously, synchronized shifting of bits within the register is assured. Data enter a serial input register through the J-input of the MSB and, through an inverter, at the corresponding K-input. This guarantees opposite signals at J and K. In a 4-bit register the data input terminals are J_D and, through the inverter, K_D. It is possible to eliminate the inverter by connecting J_D to the Q output of the data source and K_D to the $\bar{Q}$ output. This requires two data lines and removes the economic advantage of a single-rail data line.

Data entering a serial input register must be in ascending order, beginning with the LSB. Thus, if the $DCBA$ word is 1010, it must enter a serial input register in $ABCD$ form as 0101. On the first clock pulse 0 enters J_D, which means K_D is 1. After the time required for propagation delay, Q_D is 0. In this case the $DCBA$ output of the register after the first clock pulse is 0000. Because of the specific data word this happens to be the same as the output after resetting; Q_D is 0 and $\bar{Q}_D$ is 1. Thus, after the first clock pulse J_C is 0 and K_C is 1.

On the second clock pulse 2 data bits are moved. The 0 at the Q_D is shifted into flip-flop C, and the 1 on the data input line is simultaneously shifted into flip-flop D. Therefore the register output after the second clock pulse is 1000. As each successive bit is clocked in, one more bit enters the shift register and each bit in the register is shifted one flip-flop to the right. On the third clock pulse three data transfers occur, the 0 at Q_C shifts into flip-flop B, the 1 at Q_D shifts into flip-flop C, and the 0 on the rail shifts into flip-flop D. The register output is 0100 after the third clock pulse. Finally, on the fourth clock pulse the last data from the rail, in this case a 1, enters flip-flop D. At the same time each of the 3 bits already in the register is shifted down. The $DCBA$ output is 1010 and transfer of the word into the register is complete. Although the serial input shift register is based on the synchronous counter and contains features of a ripple counter, it is neither a counter nor a frequency divider. A shift register accepts and stores data.

Figure 5.18 shows flip-flop status as each bit enters the serial-input register. Data bits should be wider than clock pulses to avoid ambiguity.

Figure 5.18
Serial-input register characteristics. (*a*) Truth table; (*b*) timing diagram.

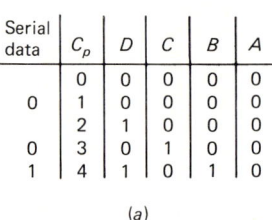

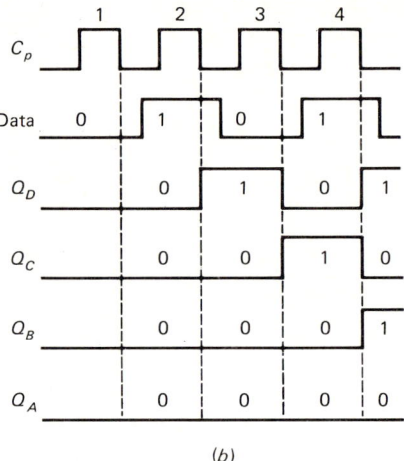

(a) (b)

Once the entire word has been entered into the register, it can be transferred out on a single clock pulse. In this case the circuit is a serial input–parallel output (SIPO) shift register. SIPOs are a convenient way of transmitting computer data along telephone and teletype lines because only one data line is required.

Data transmission is expressed in bits per second, or *bauds*. Originally the term baud had a slightly different meaning, but through usage it has become synonymous with bits per second. Coding is required to separate data words on a transmission line. Even before computers, binary data were transmitted between teletype machines along a single line. The standard teletype rate is 110 bauds and the standard teletype word contains 11 bits per character. Each teletype keystroke is converted into a different 11-bit binary code. In the 11-bit code, 7 bits contain data, the remaining bits being used for initializing and parity checking. With 11 bits per keystroke, 110 bauds represents 11 characters per second, while, including spaces, the average English word contains 5 characters. Therefore 110 bauds corresponds to 132 words per minute.

When a SIPO receives data from a transmission line, synchronizing signals are sent to initialize control logic. As each bit enters the SIPO, it is counted. After the appropriate count has been reached, parallel data is removed. The counter resets and clears the register.

Example 5.11   A 1-byte SIPO receives data at 1000 bauds. How much time is required to load 1 byte?

Solution   Since $f = 1000$ bits/s

$$T_{bit} = 1/1000 = 0.001 \text{ s}$$

Figure 5.19 Four-bit SIPO with D flip-flops.

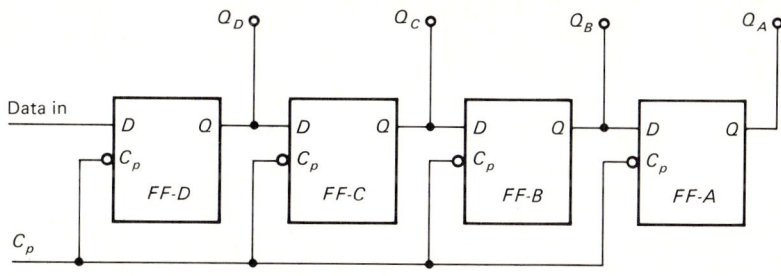

A byte is 8 bits long. Therefore

$$T_{\text{byte}} = 8 \times 0.001 \text{ s} = 8 \text{ ms}$$

This particular SIPO uses J-K flip-flops, but it is possible to use the less expensive D flip-flops. Figure 5.19 shows a 4-bit SIPO with D flip-flops. The timing diagram and truth table are the same as for J-K flip-flops.

The other type of serial input register is the serial in–serial out (SISO). Data is received and shifted in the same manner, 1 bit at a time controlled by synchronous clocking. The SISO has a single output terminal, which is the Q terminal of the LSB. Data enters a SISO 1 bit at a time and leaves 1 bit at a time in the same sequence. The SISO is an example of a first in–first out (FIFO) circuit. The time required for data to pass through a SISO is determined by the number of flip-flops in the register.

$$T_{\text{SISO}} = n \cdot C_p \tag{5.8}$$

where n is the number of flip-flops and C_p is the clocking period. SISOs are used to delay data for a specified time. C_p is fixed during computer design, and n can be selected to achieve the desired delay.

Example 5.12 A computer has a C_p which occurs every 250 ns. How long must a SISO circuit be to obtain a 1 μs delay?

Solution Equation (5.8) is applicable

$$1 \times 10^{-6} = n \times 250 \times 10^{-9}$$

$$n = 4 \text{ stages}$$

Figure 5.20 shows a 4-bit SISO.

Shift-right SIPOs and SISOs have been described. Data enters the leftmost flip-flop and is clocked to the right. If required, registers can be rewired to obtain shift-left registers. In a SISO, conversion to shift-left is accomplished by interchanging the input and output terminals. A similar change converts a shift-right SIPO into a shift-left.

Some situations require the ability to shift in either direction. The same type of AND/OR logic used in up/down counters is applicable to right/left

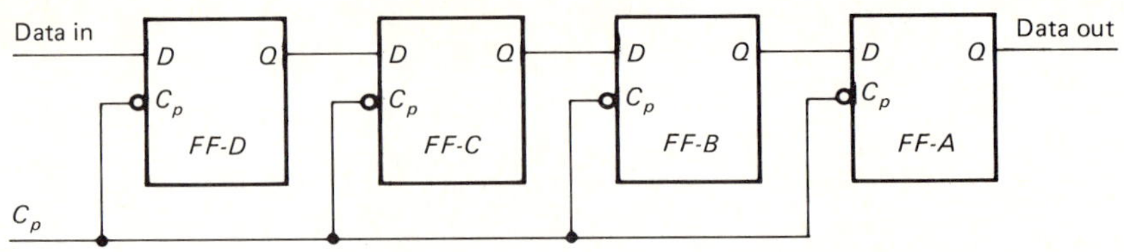

Figure 5.20 Four-bit SISO.

shift registers. Figure 5.21 shows such a 3-bit SIPO. A 1 on the control line enables the shift-right AND gates, labeled R on the schematic. Similarly, when the control line is 0, shift-left AND gates, labeled L, are enabled.

When the control line is high, serial data enters the leftmost R gate and activates the OR gate, which in turn triggers the input to flip-flop C. Other flip-flops are also activated from the Q outputs on the left in the same AND/OR circuit. When the control line is low, the L AND gates are enabled and serial data must be entered at the rightmost AND gate. With shift-left enabled, the output of each flip-flop activates the flip-flop to the left through the AND/OR logic. The next section contains an interesting application of a shift right/left register.

SISO and SIPO IC registers are available, some with shift right/left capability, in both TTL and MOS. In particular, MOS SISOs exist from 256 up to 4096 flip-flops. These registers are the delay element in circuits which furnish "reverb" for musical systems. The separation time between a sound and its echo, or reverberation, is controlled by varying the clock frequency.

5.7 PARALLEL INPUT REGISTERS

Because each bit requires one clock pulse, longer words require more time to load when serial input registers are used. Instead, all bits can be loaded

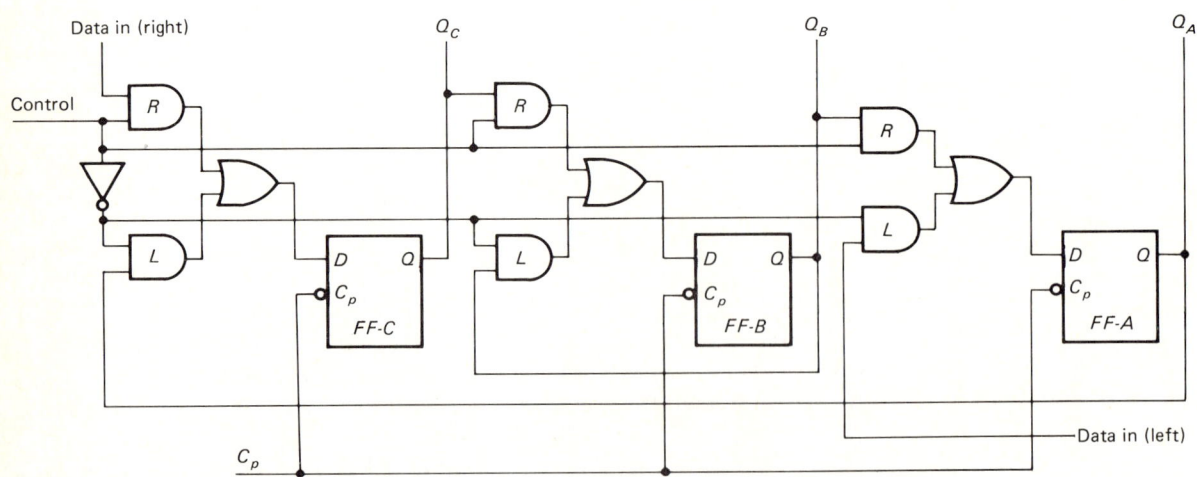

Figure 5.21 Three-bit shift right/left SIPO.

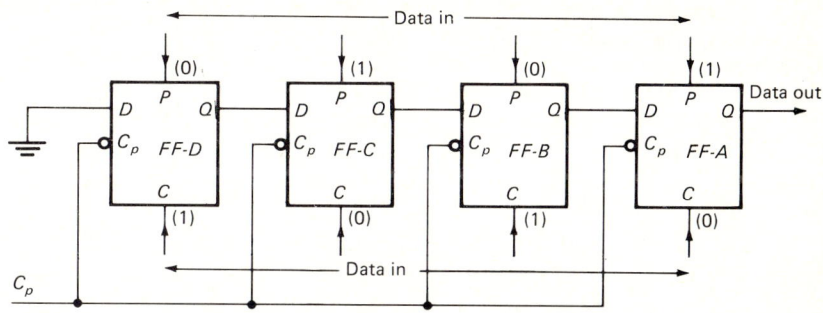

Figure 5.22 Four-bit PISO shift register.

during a single clock pulse by using a register which loads all bits of an entire word simultaneously. Such registers are called *parallel input* registers. Parallel loading requires a separate line for each bit. The penalty for faster loading is a more intricate circuit.

A 4-bit parallel input–serial output (PISO) shift register using D flip-flops is shown in Fig. 5.22. PISO clocking of parallel input registers is synchronous. Bits enter at the flip-flop Preset and Clear inputs. If a particular flip-flop type has only Clear inputs, then setting is required before data are entered. Setting is not required when P and C inputs both exist. For comparison, the same word, 1010, is used to describe PISO operation. Observe that in a PISO the least significant data bit appears at the output terminal even before data transmission begins.

During the first clock pulse data shifts onto the transmission line from Q_A. In this case the LSB happens to be 0. When the LSB shifts out of the register, data at Q_B shifts into flip-flop A. Also at this time Q_C data shifts into flip-flop B and Q_D data shifts into flip-flop C. At each successive clock pulse, the next most significant shifts out and each remaining bit shifts one flip-flop to the right. Data leaves the register in $ABCD$ form. After four clock pulses the entire word has been shifted out.

When J-K flip-flops are used, the Q output of each flip-flop is connected to the next J input and the $\bar{Q}$ output is connected to the next K input. Whether D or J-K flip-flops are used, the status of each bit on each clock pulse is the same. Figure 5.23b shows a PISO timing diagram. Data is simultaneously input and serially output. Thus the time required to transfer data into and out of a PISO is the same as for a SIPO. Also, since data leaves a PISO in ascending order, the PISO is compatible with the SIPO, which accepts bits in ascending order. In fact, when data is sent between devices, a PISO and SIPO are at opposite ends of the transmission line. Typically, computer data is processed in parallel and output through a PISO. Serial data is sent on a single rail and enters a SIPO at the receiving end.

A 1-byte data transmission system is depicted in Fig. 5.24. Bits are numbered in standard sequence. D_0 is the least significant data bit and represents the 2^0 position. In an 8-bit word, D_7 is the MSB and represents the

Figure 5.23
PISO characteristics. (*a*) Truth table; (*b*) timing diagram.

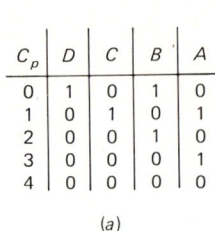
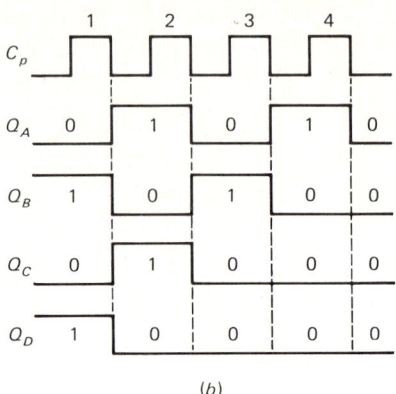

(a)

(b)

2^7 position. A shift-left PISO can be constructed by connecting the output of each flip-flop to the input of the flip-flop on the left. The same type of logic used in up/down counters and shift right/left SIPOs can also be used to construct shift right/left PISOs. Registers which shift in either direction can perform mathematical operations.

Example 5.13 The decimal number 12 is stored in binary form in an 8-bit shift right/left register. Determine the result of shifting this number, (*a*) 1 bit to the left and; (*b*) 1 bit to the right.

Solution As an 8-bit binary number 12_{10} is 0000 1100.
(*a*) When this number is shifted 1 bit to the left, the binary number is 0001 1000. This is the binary form of 24_{10}.
(*b*) Similarly when 12_{10} is shifted 1 bit to the right, the binary number is 0000 0110. This is the binary form of 6_{10}.

Shifting left is equivalent to multiplication and shifting right to division. Therefore, shifting left or right in a sufficiently long shift register is one method of multiplying or dividing by any power of 2. Since a shift right/left register can shift in either direction, multiplication and division are performed in the same circuit. The time required to perform multiplication or division is determined by the required number of shifts. It takes as long to perform mathematical operations as to shift serially.

Figure 5.24
Single-rail data transmission.

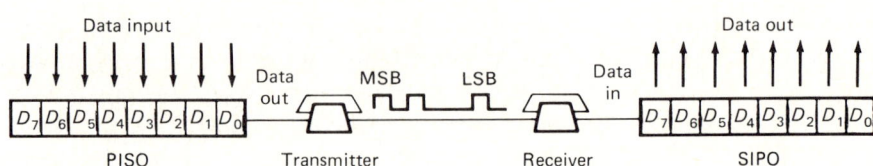

Example 5.14 A computer with a 1-MHz clock rate performs multiplication and division using a shift right/left register. How long does it take to, (a) multiply a number by 4; (b) divide a number by 1,048,576?

Solution $T = 1/f$ and therefore a clock pulse occurs every microsecond.

(a) Since $4 = 2^2$, two shift-left pulses are needed to multiply by 4. Therefore

$$T = 2 \times 1 \ \mu s = 2 \ \mu s$$

(b) Since $1,048,576 = 2^{20}$, 20 shift-right pulses are needed to divide by 1,048,576. Therefore

$$T = 1 \ \mu s \times 20 = 20 \ \mu s$$

It is difficult to multiply and divide by numbers which are not powers of 2. However, digital computers do use shift registers in performing multiplication and division. Circuit details will be described in Chap. 6.

Registers are empty after data is transmitted, and the next word can be loaded. However, loss of data at the sending end may cause problems. Many situations require retaining a word after transmission. Data can be restored by returning each bit to the input as it is transmitted. This type of circuit is called a *recirculating* register. As shown in Fig. 5.25, recirculation is another example of feedback. In this case the circuit is a PISO. The word is originally loaded in parallel. However, as each bit leaves, it is serially returned to the MSB position. As the second bit leaves the register, it is returned to the MSB and the first bit shifts one position to the right. The process continues with each bit being returned as it is transmitted. After four clock pulses the entire word is transmitted and also returned to the register in correct sequence. If the transmitted *DCBA* word is 1010, then 1010 is restored.

Recirculating registers are similar to ring counters. *D* flip-flop recirculating registers have a single return line. When *J-K* flip-flops are used, two return lines are required, just as in a twisted ring counter. However, in a *J-K* recirculating register the return lines are not cross-coupled. In the *J-K*

Figure 5.25 Recirculating PISO register.

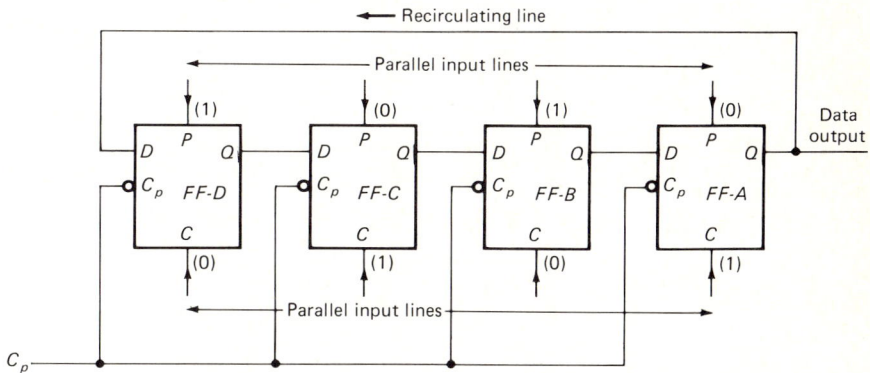

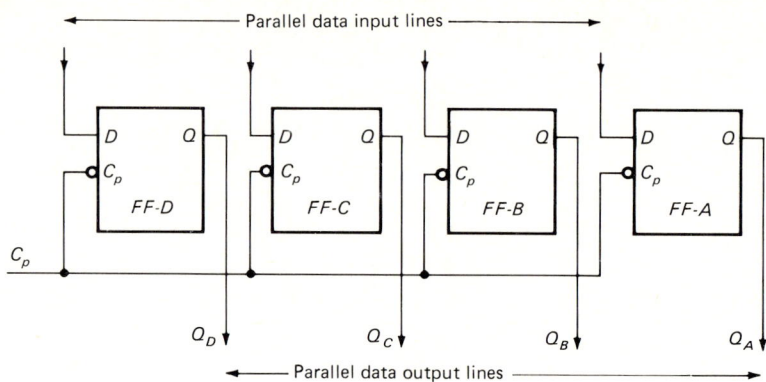

Figure 5.26 Direct-output PIPO register.

version of this PISO recirculating register, the Q_A output is fed back to the J_D input and the $\bar{Q}_A$ output is fed back to the K_D input.

The time for loading a parallel register is independent of word length. All bits are loaded in a single clock pulse. This is also true for a parallel output register. Therefore a parallel input–parallel output (PIPO) is the fastest shift register. One PIPO version, shown in Fig. 5.26, inputs and outputs data on the same clock pulse. Inputs are loaded at the D inputs of each flip-flop during the same clock pulse. The only time difference between input and output is flip-flop propagation delay. A PIPO which completes operation during a single clock pulse can also be constructed with J-K flip-flops.

Other PIPO applications require loading on one pulse and removing on a different clock pulse. Independent control of input and output is accomplished by gating. As shown in Fig. 5.27, input data are loaded into the Preset input of each flip-flop through individual AND gates. However data cannot be transferred to the Preset inputs until the AND gates are enabled. All AND gates are simultaneously enabled when the shift pulse line goes high. Then data appear at the P inputs and, after propagation delay, at the output.

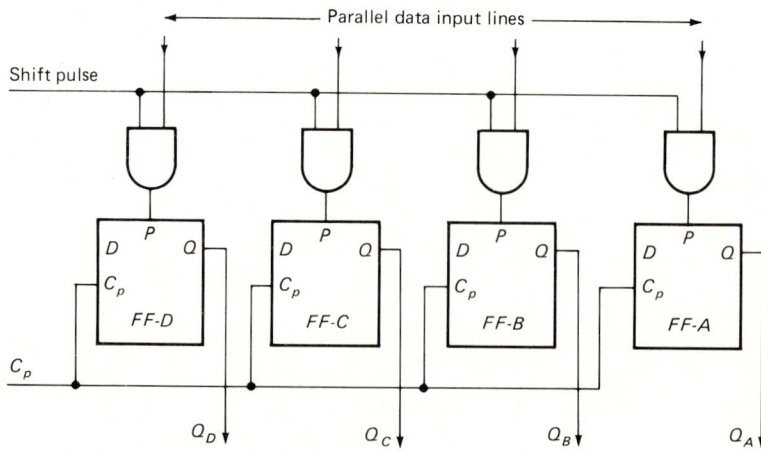

Figure 5.27 Controlled-output PIPO register.

The D inputs are not used in a PIPO shift register. P terminals accept input data and Q terminals accept output data. Similarly, if J-K flip-flops were used, J and K terminals would not be used.

Besides temporary storage and rapid data transfer, PIPOs can perform arithmetic operations. This arithmetic is an extension of the basic shifting method. Completely self-contained IC shift registers are available in TTL and MOS. Several versions of SISO, SIPO, PISO, and PIPO exist in a variety of word lengths. Many chips provide the option of changing to serial or parallel by applying control signals. Some chips can also be converted from shift-left to shift-right. Manufacturers' data books contain all the details.

5.8 COUNTER AND REGISTER DISPLAYS

Troubleshooting and other operations require displaying register and counter outputs. Two display techniques are compatible with low-voltage computer circuits, namely, the *light-emitting diode* (LED), and the *liquid crystal display* (LCD).

The LED is a modification of a conventional silicon diode. LEDs have low forward-bias resistance and high reverse-bias resistance. In an ordinary diode the dopants are phosphorus and boron. With these dopants, energy expended when current carriers flow is dissipated as heat. In LEDs additional dopants are used. The result is emission of light as well as heat when current flows. If a junction is doped with phosphorus, gallium, and arsenic, the diode emits red light on forward bias. Doping with only phosphorus and gallium causes green light to be emitted. If the compounds which cause red and green light are combined, the diode glows yellow, a color between red and green in the spectrum. The voltage drop across a forward-biased LED depends on the materials. Typical voltage drops are three times greater than across conventional silicon diodes. Series resistors are required to present excessive current.

When an LED is driven from a TTL circuit, modification of the basic circuit shown in Fig. 5.28a is needed. As described, TTL circuits are capable of sinking milliamperes but can only source microamperes. Therefore, LEDs must be driven from logic low rather than logic high. This is accomplished by inverting the output of the drive circuit, as shown in Fig. 5.28b.

Figure 5.28 Activating an LED. (*a*) Conventional method; (*b*) modification for TTL.

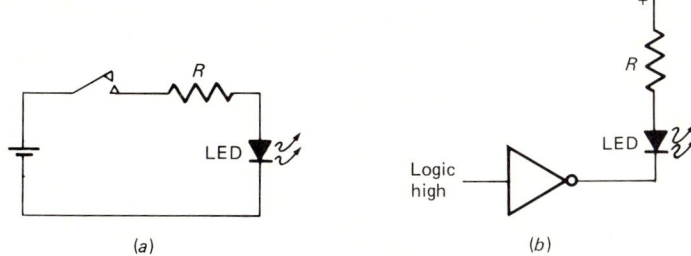

Example 5.15 An LED with a 10-mA maximum current is operated from a 5-V source. Determine the current-limiting resistor.

Solution The voltage drop across a forward-biased LED is about

$$3 \times 0.7 = 2.1 \text{ V}$$

Therefore the voltage drop across the resistor is

$$5 - 2.1 = 2.9 \text{ V}$$

Since

$$R = V/I$$

$$R = \frac{2.9}{10 \times 10^{-3}} = 290 \text{ }\Omega$$

By contrast an LCD is not a semiconductor. Liquid crystals are special long-chain molecules with interesting electrical and optical properties. LCDs are formed by enclosing the liquid crystal material between glass plates. A conductive but transparent film is attached to the rear glass plate, as shown in Fig. 5.29. With no voltage applied, the liquid crystals are vertically aligned between the plates and are rather transparent. When voltage is applied, liquid crystals react to the electric field by forming completely random orientations. This results in the scattering of incident illumination as white light from a dark background. Using an LCD in this manner is called *dynamic scattering*.

The more popular method of using LCDs is the *field effect*. In addition to the basic components, a field-effect LCD contains a light polarizer attached to the front glass plate. When voltage is applied, the polarizer prevents the scattered liquid crystals from reflecting light and the LCD is black. In either the dynamic or field-effect configuration, an LCD does not emit light. Liquid crystal displays only change the method in which incident light is reflected.

Since direct current destroys LCD crystal structure, LCDs are driven from an ac source. The frequency should be above 30 Hz to prevent the display from flickering. LCDs can be operated at voltages as low as $3V_{\text{rms}}$, and a single segment of an LCD draws less than 1 µA.

LEDs or LCDs can be connected to the output bits of a register or

Figure 5.29 Activating an LCD. (*a*) No voltage; (*b*) voltage.

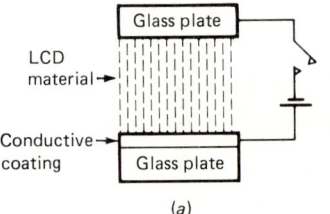
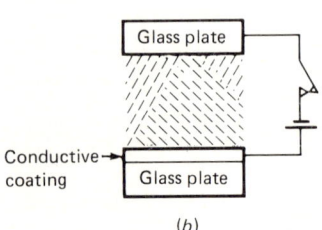

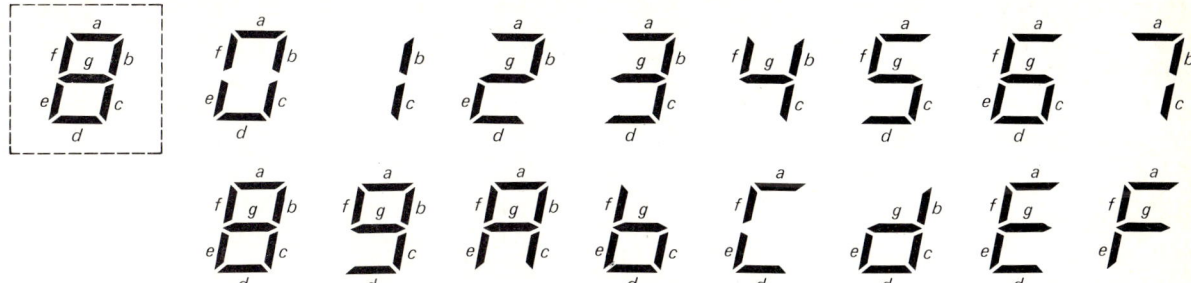

Figure 5.30 Seven-segment display capability.

counter. In this case the display is binary. When the LED is on, the bit is high, and a binary 1 is displayed. When the LED is off, a binary 0 is displayed. However, it is tedious to read binary outputs. A popular solution is the 7-segment display, which contains 7 electrically independent segments. As shown in Fig. 5.30, the segments are capable of presenting numbers in any base. Seven-segment displays of *B* and *D* in hex are lowercase because uppercase *B* cannot be distinguished from 8 and uppercase *D* cannot be distinguished from 0.

If the 7-segment display is LED, each segment contains a plastic light pipe with an LED in back. When the LED is activated, the entire pipe appears to emit light. In an LCD 7-segment display, the polarizer for each segment is attached to a separate input lead, and when voltage is applied, only the activated segments block light. This results in black numbers on a transparent background. Seven-segment displays are usually attached to registers which output in *DCBA*. A 4-bit output contains 16 possible combinations and is compatible with hex or any lower base.

Regardless of whether LCDs or LEDs are used for the 7-segment display, a decoder is needed to convert from *DCBA* to whatever number system is used. Figure 5.31 shows a block diagram and truth table for a decimal display using the same format shown in Fig. 5.30. The binary output of the source is fed to the decoder circuit, which contains the gates needed to activate each of the $g - a$ segments of the display. Current-limiting resistors are connected from decoder to the segments. Decoder circuits can be determined from truth tables or Karnaugh maps.

Example 5.16 Design a decoder to activate the *a* bar of a 7-segment display in the decimal number system.

Solution As shown in Fig. 5.31b, the *a* bar is on for every decimal number except 1 and 4. Therefore:

$$a = \Sigma(0, 2, 3, 5, 6, 7, 8, 9)$$

Figure 5.31
Output display circuit. (*a*) Block diagram; (*b*) decimal display truth table.

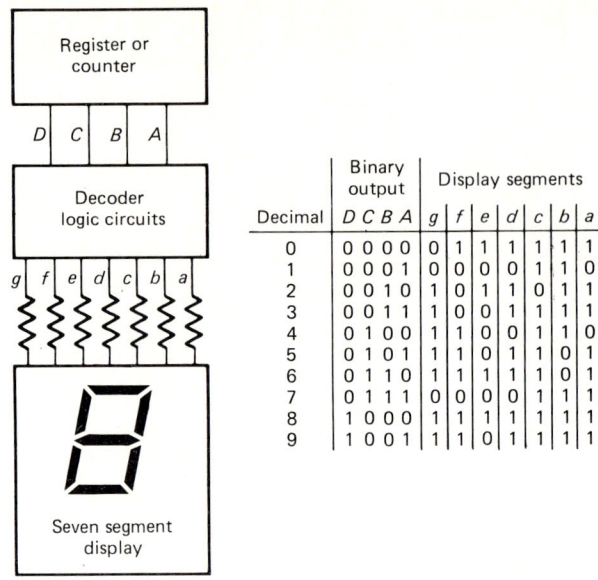

Decimal	Binary output D C B A	Display segments g f e d c b a
0	0 0 0 0	0 1 1 1 1 1 1
1	0 0 0 1	0 0 0 0 1 1 0
2	0 0 1 0	1 0 1 1 0 1 1
3	0 0 1 1	1 0 0 1 1 1 1
4	0 1 0 0	1 1 0 0 1 1 0
5	0 1 0 1	1 1 0 1 1 0 1
6	0 1 1 0	1 1 1 1 1 0 1
7	0 1 1 1	0 0 0 0 1 1 1
8	1 0 0 0	1 1 1 1 1 1 1
9	1 0 0 1	1 1 0 1 1 1 1

Each of these Karnaugh squares must contain a 1. Squares numbered 10 through 15 cannot occur and represent don't cares. Figure 5.32 shows the Karnaugh map and logic circuit. If the particular register contains Q and $\bar{Q}$, inverters are not needed.

In rare cases it is not possible to use IC decoders. When such instances arise, it is worth considering octal rather than decimal displays. An octal display is driven from a 3-bit rather than a 4-bit output, and hence an octal decoder is simpler than a decimal decoder. Another small advantage of octal is its slightly reduced power requirement; this is explained by the fact that

Figure 5.32
Logic for *a* in decimal display. (*a*) Karnaugh map; (*b*) logic diagram.

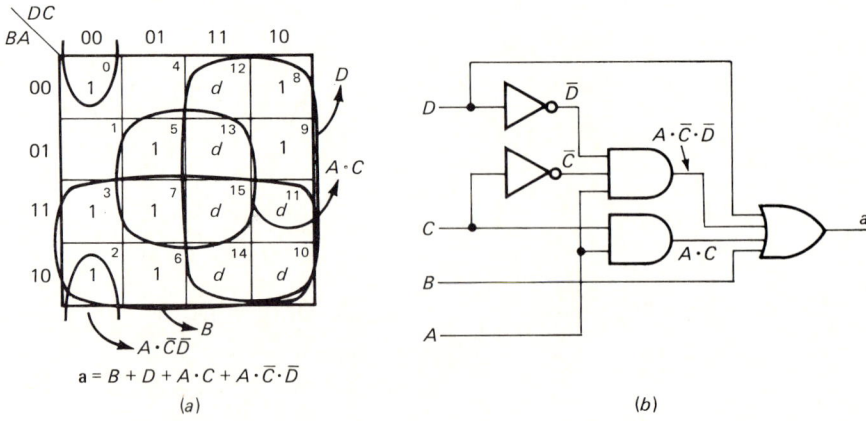

$a = B + D + A \cdot C + A \cdot \bar{C} \cdot \bar{D}$

(a) (b)

7-segment display requires maximum power when all segments are on, a situation that only arises when the numeral 8 is displayed, but 8 is not an octal number. However, these two advantages of octal are usually overshadowed by the convenience of working with IC decimal circuits.

The popular 7447 is a 16-pin TTL 7-segment decoder, which contains all logic necessary for converting $DCBA$ binary to decimal. This chip contains open-collector outputs, which can be connected to current-limiting resistors. Thus the 7447 is a combination decoder/driver. Built-in testing is another feature. In an IC decoder the format of the numerical display is determined by chip design.

There is one slight difference between the 0 through 9 portion of the display shown in Fig. 5.30 and the 7447 decoder logic: the d is not activated when 9 is displayed, but all other numerals are exactly the same. The number format shown in Fig. 5.30 is used in most electronic calculators. Other TTL chips such as the 9368 are decoder/drivers for presenting hexadecimal on a 7-segment display.

Decoder/driver chips are also available in MOS. The 4543 and the 4743 are MOS decimal decoders. Since MOS circuits draw much less current than TTL circuits, MOS decoder/drivers are excellent for activating a 7-segment LCD display. Typically, LCD displays draw microamperes per segment.

Example 5.17 The typical 0.5-in LED 7-segment display draws 10 mA per segment from a 5-V supply. The same size LCD draws 1 μA per segment. Compare maximum power dissipations.

Solution Maximum power is used when all seven segments are activated. For the LED display

$$P_{max} = 5 \text{ V} \times 7 \text{ segments} \times \frac{10 \text{ mA}}{\text{segment}} = 350 \text{ mW}$$

For the LCD display

$$P_{max} = 5 \text{ V} \times 7 \text{ segments} \times \frac{1 \text{ }\mu\text{A}}{\text{segment}} = 35 \text{ }\mu\text{W}$$

Thus the foregoing example shows that for the same size display, an LED draws 10,000 times as much power as an LCD. While this is a tremendous difference, the extra power is available if the display is operated from power lines. The increase in power is largely offset by the ease of viewing a LED display. However, when the display is part of a battery-operated circuit, lower LCD power dissipation is a distinct advantage.

Although called 7-segment displays, most packages really contain an extra LED or LCD segment which is a decimal point. Some displays contain colons for digital clock displays and others have plus and minus signs for digital multimeter displays. Manufacturers of displays provide packages with 3, 4, or more digits. Such displays are rapidly replacing traditional meter movements in most test equipment.

Previous chapters have discussed basic computer principles and circuits to implement these principles. Simpler circuits can be constructed from SSI ICs which contain up to 12 gates on a single chip. These chips contain one gate with many inputs, several gates with fewer inputs, or two flip-flops. MSI circuits contain more than 12 but less than 100 gates; completely packaged 4-bit counters and registers are examples of MSI circuits. The following chapters extend basic principles to obtain complete computer circuits containing the more complex MSI and LSI circuits. Some LSI circuits contain thousands of gates on a single chip. Not very many years ago SSI was a great advance compared with discrete transistor circuits. The transition from SSI to MSI to LSI is a matter of decreasing gate area and interconnections. While simple in principle, the reductions required to produce LSI required production techniques which are as different from those used for SSI as the latter are from discrete transistor techniques.

SUMMARY

1. The T flip-flop changes state once for each input and is the basis of electronic counting. In ripple counters the output of each stage triggers the next higher stage.

2. It is as easy to count down as to count up. Propagation delays in a ripple counter are cumulative and problems occur at high counting rates.

3. All flip-flops in a synchronous counter are triggered simultaneously to eliminate cumulative propagation delays. J and K inputs control the activation sequence of bits in a synchronous counter.

4. Ripple and synchronous counters can be designed to count in any modulus. A large selection of packaged counters is available in TTL and MOS.

5. Some counters do not have the ordinary base 2 counting sequence. While ring counters are self-decoding, Johnson counters require some decoding.

6. Registers are used to temporarily store and then transmit data. Serial input registers are loaded 1 bit at a time, and data can be output either serially or simultaneously.

7. All bits of a parallel input shift register are loaded at the same time. Depending on the application, parallel input registers are available in both serial and parallel output modes.

8. LEDs and LCDs are low-voltage circuits for displaying counter and register outputs. LEDs emit light, but LCDs reflect incident light.

PROBLEMS

5.1 Find the (a) maximum count and (b) maximum number of events which can be processed by a 5-bit ripple counter.

Answer: (a) 31; (b) 32.

5.2 Find the maximum possible count for the circuit in Example 5.1.

5.3 A designer of an electronic music synthesizer has determined that 55 Hz is the lowest A and that 8 octaves are required. With what frequency should the designer begin?

Answer: 14,080 Hz.

5.4 A digital wristwatch uses a 32,768-Hz quartz crystal. How many stages must a binary counter have to obtain one pulse per second?

5.5 Two 7473 chips are used to construct a 4-bit counter. Find: (a) the maximum propagation delay; and (b) the maximum counting rate.

Answer: (a) 160 ns; (b) 6.25 MHz.

5.6 Show a Karnaugh map for Eq. (5.3). Determine if any reduction is possible.

5.7 Using a 7493 chip determine the maximum counting rate when the circuit is used as: (a) a 3-bit counter; (b) a 4-bit counter.

Answer: (a) 19.6 MHz; (b) 14.5 MHz.

5.8 Two 7493 chips are used to construct a divide-by-12 circuit. Find the maximum rate at which this circuit can operate.

5.9 Start with a truth table and develop the control equations for a 3-bit synchronous down counter.

Answer: $J_A = K_A = 1$; $J_B = K_B = \bar{A}$; $J_C = K_C = \bar{A} \cdot B$.

5.10 Use a truth table to develop the control equations for a 4-bit synchronous up counter.

5.11 Compare the maximum counting rate of (a) a 3-bit ripple and (b) a 3-bit synchronous counter. Assume a propagation delay of 40 ns for the flip-flops and 20 ns for the AND gates.

Answer: (a) 8.33 MHz; (b) 16.7 MHz.

5.12 Show a schematic for a 3-bit synchronous up/down counter.

5.13 Find the J and K equations for a synchronous mod 6 counter.

Answer: $J_A = K_A = 1$; $J_B = A \cdot \bar{C}$; $K_B = A$; $J_C = A \cdot B$; $K_C = A$.

5.14 Show the schematic for a mod 6 ripple counter using J-K flip-flops which require logic lows for reset.

5.15 What is the Preset count for a 4-bit presettable counter which counts to 6?

Answer: 1010.

5.16 Find the J and K equations for a synchronous divide-by-5 counter.

5.17 Compare the flip-flop counts of a 1-MHz frequency counter using synchronous and ring counters.

Answer: 24 for synchronous, 60 for ring.

5.18 Show the schematic of a gateless divide-by-12 counter which counts in conventional binary.

5.19 Compare the maximum counting rates of (a) ripple, (b) synchronous, and (c) ring counters which use 25-MHz flip-flops and are 5 bits long.

Answer: (a) 5 MHz; (b) 25 MHz; (c) 125 MHz.

5.20 Show truth table, schematic, and timing diagram for a 5-bit ring counter with D flip-flops.

5.21 Write the Boolean equations for each J and K of a mod 5 twisted ring counter.

Answer: $J_A = \bar{Q}_C$; $K_A = Q_B$; $J_B = Q_A$; $K_B = \bar{Q}_A$; $J_C = Q_B$; $K_C = \bar{Q}_B$.

5.22 The clock pulses used to load a serial input shift register are high for 30 ns and low for 70 ns. How long does it take to load (a) a 4-bit register, (b) a 8-bit register?

5.23 A circuit requires a 1-ms delay. Find C_p if a 5-stage SISO is available.

Answer: 0.2 ms.

5.24 A single-rail data transmission line transmits 100 bauds. Data words are byte size. If each byte represents an alphabetic character, how many words per minute can be transmitted?

5.25 A computer has a 0.5-MHz clock rate. It is designed to perform multiplication and division by left/right shifting in a register. How long does it take to (a) divide by 64, (b) divide by 32,768?

Answer: (a) 16 μs; (b) 30 μs.

5.26 Show the schematic of a 4-bit PISO shift register with J-K flip-flops.

5.27 What value of resistor is necessary to limit the current through an LED to 5 mA if the voltage source is 5 V?

Answer: 580 Ω.

5.28 Find the logic equation for decoding g in a seven-segment decimal display.

5.29 An octal display is driven from a CBA register output. Find the Boolean equation for the a segment of a 7-segment display which is in the octal system.

Answer: $a = A + B + \bar{C}$.

5.30 Compare the power dissipations of operating 4-digit 7-segment displays. The LCD version draws 0.7 μA per segment from a 5-V supply and the LED version draws 20 mA per segment from a 5-V supply.

6
MATHEMATICAL OPERATIONS

6.0 INTRODUCTION

A text describing computers must cover a wide range of material. Accordingly, the first three chapters have been devoted to Boolean algebra, binary numbers, and gates, and the next two chapters to flip-flops and counters. We now turn attention to how circuits are combined to perform useful functions.

Regardless of size or complexity, the main function of a computer is high-speed arithmetic. Mathematical operations will be discussed in this chapter. The first sections describe a circuit which adds 2 bits in accordance with the laws of binary addition; this circuit is the basis of all computer arithmetic. Serial adders add numbers bit by bit, and parallel adders add all bits simultaneously. An adder circuit can perform subtraction by using the complement of a number. The adder circuit is also the basic component for performing multiplication and division.

Operating speed is a consideration in selecting a specific adder. Fast adders are more complex than slow adders. Generally, faster adders are used in large computers and relatively slower but less complex adders in small computers. Besides the ability to perform arithmetic operations, the number range is an important factor. Floating-point arithmetic, the computer equivalent of scientific notation, is the typical method of obtaining a practical number range.

Finally, this chapter concludes with a discussion of binary-coded decimal, which is a useful number system for pocket calculators and test equipment with digital outputs.

6.1 HALF-ADDERS

The simplest form of electronic binary arithmetic is performed in a *half-adder* circuit. A half-adder adds 2 bits and generates sum and carry-out bits. Although Boolean gates are used to perform binary addition, the half-adder is designed to perform true binary rather than logical addition. Half-adder design follows the same sequence as other digital circuits: problem definition,

Figure 6.1
Half-adder requirements.
(a) Binary addition table;
(b) corresponding truth table.

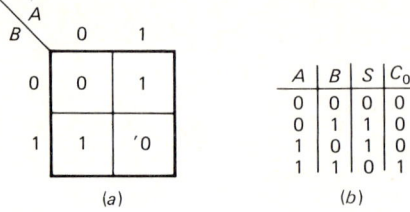

truth table, conversion to Boolean algebra, attempted reduction of the Boolean equation, and finally construction of the circuit. A half-adder is the electronic version of the binary addition table originally presented in Chap. 2. This table is shown again in Fig. 6.1a, and as before the prime symbol represents a carry-out. The binary addition table defines the problem. Figure 6.1b represents the next step in designing a half-adder, conversion into a truth table. The truth table has four columns: A and B are the 2 input bits; and S, which represents the sum, and C_0, which represents carry-out, are the multiple outputs.

The next step in the design sequence is writing Boolean equations from the truth table. This particular truth table shows that

- S is 1 only if the inputs are different.

- C_0 is 1 only when both inputs are 1.

Boolean equations, in SOP form, are written by ORing the combination of variables which result in 1s. Thus

$$S = \bar{A} \cdot B + A \cdot \bar{B} \qquad (6.1a)$$

and similarly $\quad C_0 = A \cdot B \qquad (6.1b)$

Observe that the sum equation for the half-adder is exactly the same as the equation developed in Chap. 1 for Andy and Bruce going to the movies.

The next step is reduction, if possible, of these Boolean equations. Figure 6.2a and b shows the S and C_0 Karnaugh maps. The S map contains two 1s which are not adjacent; thus S cannot be simplified. The map for C_0 has only a single 1, and no simplification of C_0 is possible. Since no simplification is possible, the final step is construction of the half-adder circuit. Figure 6.2 also shows the block diagram and logic circuit. As shown in Fig. 6.2d, a half-adder logic diagram requires six gates, namely, one OR, three AND, and two inverters.

It is not quite true that six is the minimum number of gates required to construct a half-adder circuit. This particular circuit is the best in terms of Karnaugh map reduction. Recall that map construction requires a single change of state between adjacent squares. Thus, as mentioned in Chap. 1, a Karnaugh map lacks the capability of displaying terms with extended inverts. Maps are useful for simplifying minterm expressions but cannot be used for

Figure 6.2 Half-adder circuit development. (a) Sum map; (b) carry-out map; (c) block diagram; (d) logic diagram.

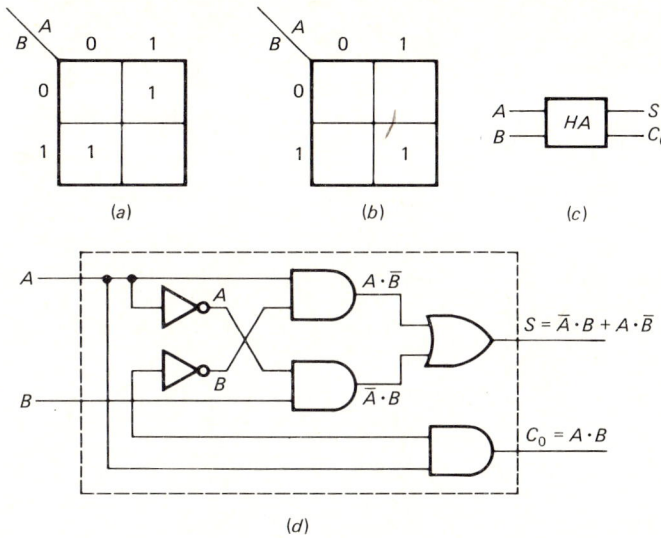

the De Morgan equivalent of an equation. The half-adder happens to be a circuit which can be reduced by applying De Morgan's theorem.

Using the half-adder truth table, begin by writing the sum-of-products form of the invert of S. This is accomplished by ORing the zero combinations.

$$\bar{S} = \bar{A} \cdot \bar{B} + A \cdot B \tag{6.2a}$$

Next invert both sides of the equation

$$\bar{\bar{S}} = \overline{\bar{A} \cdot \bar{B} + A \cdot B} \tag{6.2b}$$

$\bar{\bar{S}}$ is the same as S, and the right-hand side of this equation can be De Morganized:

$$S = (A + B) \cdot (\overline{A \cdot B}) \tag{6.2c}$$

This version of the sum equation combined with the previous carry-out equation is shown in Fig. 6.3. Only four gates, namely, one inverter, one OR gate, and two AND gates, are required. In this instance, strategic use of De Morgan's theorem results in significant circuit reduction.

Table 6.1 shows the truth table for the sum output of a half-adder and also for an OR gate. The outputs are similar. S and OR have the same outputs for the first three combinations. Of course, similarity does not indicate equality between S and OR, but it does indicate a close correspondence.

Figure 6.3 De Morganized half-adder.

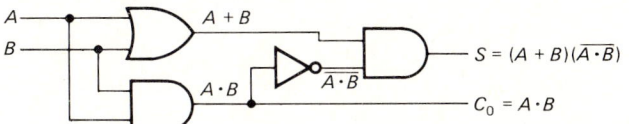

TABLE 6.1
Sum/OR Comparison

A	B	S	OR
0	0	0	0
0	1	1	1
1	0	1	1
1	1	0	1

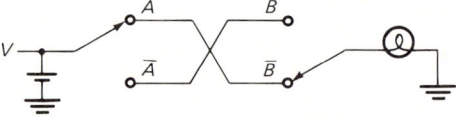

Figure 6.4 EOR using switches.

In particular, S outputs a 1 when the inputs are different and a 0 when the inputs are the same. For this reason the S truth table is called an *Exclusive-OR*, which is usually abbreviated as EOR.

The mechanical switch version of an EOR gate is a very common circuit. As shown in Fig. 6.4, the lamp is off when both switches are in the same position. Similarly, the lamp is on when the switches are in opposite positions. This circuit controls the same lamp from two different positions, such as the top and bottom of a staircase or the opposite sides of a room.

In Boolean algebra the EOR symbol is formed by encircling the OR symbol. Thus

$$S = A \oplus B \tag{6.3a}$$

has the same meaning as

$$S = \bar{A} \cdot B + A \cdot \bar{B} \tag{6.3b}$$

In both cases S is high when A and B are different and low when A and B are at the same logic level. Thus

$$A \oplus B = \bar{A} \cdot B + A \cdot \bar{B} \tag{6.3c}$$

The electronic symbol for an EOR gate is shown in Fig. 6.5a, and Fig. 6.5b shows the two-gate half-adder which results when an EOR is used to perform the sum function. SSI EOR gates exist in both TTL and MOS. The 7486 is a

Figure 6.5
EOR circuits. (*a*) EOR gate; (*b*) EOR half-adder.

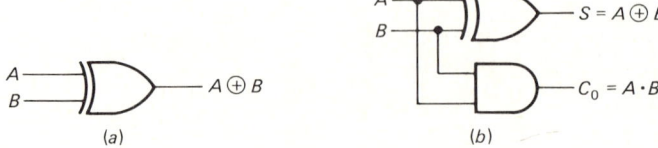

Figure 6.6 True/complement generator.

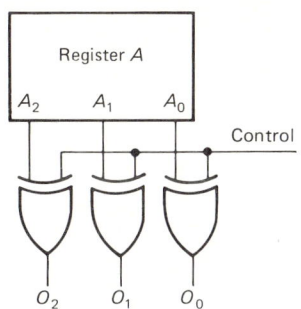

14-pin TTL quad 2-input EOR, and the 4030B and 4070B are MOS versions of the same circuit.

Using an EOR reduces the number of gates required for a half-adder to only two. However, the EOR is constructed by packaging the sum section of a half-adder as a single unit. Thus the EOR gate is more complex and therefore more expensive than ordinary 2-input SSI gates. The decision to construct a half-adder using either an EOR gate or else to construct the equivalent S circuit is a trade-off. A two-gate half-adder uses fewer gates which are costlier, while a four-gate half-adder uses more gates which are less expensive. Factors such as circuit board complexity and the quantity of circuits to be constructed must be considered before the "best" half-adder version is chosen.

The EOR has other applications besides reducing the number of gates in a half-adder. Consider the circuit shown in Fig. 6.6. The outputs of each bit of a parallel output register, in this case A_2, A_1, and A_0, are connected to separate EOR gates. The EOR gate outputs are labeled O_2, O_1, O_0 to correspond with register outputs. One input of each EOR gate is tied to the common control line. Control line inputs can be either high or low.

Example 6.1 For the circuit shown in Fig. 6.6 the $A_2 A_1 A_0$ sequence is 101. Determine the output sequence $O_2 O_1 O_0$ when: (a) the control line is low, (b) the control line is high.

Solution (a) The output of an EOR gate is high when the inputs are different and low when the inputs are the same. Thus when the control line is low

$$\text{Bit} \oplus \text{Control} = \text{output}$$

$$
\begin{array}{lrcccl}
A_2 & 1 \oplus & 0 & = & 1 & O_2 \\
A_1 & 0 \oplus & 0 & = & 0 & O_1 \\
A_0 & 1 \oplus & 0 & = & 1 & O_0 \\
\end{array}
$$

The EOR outputs are the same as the register outputs.

(b) When the control line is high

$$\begin{array}{llllll} & \text{Bit} & \oplus & \text{Control} & = & \text{output} \\ A_2 & 1 & \oplus & 1 & = & 0 \quad O_2 \\ A_1 & 0 & \oplus & 1 & = & 1 \quad O_1 \\ A_0 & 1 & \oplus & 1 & = & 0 \quad O_0 \end{array}$$

The EOR outputs are the inverse of the register outputs.

The EOR outputs which are the same as the register are called true. Since complement is another word for inverse, the device is called a *true/complement* generator. The condition of the control line determines whether the true or complementary outputs appear. In Chap. 2 the use of binary complements to simplify subtraction was introduced. The true/complement circuit is a convenient method of obtaining the data or their complements when a register does not output both forms. Packaged true/complement IC circuits exist. For example, the 7487 is a self-contained 4-bit true/complement generator. This IC has four possible output conditions depending on the conditions on the control lines: true, complement, all highs, all lows.

Another EOR application is checking the *parity* of a data word. Regardless of bit length, the number of 1s in a word can only be odd or even. If there are an even number of 1s, the word has *even* parity. Similarly if there are an odd number of 1s, the word has *odd* parity. For example, 1001 has even parity while 1110 has odd parity. A change in any bit from 1 to 0 or vice versa changes the parity of the word. Thus a parity check can be used to determine if a data word has been transferred incorrectly. Typically, parity checking is used for input or output data, especially when stored on magnetic tape. It is also possible to parity-check for more than a single bit change.

A parity-checking circuit can be constructed by cascading EOR gates. Figure 6.7 shows a 4-bit parity checker. The output of this circuit is:

$$L = A_3 \oplus A_2 \oplus A_1 \oplus A_0 \tag{6.4}$$

In particular, the output will be 0 if all inputs are 0 or if the word contains an even number of 1s. Similarly, the output of a parity checker will be 1 if the word contains an odd number of 1s. Thus a change in parity indicates that a bit has changed state and the word has been altered in transmission. TTL and MOS MSI parity checkers exist in a variety of word lengths.

A digital *comparator* is a circuit which compares the value of two words of equal length. EORs can be used to construct a comparator, and a 3-bit

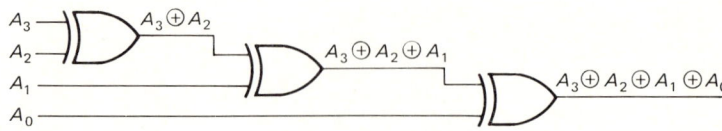

Figure 6.7 Four-bit parity checker.

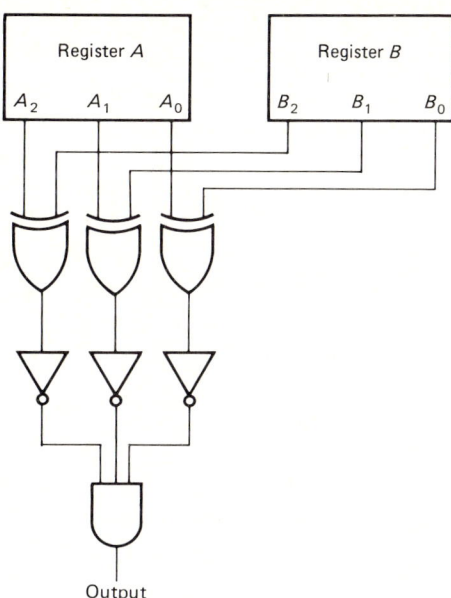

Figure 6.8 Three-bit comparator.

comparator is shown in Fig. 6.8. Corresponding bits of two parallel-output registers are connected to the same EOR. The EOR outputs are connected to inverters, which are in turn connected to the inputs of an AND gate. The AND gate is the comparator output. Each EOR will output a 0 only if the corresponding bits of both words are equal, and the inverter complements this 0 to a 1. Similarly, if the corresponding bits are not the same, the EOR will output a 1, which will be inverted to a 0.

The output of an AND gate is high only when all the inputs are high. For the circuit shown in Fig. 6.8 this can only occur if registers A and B both contain the same word. Thus, a high comparator output indicates that equal words are being stored. Use of a digital comparator is not limited to comparing numerical data. Alphabetic and special characters are also sequences of 1s and 0s and can be compared with a digital comparator. Comparators can also indicate which word is larger if the words are not equal. Typically MSI digital comparators have 3 outputs, 2 low and 1 high. Thus for words A and B, the position of the high indicates whether $A = B$, $A > B$, or $B > A$.

Thus, while the EOR is a component in a minimum-gate half-adder, it has other important applications. The next section describes how the capability of the basic half-adder can be extended.

6.2 FULL-ADDERS

Adders described in the previous section fulfill design requirements; addend and augend bits are added to produce sum and carry-out bits. Useful as it is, a

Figure 6.9
Full-adder requirements.
(a) Block diagram; (b) truth table.

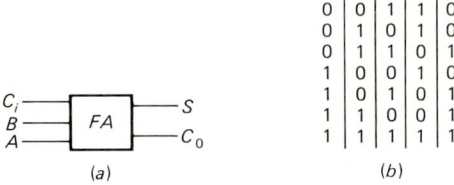

half-adder does not perform binary addition for all possible conditions. When binary addition was introduced in Chap. 2, combinations arose for which a carry-in was transferred to the next higher significant bit column. For example:

$$\begin{array}{r} 1\;'0\;'1\;\;1 \\ +\,0\;\;\;0\;\;\;1\;\;\;1 \\ \hline 1\;\;\;1\;\;\;1\;\;\;0 \end{array}$$

As indicated by primes, the two middle columns must accept a carry-in C_i from the previous column.

Carry-ins are, of course, carry-outs from previous columns and demonstrate the limitation of half-adders. A half-adder generates a carry-out, but there is no provision for accepting a carry-in. A circuit which performs complete binary addition is called a *full-adder*. The full-adder accepts three separate input bits, namely, addend, augend, and carry-in. Full-adder inputs are A, B, and C_i, while outputs are S and C_0. Since a full-adder has three independent inputs, there are eight possible input combinations. Figure 6.9 summarizes the full-adder conditions.

The truth table of Fig. 6.9b is also a true binary addition table and does not represent logical sums. A binary sum of 1 is generated whenever the addition of 3 bits results in a 1. The same condition applies to the carry-out column of the truth table. SOP equations for sum and carry-out are obtained from the truth table by ORing the terms which result in outputs of 1.

$$S = \bar{C}_i \cdot \bar{B} \cdot A + \bar{C}_i \cdot B \cdot \bar{A} + C_i \cdot \bar{B} \cdot \bar{A} + C_i \cdot B \cdot A \tag{6.5a}$$

and $$C_0 = \bar{C}_i \cdot B \cdot A + C_i \cdot \bar{B} \cdot A + C_i \cdot B \cdot \bar{A} + C_i \cdot B \cdot A \tag{6.5b}$$

Equation (6.5a) shows that sum is 1 for all combinations of the 3 inputs which contain 2 inverts and the input combination which contains no inverts. Similarly, Eq. (6.5b) shows that carry-out is 1 for each input combination which contains a single invert and the combination which contains no inverts. The next step in designing a full-adder is an attempt to reduce the Boolean equations. Figures 6.10a and b show the Karnaugh maps. At least in SOP form, the sum equation cannot be reduced. The carry-out equation reduces to

$$C_0 = B \cdot A + C_i \cdot B + C_i \cdot A \tag{6.5c}$$

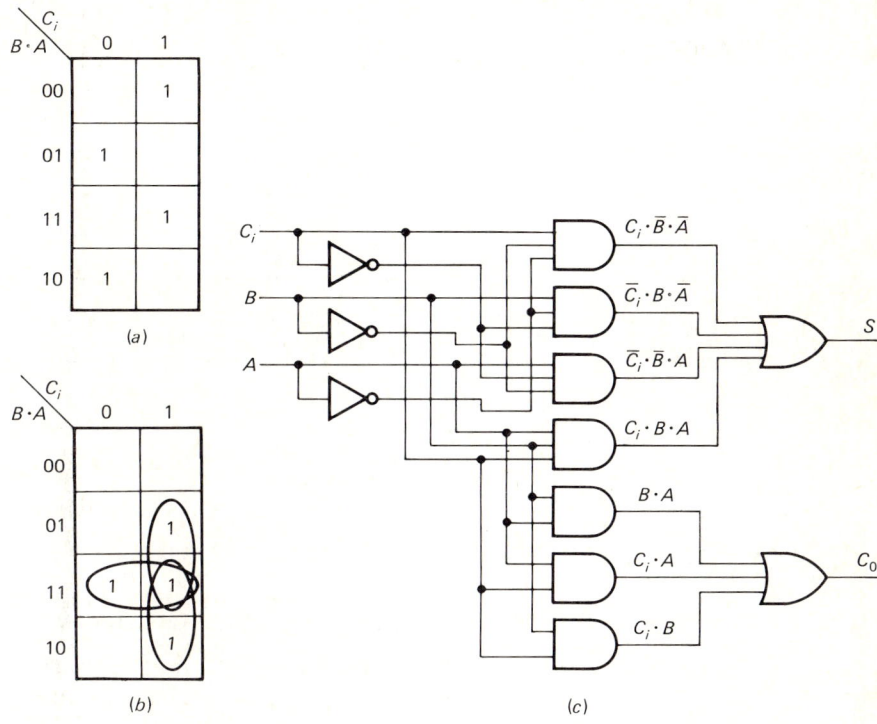

Figure 6.10 Basic full-adder. (*a*) Sum map; (*b*) carry-out map; (*c*) logic diagram.

In words this is all possible combinations of 2 of the 3 inputs. The resultant full-adder is shown in Fig. 6.10*c*.

This full-adder requires 12 gates. In order of increasing complexity these gates are: three inverters, three 2-input ANDs, one 3-input OR, four 3-input ANDs, and one 4-input OR. As with the half-adder, use of De Morgan's theorem reduces the number of gates. Another method of circuit reduction uses the EOR gate. Consider the circuit shown in Fig. 6.11*a*. If this circuit had been presented in the previous section, it would have been called a 3-bit parity checker. The equation is

$$L = C_i \oplus B \oplus A \tag{6.6}$$

and the truth table is shown in Fig. 6.11*b*. The output of an EOR is high when the inputs are different. Thus the output of this circuit is high whenever 2 of the 3 inputs are low and also when all 3 inputs are high. A 3-bit parity checker has the same truth table as the sum section of a full adder. Figure 6.11*c* shows a full-adder using the Karnaugh version of the carry-out section combined with a 3-bit EOR parity checker.

In this version there is no need for any inverters, and the number of gates is reduced from 12 to 6: three 2-input ANDs, two 2-input EORs, and one 3-input OR. The EOR-gate full-adder reduces the number of gates by half; however, it is not automatically less expensive since, as has been discussed,

Figure 6.11 EOR-gate full-adder. (a) Sum section; (b) sum truth table; (c) logic diagram.

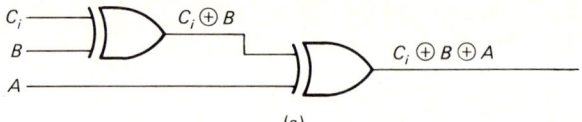

(a)

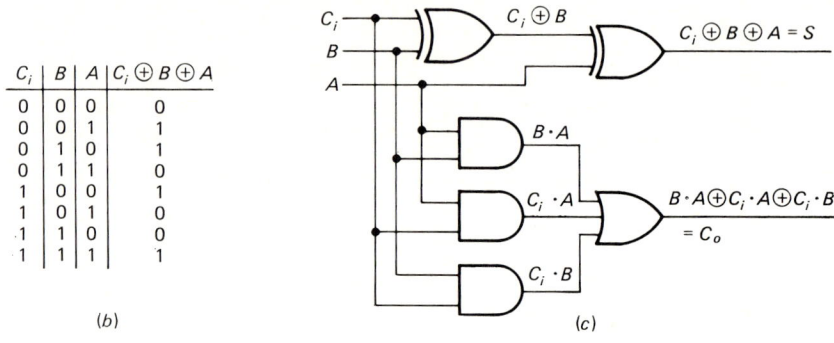

C_i	B	A	$C_i \oplus B \oplus A$
0	0	0	0
0	0	1	1
0	1	0	1
0	1	1	0
1	0	0	1
1	0	1	0
1	1	0	0
1	1	1	1

(b)   (c)

the EOR is more complex and therefore more expensive than conventional 2-input gates.

An interesting version of the full-adder is the circuit shown in Fig. 6.12, which consists of two half-adders and one 2-input OR gate. The outputs of half-adder I with B and A as inputs are

$$S_\text{I} = \bar{B} \cdot A + B \cdot \bar{A} \tag{6.7a}$$

and $\quad C_\text{oI} = B \cdot A \tag{6.7b}$

These are the same half-adder equations derived in the previous section. Since half-adder II has C_i and S_I as inputs, the outputs of this section are

$$S_\text{II} = \bar{C}_i \cdot S_\text{I} + C_i \cdot \bar{S}_\text{I} \tag{6.8a}$$

and $\quad C_\text{oII} = C_i \cdot S_\text{I} \tag{6.8b}$

S_II is the sum output of the complete full adder. Therefore, we must demonstrate that S_II has the same Boolean equation as S of the previous full-adders.

When S_I of Eq. (6.7a) is substituted into Eq. (6.8a), the result is

$$S_\text{II} = \bar{C}_i(\bar{B} \cdot A + B \cdot \bar{A}) + C_i \overline{(\bar{B} \cdot A + B \cdot \bar{A})}$$

Expanding the first term and De Morganizing the second yields

$$S_\text{II} = \bar{C}_i \cdot \bar{B} \cdot A + \bar{C}_i \cdot B \cdot \bar{A} + C_i(\bar{\bar{B}} + \bar{A}) \cdot (\bar{B} + \bar{\bar{A}})$$

Figure 6.12 Generalized full-adder.

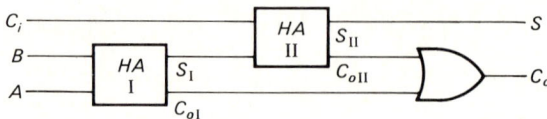

Since a double inverted quantity is the same as the uninverted quantity, the equation can be rewritten as

$$S_{II} = \bar{C}_i \cdot \bar{B} \cdot A + \bar{C}_i \cdot B \cdot \bar{A} + C_i(B + \bar{A}) \cdot (\bar{B} + A) \tag{6.8c}$$

Consider for a moment the quantities in the parentheses. Expansion of these terms results in

$$B \cdot \bar{B} + B \cdot A + \bar{B} \cdot \bar{A} + A \cdot \bar{A}$$

Since any quantity ANDed with its invert is 0, this expansion reduces to

$$B \cdot A + \bar{B} \cdot \bar{A}$$

Therefore Eq. (6.8c) can be rewritten

$$S_{II} = \bar{C}_i \cdot \bar{B} \cdot A + \bar{C}_i \cdot B \cdot \bar{A} + C_i(B \cdot A + \bar{B} \cdot \bar{A})$$

which upon expansion becomes

$$S_{II} = \bar{C}_i \cdot \bar{B} \cdot A + \bar{C}_i \cdot B \cdot \bar{A} + C_i \cdot B \cdot A + C_i \cdot \bar{B} \cdot \bar{A} \tag{6.8d}$$

Equation (6.8d) consists of all combinations of the three inputs with 2 inverts and the combination with no inverts. It is identical to Eq. (6.5a), which demonstrates that the sum section of the full-adder shown in Fig. 6.12 is correct.

Boolean algebra is also used to determine the equation for the carry-out section shown in Fig. 6.12:

$$C_o = C_{oI} + C_{oII} \tag{6.9a}$$

The value of C_{oI} was determined in Eq. (6.7b), so that

$$C_o = B \cdot A + C_{oII}$$

Similarly, substituting the C_{oII} from Eq. (6.8b) yields

$$C_o = B \cdot A + C_i \cdot S_I$$

and when the S_I value of Eq. (6.7a) is substituted, the result is

$$C_o = B \cdot A + C_i(\bar{B} \cdot A + B \cdot \bar{A}) \tag{6.9b}$$

This version of the carry-out equation can be expanded as follows:

$$C_o = B \cdot A + C_i \cdot \bar{B} \cdot A + C_i \cdot B \cdot \bar{A} \tag{6.9c}$$

In this form the carry-out equation cannot readily be compared with Eq. (6.5b). However, ANDing with 1 is a permissible operation. In particular, if the first term in Eq. (6.9c) is ANDed with 1 in the form of $(C_i + \bar{C}_i)$, the result is

$$C_o = B \cdot A(C_i + \bar{C}_i) + C_i \cdot \bar{B} \cdot A + C_i \cdot B \cdot \bar{A}$$

Upon expansion

$$C_0 = B \cdot A \cdot C_i + B \cdot A \cdot \bar{C}_i + C_i \cdot \bar{B} \cdot A + C_i \cdot B \cdot \bar{A} \quad (6.9d)$$

Equation (6.9d) is identical to Eq. (6.5b). All input combinations containing one invert and the input combination with no inverts are included. Both outputs of the full-adder consisting of 2 half-adders with an OR gate are exactly the same as any other full-adder.

Thus, the most efficient half-adder can be used to construct the most efficient full-adder by using the circuit shown in Fig. 6.12. For example, an application may require a minimum-gate full-adder. In this case, we select the 2-gate half-adder shown in Fig. 6.5b. Thus a minimum-gate full-adder consists of 5 gates, namely, 2 EORs, 2 ANDs, and 1 OR. Observe that all gates are 2-input. By way of contrast, the full-adder circuit shown in Fig. 6.11 requires 6 gates, and one of these is a 3-input gate. Since EOR gates are relatively expensive, minimum gate count is not the only consideration. However, an optimum full-adder can be constructed using 2 half-adders plus an OR gate regardless of the optimization criteria.

6.3 SERIAL ADDERS

The full-adder described in the previous section outputs a single sum bit and a single carry-out bit. Nevertheless, a single-bit full-adder can compute the sum of numbers of any bit length. Consider the circuit shown in Fig. 6.13a. Registers B and A are the input registers; in this case, they are 4-bit serial output shift-right registers. The input registers are connected to the addend and augend input leads of a 1-bit full-adder, while the sum output is connected to the serial input of the sum register. All three registers and the flip-flop are clocked simultaneously.

Recall that data leave a serial output register in ascending order and enter a serial input register in the same LSB-to-MSB sequence. As soon as the numbers to be added enter the input registers, the correct values of S and C_0 for the LSB appear at the full-adder outputs. However, this value of S cannot enter the sum register until the first clock pulse. When the first clock pulse occurs, several operations take place simultaneously:

1. The LSB sum bit is input to the MSB of the sum register.
2. The flip-flop transfers C_0 to C_i.
3. The next higher-order bits are shifted out of registers B and A.
4. C_i, B, and A inputs to the full-adder are combined to obtain S and C_0.

The flip-flop performs an important function. Thus, let us assume the C_0 bit to be a 1. This bit cannot be shifted into C_i until the following clock pulse activates the flip-flop. Thus, the clock pulse which transfers C_0 to C_i is the same pulse which inputs the next higher sum and augend bits. C_i is therefore in the correct time sequence to be added to the B and A full-adder inputs. The carry

Figure 6.13
Four-bit serial adder. (*a*) Block diagram; (*b*) addition sequence.

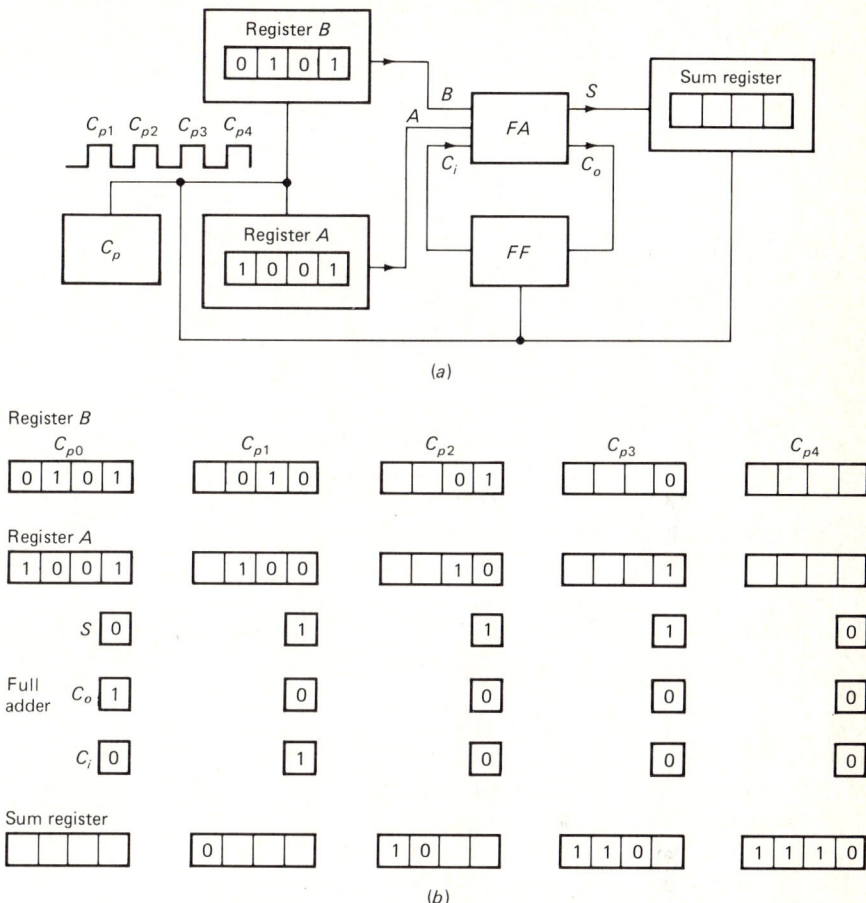

flip-flop provides a 1-bit delay to synchronize C_i with the incoming bits from registers A and B. Figure 6.13*b* shows the bit status for each clock pulse. Empty bits are used for clarity. There are, of course, no empty bits in an actual computer. As a result of adding the two LSBs shown in Fig. 6.13*b*, S is 0 and C_0 is 1. On the first clock pulse this sum is transferred into the MSB of the sum register, and carry-out is transferred into carry-in. Serial, or bit-by-bit, addition continues. On the fourth clock pulse the correct 4-bit sum is stored in the sum register.

Serial addition works well for two numbers but is relatively inconvenient for adding a column of numbers. We might have to add a third number to the result in the sum register. In this case the number in the sum register has to be transferred into one of the input registers. At the same time the third number enters the other input register. Similarly, the addition of a fourth number would again require shifting the sum-register word into an input register, and so forth.

Figure 6.14
Accumulator serial adder.

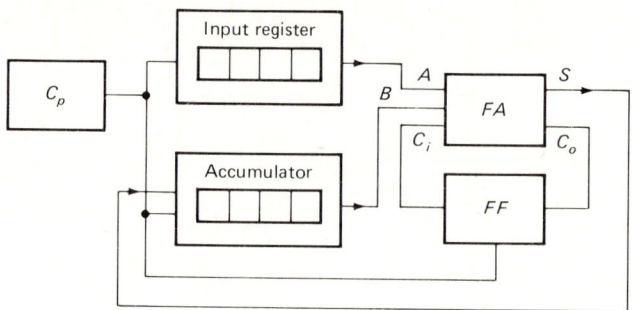

There is a clever technique for adding a column which actually reduces the number of registers required. Figure 6.14 shows the improved circuit. One of the input registers is deleted and the sum register becomes an *accumulator*. In this version, the accumulator is a circulating register. The full-adder with carry flip-flop and the clocking sequence are the same as for a 3-register serial adder.

Addition proceeds 1 bit at a time as in the previous case. Since the accumulator is a circulating register, the result of adding the LSBs appears at the MSB position of the accumulator. On the next pulse, the next 2 bits are added and enter the MSB position. At this time the previous sum bit moves down 1 bit position in the accumulator. After all bits are added, the sum is in correct sequence in the accumulator.

An accumulator is reset prior to use, and the first number is summed with 0s from the reset. After 4 clock pulses, this first partial sum is in the accumulator. Then the next number to be added enters the input register. This second number is added to the previous partial sum, and so on, until all numbers have been added. The accumulator contains the correct partial sum at the end of each 4-bit addition cycle and the correct total sum after the last number is added. An accumulator serial adder is simple and practical. Slow speed is the chief difficulty with serial addition. An *n*-bit serial adder requires *n* clock pulses to complete the addition of two numbers.

$$T_{\text{serial}} = nT_p \qquad (6.10)$$

where T_p, the complete clock pulse period, must be longer than the longest propagation delay in the serial adder circuit.

Example 6.2 A serial adder uses a square-wave clock pulse which is on for 0.5 µs. Determine the time needed to add two 8-bit numbers.

Solution Since the clock pulse is a square wave, it is off for as long as it is on:

$$T_p = t_{\text{on}} + t_{\text{off}}$$
$$= 0.5 \ \mu s + 0.5 \ \mu s$$
$$T_p = 1 \ \mu s$$

The time required to serially add two 8-bit numbers is determined by Eq. (6.10):

$$T_{\text{serial}} = nT_p$$
$$= 8 \times 1 \ \mu s$$
$$T_{\text{serial}} = 8 \ \mu s$$

Eight microseconds may seem a relatively short time to complete the addition of two 8-bit numbers. However, the time to add a column of numbers is cumulative. Thus, addition of a third number requires another 8 μs, which increases the time to 16 μs. Similarly, adding four numbers requires 24 μs, etc.

Although an accumulator continues to store the result of successive additions, there is a fundamental limitation to the addition performed by any computer. Maximum accumulator capacity is determined by the number of bits. For example, a 4-bit accumulator has a maximum capacity of 1111_2, which is 15_{10}. Similarly, the maximum number in an 8-bit accumulator cannot exceed $1111\ 1111_2$, or 255_{10}. In general, the maximum number N_{max} is given by

$$N_{\text{max}} = 2^n - 1 \tag{6.11}$$

where n is the number of bits in the accumulator.

Example 6.3 Find the size of an accumulator which must contain 1000_{10}.

Solution Since
$$N_{\text{max}} = 1000$$
$$1000 = 2^n - 1$$
$$2^n = 1001$$
$$n \cdot \log 2 = \log 1001$$
$$n = \log 1001 / \log 2 \simeq 3/0.301$$
$$n = 9.96 \text{ bits}$$

Therefore the accumulator must be 10 bits long.

A carry-out from the MSB of an accumulator indicates an overflow. The computer is incapable of handling such a number. Since an overflow is always possible, logic circuits must be included to halt operation whenever an overflow occurs. An overflow detector for a 4-bit serial adder is shown in Fig. 6.15. A program counter has been added. As each clock pulse sequences the serial adder, it also increases program count by 1. Carry-outs can occur for

Figure 6.15 Overflow detector.

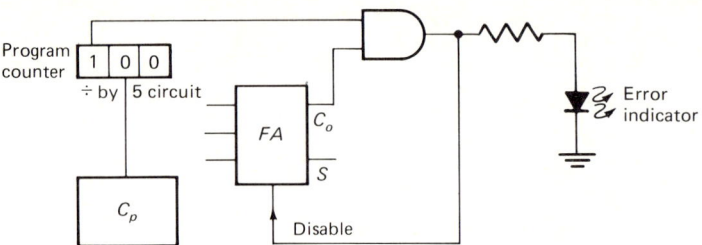

any count except the last. On the last count the program counter delivers a high to the AND gate. If carry-out occurs at the same time, the AND gate output goes high. This disables the full-adder and also signals an overflow by turning on the LED.

One method of extending the range of an accumulator is to break the numbers to be added into a size which the accumulator can process. A 4-bit accumulator can add 16-bit numbers if the 16 bits are processed in four 4-bit segments. The result of each 4-bit addition is stored in a separate register. Early hand-held calculators operated in just this way because 4-bit ICs were the only available size at the time.

Single-bit full-adder ICs are available and are used to construct n-bit serial adders. The 7480 is a 14-pin TTL single-bit full-adder. This chip contains the three inputs and two outputs required to construct a full-adder. The 7480 also contains $\bar{S}$ and $\bar{C}_0$.

Mathematical operations such as multiplication and division are performed by repetitions of the basic addition process. Thus, the time required to perform serial arithmetic becomes excessive. Serial addition is used when low cost rather than speed is the prime consideration.

6.4 PARALLEL ADDERS

The relatively simple serial adder, which performs addition 1 bit at a time, can be replaced by a *parallel* adder. This circuit adds all bits during a single clock pulse. The penalty for faster addition is a more complex circuit. In fact, a parallel adder requires one full-adder for each bit. Figure 6.16 shows a 4-bit parallel adder. Since the LSB cannot have a carry-in, full-adder FA_0 can be replaced with a half-adder. However, this component reduction is negligible, and most parallel adders use full-adders in each bit position.

Input registers A and B have parallel outputs. Corresponding bits are connected to the same full-adder, A_3 and B_3 to $FA_3, \ldots,$ etc. Sum outputs of each full-adder are stored in corresponding bits of the parallel input sum register. The carry-out bit of each full-adder is the carry-in bit to the next higher-order bit. Numbers to be added enter the input registers, and the complete sum appears in the sum register on the next clock pulse. Parallel addition is considerably faster than serial addition.

Figure 6.16 Basic parallel adder.

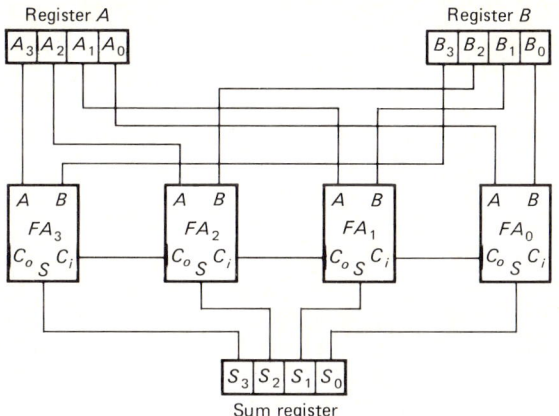

A 4-bit parallel adder is four times as fast as a 4-bit serial adder, but four times as many full-adders are required. This ratio remains constant. An 8-bit parallel adder is eight times as fast as an 8-bit serial adder at a cost of eight times as many full-adders. Whether serial or parallel addition circuits are used, maximum number limitation is the same. The largest number occurs when each bit in the sum register is a 1, and a carry-out from the MSB indicates an overflow.

As a practical matter this basic parallel adder, although more complex, may not be faster than a serial adder. Consider the following two examples of binary addition.

$$\begin{array}{r} 1\ 0\ 1\ 0 \\ +0\ 1\ 0\ 0 \\ \hline 1\ 1\ 1\ 0 \end{array}$$

In this case no carry-ins are required and addition is performed during a single clock pulse. Now consider

$$\begin{array}{r} '0\ '1\ '1\ 1 \\ +\ 0\ 1\ 1\ 1 \\ \hline 1\ 1\ 1\ 0 \end{array}$$

Three carry-ins are required. On the first clock pulse the parallel adder performs the partial sum for each bit position but without any carry-ins. On the next clock pulse the carry-in from the LSB adjusts the sum of the next higher bit, and successive clock pulses complete the carry-in additions. In this case parallel addition requires four clock pulses, one for the initial partial sum and one for each of the three carry-ins.

Carry-ins "ripple up" through individual full-adders just as counts ripple up through a ripple counter. As a result, the more complicated parallel adder can be just as slow as a serial adder. The time delay caused by carry ripple propagation can be eliminated but only by increasing circuit complexity. Inclusion of a *look ahead*, or carry anticipation circuit, to the basic parallel adder results in a *fast adder*. The fast adder performs parallel addition during a single clock pulse regardless of how many carry-ins are required. Carry anticipation circuits determine which bit combinations result in carry-ins at the same time as each full-adder obtains a sum.

Example 6.4 Design C_{A2}, the carry anticipation circuit, for the third bit in a parallel adder.

Solution A carry-in to the third bit is due to carry-outs generated by the two lower-order full-adders. For example if $B_1 B_0$ is 01 and $A_1 A_0$ is 11, the result is:

$$\begin{array}{r} 0\;1 \\ +\;1\;1 \\ \hline 1\;0\;0 \end{array}$$

which is a carry-in to the third bit.

Since each lower-order full-adder bit contains two inputs, there are 4 bits, B_1, A_1, B_0, and A_0, which can cause a carry-in to FA_2. As shown in Table 6.2,

TABLE 6.2
C_{A2} **Truth Table**

B_1	A_1	B_0	A_0	C_{A2}
0	0	0	0	0
0	0	0	1	0
0	0	1	0	0
0	0	1	1	0
0	1	0	0	0
0	1	0	1	0
0	1	1	0	0
0	1	1	1	1
1	0	0	0	0
1	0	0	1	0
1	0	1	0	0
1	0	1	1	1
1	1	0	0	1
1	1	0	1	1
1	1	1	0	1
1	1	1	1	1

Figure 6.17 Carry anticipation reduction. (*a*) Karnaugh map; (*b*) logic diagram.

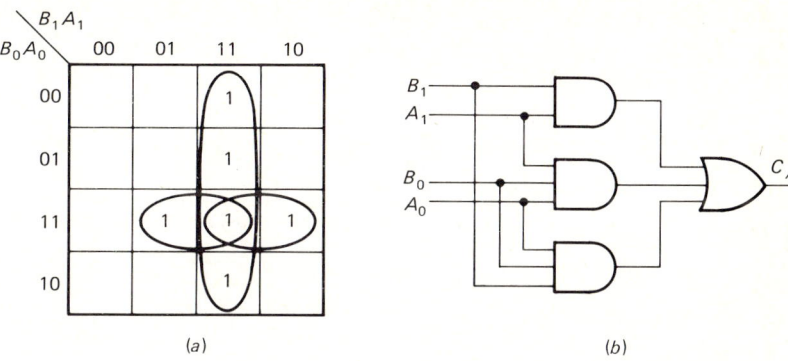

the carry anticipation truth table contains 16 possible combinations. For this truth table, the SOP form of the equation of C_{A2} is

$$C_{A2} = \bar{B}_1 \cdot A_1 \cdot B_0 \cdot A_0 + B_1 \cdot \bar{A}_1 \cdot B_0 \cdot A_0 + B_1 \cdot A_1 \cdot \bar{B}_0 \cdot \bar{A}_0 + B_1 \cdot A_1 \cdot \bar{B}_0 \cdot A_0$$
$$+ B_1 \cdot A_1 \cdot B_0 \cdot \bar{A}_0 + B_1 \cdot A_1 \cdot B_0 \cdot A_0$$

Constructing this version of C_{A2} requires 11 gates: four inverters, six 4-input AND gates, and one 6-input OR gate. This is possible, but the Karnaugh map shown in Fig. 6.17*a* yields a much simpler circuit. In this reduced version, gates are simpler and fewer: two 3-input AND gates, one 2-input AND gate, and one 3-input OR gate for a total of four gates.

Each bit of a fast adder requires a separate carry anticipation analysis. For example, C_{A3} is the result of carry-ins from the three lower-order full-adders. This means 6 bits can cause carry-ins to the fourth full-adder. Therefore, the starting point for the design of C_{A3} is a 64-combination truth table. Similarly, the fast-carry circuit for C_{A4} is due to 4 lower-order bits. Thus the design for C_{A4} begins with a 256-combination truth table, etc.

Such analyses are necessary and have been performed. IC parallel adders with self-contained carry anticipation circuits for each bit exist in both TTL and MOS. The 7483 is a 16-pin 4-bit TTL parallel fast adder, and the 4008 is the MOS equivalent. Both versions contain carry-in and carry-out leads. These leads make it possible to cascade adder chips to increase word length. Figure 6.18b shows two 4-bit adders cascaded to form an 8-bit adder.

Regardless of how many fast adders are cascaded, complete addition of two numbers is completed during one clock pulse. However, a two-input register parallel adder suffers from the same problem as a two-input register serial adder. Addition of two numbers is straightforward, but addition of a column of numbers is complicated. Use of an accumulator rather than a two-input register circuit simplifies columnar parallel addition just as in serial addition.

A simplified diagram of an accumulator-type 4-bit parallel adder with control logic is shown in Fig. 6.19. For clarity, reset lines are omitted, but the accumulator must be reset prior to use. During the low portion of the first

248 Computer Circuit Concepts

Figure 6.18
Fast parallel adder. (*a*) IC block diagram; (*b*) cascaded 8-bit adder.

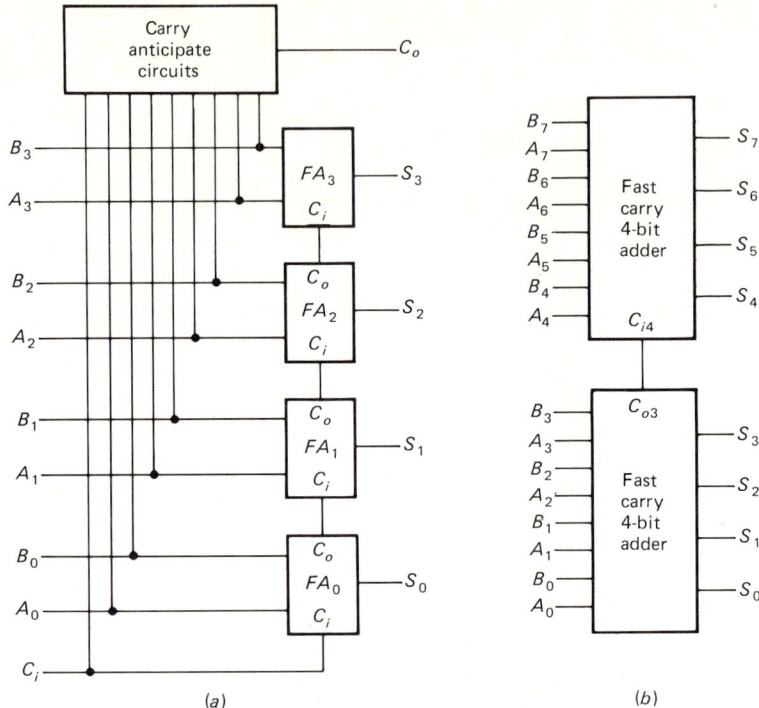

Figure 6.19
Parallel addition using an accumulator.

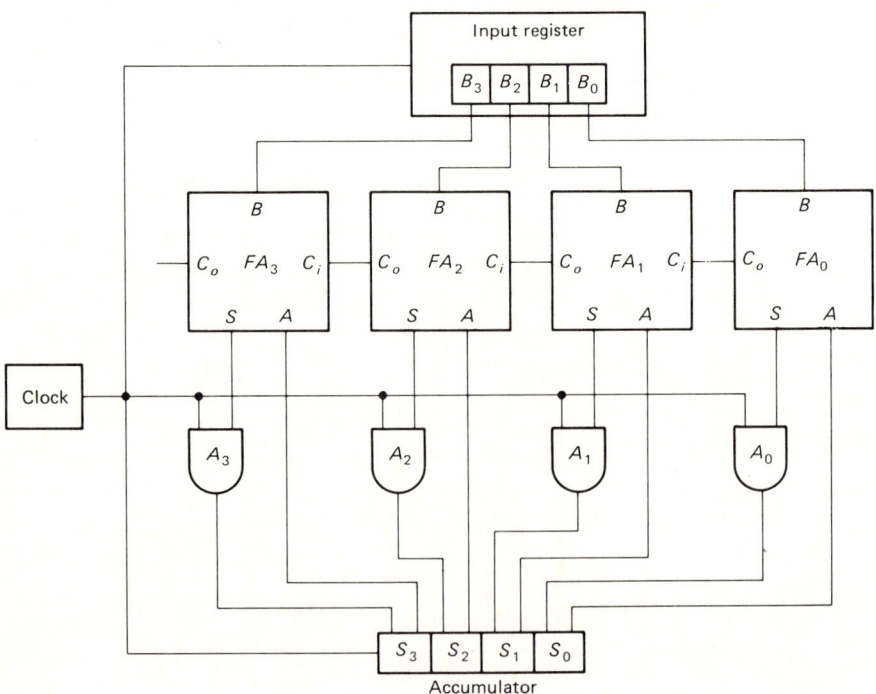

clock pulse, the first number to be added enters the input register. While the clock is low, the first number is added to the 0s in the previously reset accumulator. Each sum bit appears as an input to the respective AND gate, S_3 to A_3, etc. On the alternative half of the same clock cycle, C_p goes high, enabling the AND gates. Sum bits enter the accumulator when the clock is high. At the end of this initial clock cycle the accumulator contains the first number.

During the low portion of the second clock pulse, the second number in the column enters the input register and is summed with the first number now in the accumulator. Since C_p is still low, the partial sum of the first two numbers exists at the full-adder outputs but is not yet in the accumulator. Transfer into the accumulator occurs during the high portion of the second clock cycle. The process continues. Successive numbers enter the input register while C_p is low, and the sum enters the accumulator while C_p is high. On each successive clock pulse, the correct partial sum is in the accumulator. When the last number enters the input register, the result is the sum for an entire column of numbers. An accumulator-type parallel adder sums n numbers in n clock pulses.

6.5 SIGNED ARITHMETIC

Serial and parallel adders both add columns of numbers. However a general-purpose adder must add negative as well as positive numbers, and subtraction is also required. While it is possible to construct separate circuits to implement subtraction, a single circuit to perform addition and subtraction is more efficient.

As described in Chap. 2, the inclusion of negative numbers requires a bit position to indicate sign, and the MSB is the sign bit. If the MSB is a 0, the number is positive; if the MSB is 1, the number is negative. Each text uses a different convention to indicate signed numbers. In this text the MSB is underlined when the number is signed. For example, 0101 represents +5, while 0101 is the unsigned version. Of course, underlining is for reading ease. No such distinction exists in a computer.

Three methods of representing signed numbers have been found useful:

- *Signed magnitude*   Positive numbers begin with a 0 and negative numbers begin with a 1. For example, as a 4-bit number +6 is 0110 while −6 is 1110.

- *1s complement*   Positive numbers are the same as in signed magnitude. Negative numbers begin with a 1 and the invert of the magnitude represents the negative value. In 1s complement +6 is 0110 while −6 is 1001.

- *2s complement*   Positive numbers are the same as in signed magnitude. Negative numbers are obtained by adding a 1 to the negative 1s complement of the number. In 2s complement +6 is 0110 and −6 is 1010.

TABLE 6.3 4-Bit Number Comparison

	Unsigned	Signed Magnitude	1s Complement	2s Complement
Positive form	$DCBA$	$0CBA$	$0CBA$	$0CBA$
Negative form	—	$1CBA$	$1\overline{CBA}$	$1\overline{CBA} + 1$
Highest number	1111 = 15	0111 = +7	0111 = +7	0111 = +7
Lowest number	0000 = 0	1111 = −7	1000 = −7	1000 = −8
Zero representation	0000 = 0	0000 = +0	0000 = +0	0000 = +0
		1000 = −0	1111 = −0	
Range	16 numbers	15 numbers	15 numbers	16 numbers
	0 through 15	−7 through +7	−7 through +7	−8 through +7

Table 6.3 compares features of unsigned numbers in these three signed number systems. Allocating a bit for a sign decreases the largest number which can be represented, but this does not necessarily decrease total number range. For example, 4-bit unsigned and 2s complement numbers both have a range of 16 numbers. In signed magnitude and 1s complement, there are two versions of 0. In both cases, the range is decreased by 1.

Addition with signed magnitude requires testing to determine the sum sign. When adding two signed magnitude numbers, the signs may be the same, either both $\underline{0}$ or both $\underline{1}$. In this case magnitudes are added and the common sign is retained for the sum. When the signs are opposite, one method is required when the augend magnitude is greater and another method when the addend magnitude is greater. Several different sign possibilities must also be evaluated when signed magnitude numbers are subtracted. Testing signed magnitude numbers slows a computer down. Signed magnitude arithmetic is no longer popular. Complementary arithmetic circuits are easier to implement and result in simpler circuits.

In 1s complement, three sign possibilities exist: both numbers are positive, both numbers are negative, or the numbers have opposite signs. When both numbers are positive, all bits including the sign bit are added. Both sign and magnitude of the sum are correct.

For example

$$+3 = \underline{0}011$$
$$+2 = \underline{0}010$$
$$+5 \quad \underline{0}101$$

When both numbers are negative, all bits, including the sign bit, are added. Since both sign bits are $\underline{1}$, an end-around carry is always generated. Sign and magnitude are correct after end-around carry is completed.

$$
\begin{array}{r}
-3 = \underline{1}100 \\
-2 = \underline{1}101 \\
\hline
①\ 1001 \\
\hookrightarrow 1 \\
\hline
-5 = \underline{1}010
\end{array}
$$

When numbers have opposite signs, two possibilities exist. If the magnitude of the positive number is greater than that of the negative number, an end-around carry is generated. Sign and magnitude are correct after end-around carry.

$$
\begin{array}{r}
+3 = \underline{0}011 \\
-2 = \underline{1}101 \\
\hline
①\ 0000 \\
\hookrightarrow 1 \\
\hline
+1 = \underline{0}001
\end{array}
$$

When the magnitude of the negative number is greater, no end-around carry is generated. Sign and magnitude are correct.

$$
\begin{array}{r}
+2 = \underline{0}010 \\
-3 = \underline{1}100 \\
\hline
-1 = \underline{1}110
\end{array}
$$

Thus, sign bit testing is not required in 1s complement addition. All bits, including the sign bit, are added. While an extra clock pulse may be required to perform end-around carry, sign and magnitude are correct when addition is completed. Moreover, subtraction is performed with the same circuit by adding the 1s complement of the subtrahend. Sign and magnitude of subtraction are also correct in 1s complement. Therefore 1s complement arithmetic is easier to implement than operation with signed magnitudes. Sign bits are added rather than tested. Figure 6.20 shows a serial 1s complement add/subtract circuit.

When the add/subtract line is high, the add AND gate is enabled but the subtract AND gate is disabled. In this case, bits from input register B arrive at the full-adder in original form. The contents of register B are summed with the contents of the accumulator on a bit-by-bit basis.

When the add/subtract line is low, the subtract AND gate is enabled but the add AND gate is disabled. Bits from the input register pass through an inverter before reaching the full-adder. Thus, the 1s complement of the number is processed by the full-adder, and subtraction results. The add/subtract logic consists of two inverters, two AND gates, and one OR gate.

Figure 6.20 1s complement serial add/subtract circuit.

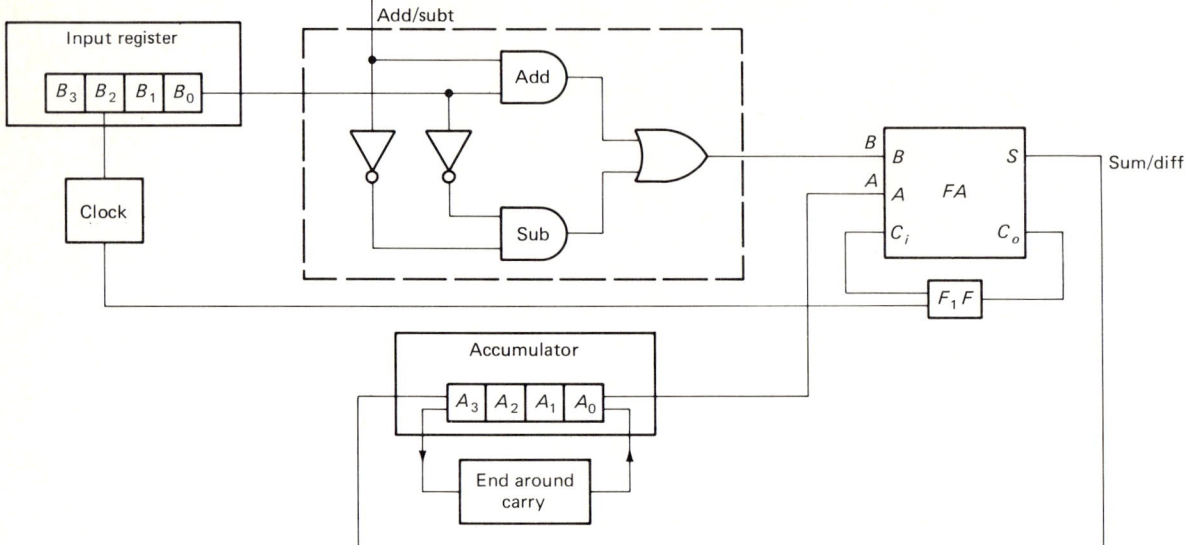

End-around carry is a problem with 1s complement arithmetic in a serial adder. End-around carry requires an additional clock pulse, and extra circuits to transfer the MSB carry-out to the LSB. Also adding 1 to the LSB can cause additional carrys to higher-order bits. Difficulty with implementing end-around carry eliminates simplicity as an advantage of serial addition.

However, end-around carry is easy in a 1s complement parallel adder, where it results from connecting C_o of the MSB to C_i of the LSB. Figure 6.21 shows a parallel 1s complement arithmetic circuit. The same logic used for serial 1s complement add/subtract could be used for the parallel version, but a more efficient solution exists. The true/complement generator, discussed in Sec. 6.1, yields either a positive or negative 1s complement number. If the add/subtract line is low, EORing outputs the original number; if this line is high, the 1s complement results. Otherwise, a 1s complement parallel add/subtract circuit is the same as that for the previously discussed parallel adder. Numbers enter the input register and are summed when the clock pulse is low. When the clock pulse is high, the result enters the accumulator.

While 1s complement arithmetic is practical, 2s complement arithmetic is even easier and is the preferred method in modern computers. The same three possibilities, namely, both positive, both negative, and opposite in sign, must be investigated for a 2s complement add/subtract circuit:

- Since positive numbers are identical in 1s and 2s complement, the situation in which both addend and augend are positive is exactly the same. Thus, 2s complement addition is the same as 1s complement addition.

- When both numbers are negative, all bits, including the sign bit, are added.

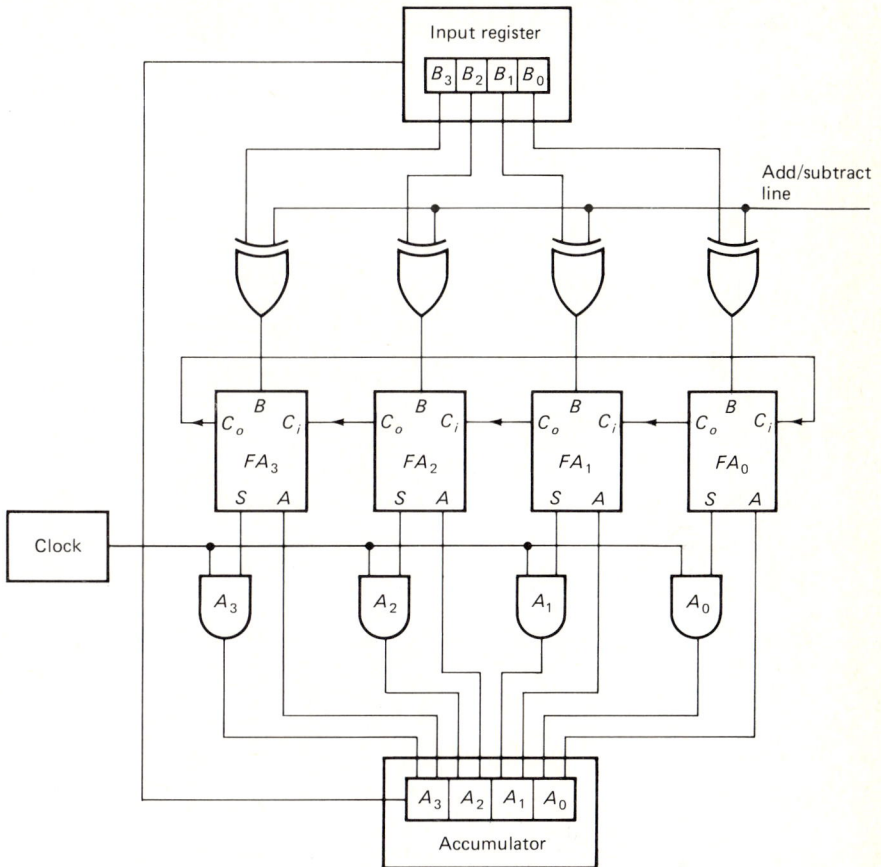

Figure 6.21
1s complement parallel add/subtract circuit.

The carry-out, which is always generated, is discarded. Both sign and magnitude are correct.

$$
\begin{array}{rl}
-3 = & \underline{1}101 \\
-2 = & \underline{1}110 \\
\hline
-5 = \text{①} & \underline{1}011 \\
\end{array}
$$
↳ Ignore carry-out

- The same two possibilities occur when adding numbers with opposite signs. If the magnitude of the positive number is greater, a carry-out is generated and discarded. Both sign and magnitude are correct.

$$
\begin{array}{rl}
+3 = & \underline{0}011 \\
-2 = & \underline{1}110 \\
\hline
+1 = \text{①} & \underline{0}001 \\
\end{array}
$$
↳ Ignore carry-out

When the negative number has greater magnitude, no carry-out is generated. Both sign and magnitude are correct.

$$-3 = \underline{1}101$$
$$+2 = \underline{0}010$$
$$-1 = \underline{1}111$$

Similarly, subtraction in 2s complement is performed by complementing the subtrahend and adding. No additional circuits are required to perform subtraction.

The advantage of 2s complement over 1s complement is clear. Carry-outs from the MSB, when and if they occur, are simply ignored. Therefore, no circuits are needed to process carry-outs. 2s complement arithmetic is also completed in one clock pulse. However, a circuit to generate 2s complement numbers is required. As described in Chap. 2, the simplest pencil-and-paper method of generating negative 2s complement numbers begins at the LSB and recopies up to and including the first 1. Then all remaining bits are inverted. This approach, which is particularly suitable for serial 2s complement arithmetic, is shown in Fig. 6.22.

The set/reset line is omitted for clarity, but the flip-flop is set prior to use. With the flip-flop set, Q is high and $\bar{Q}$ is low. This enables the AND gate connected to Q, while the AND gate connected to $\bar{Q}$ is disabled. Beginning with the LSB of the input register, data enter the input AND gate and are clocked through. The rate is 1 bit for each positive clock pulse. Lower-order 0 bits pass undisturbed through the upper AND gate until the first 1 bit reaches the input. This first 1 bit resets the flip-flop; however, it does pass undisturbed through the upper AND gate. Resetting enables the lower AND gate, while disabling the upper AND gate. Since data bits which enter the lower AND gate are first inverted, all higher-order bits after the first 1 are inverted. Observe that the flip-flop remains in the reset condition after triggering, since J is low and K is high. Thus, all bits up to and including the first 1 are unchanged, while all higher-order bits, including the sign bit, are inverted.

The circuit shown in Fig. 6.22 must be enabled to generate 2s complements. If this circuit is disabled, the original number arrives at the output, as required for a positive 2s complement number. A 2s complement serial adder

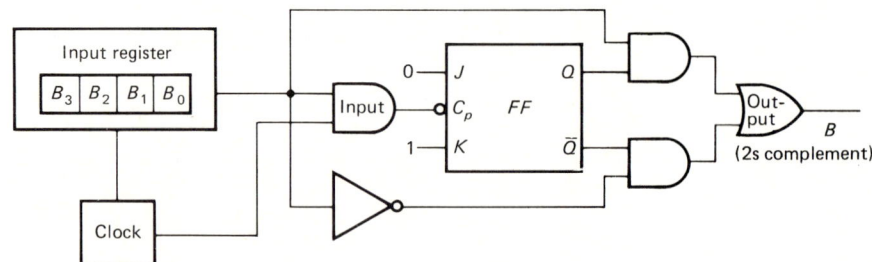

Figure 6.22 2s complement serial generator.

is obtained by replacing the add/subtract section of the circuit shown in Fig. 6.20 with the 2s complement serial generator. The end-around-carry circuit is unnecessary since carry-outs are ignored in 2s complement arithmetic.

Another method of generating 2s complement numbers is more appropriate for a parallel add/subtract circuit. The 2s complement is obtained by adding a 1 to the 1s complement, which results when a high is placed on the control line of a true/complement generator. As shown in Fig. 6.23, the 2s complement can be obtained by connecting a lead from the add/subtract line to the C_i terminal of the LSB. When the control line is high, the true/complement generator outputs the 1s complement. This same high simultaneously inputs a 1 to the LSB. This adds a 1 to the 1s complement, but when the control line is low, the true/complement generator outputs the original number from the input register, while the low at C_i of the LSB has no effect. The only difference between this 2s complement add/subtract circuit and the unsigned parallel add circuit shown in Fig. 6.19 is the true/complement generator needed to obtain negative 2s complement numbers. Neither

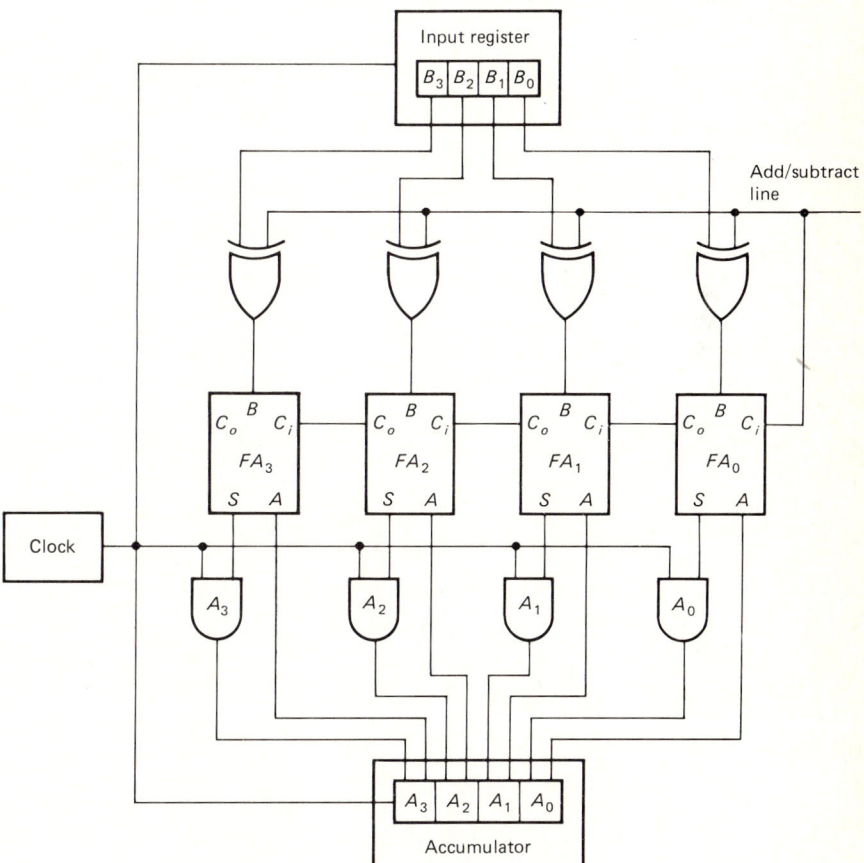

Figure 6.23 2s complement add/subtract circuit.

end-around carry nor an extra clock pulse is required. Thus, parallel 2s complement arithmetic is performed with a minimum of circuits. Although 2s complement arithmetic is farthest removed from pencil-and-paper arithmetic, it is the most efficient method for computers.

A complete description of signed arithmetic circuits should include a discussion of overflow. Since overflow in the form of end-around carry is part of 1s complement arithmetic, overflows cannot be determined until all necessary end-around carry has been completed. On the other hand, overflows in 2s complement are routinely discarded as part of any add or subtract operation. When a true overflow occurs, the result is incorrect, and the computer must be disabled.

Differences exist in detecting overflow in 1s and 2s complement arithmetic, but the principle is the same. Overflow can occur if both numbers are positive. For example:

$$+5 = \underline{0}101$$
$$+4 = \underline{0}100$$
$$+9 = \underline{1}001$$

Overflow can also occur if both numbers are negative.

$$-5 = \underline{1}011$$
$$-4 = \underline{1}100$$
$$-9 = ①\underline{0}111$$

In both cases overflow occurs when the sum sign is different from the signs of the added numbers. If both numbers are positive, a sum sign of $\underline{1}$ indicates overflow, and if both numbers are negative, a sum sign of $\underline{0}$ indicates overflow. Therefore, the Boolean equation for overflow in a 4-bit accumulator is

$$O = \bar{S}_3 \cdot B_3 \cdot A_3 + S_3 \cdot \bar{B}_3 \cdot \bar{A}_3 \tag{6.12}$$

where O = overflow,
S = sum
A = augend
B = addend

The sum of numbers with opposite signs is always less in magnitude than either number. Thus overflow is not possible if augend and addend have opposite signs.

6.6 MULTIPLICATION AND DIVISION

By including a few more control circuits, a full-adder can also perform multiplication and division. In fact, the basic computer arithmetic unit is an accumulator-type full-adder. The same accumulator contains a sum, difference, product, or quotient, depending on which control circuits are activated.

Multiplication can be performed by "brute force" repeated addition: thus to multiply 3 by 4, add 3 four times or else add 4 three times

```
  3
  3     4
  3     4
  3     4
 ──    ──
 12    12
```

Multiplication by repeated addition requires placing one of the numbers to be multiplied in the input register and the other in a down counter. Each time the down counter is stepped, the input register number is added to the accumulator. When the down counter reaches zero, the accumulator contains the product. Multiplication by repeated addition is feasible but may be inconvenient for products of large numbers.

A faster technique is related to pencil-and-paper binary multiplication described in Chap. 2. The multiplier times the multiplicand is the product.

```
      Multiplicand
   ×  Multiplier
      ──────────
      Product
```

An entire multiplicand is multiplied by each bit in the multiplier to obtain a partial product. The weight of each position of the multiplier is accounted for by shifting each successive partial product one position to the left. Summing all partial products yields the total product. Because there are fewer combinations, binary partial products are easier than decimal. In fact, binary multiplication involves only two possibilities.

When a multiplier is 1, the partial product is the multiplicand:

```
     101
  ×    1
     ───
     101
```

When a multiplier is 0 the partial product is 0:

```
     101
  ×    0
     ───
     000
```

For example, to multiply 111 by 101:

Process		Description
	111	Multiplicand
×	101	Multiplier
	111	Multiplier is 1. Partial product is multiplicand.
	000	Shift. Multiplier is 0. Partial product is 0.
	111	Shift. Multiplier is 1. Partial product is multiplicand.
	100011	Product is sum of partial products.

Binary multiplication consists of three operations:

- Either recopying or not recopying the multiplicand
- Shifting
- Adding

Circuits for implementing each of these three steps have been discussed. However, a few more details are needed to obtain a practical binary multiplier.

A full-adder only sums one addend and one augend at a time. Therefore, each partial product must be summed before the next partial product is obtained. Another important point is the bit length of an accumulator used for multiplication. For addition and subtraction an accumulator has the same length as the numbers being processed; but, as just demonstrated, multiplying two 3-bit numbers results in a 6-bit product. Since the product also contains a sign bit, an accumulator for the product of two 3-bit signed numbers must be 7 bits long, 6 for magnitude, and 1 for sign. In general, the bit length of an accumulator L_{Acc} for products must be 1 bit less than the sum of the number of bits in the signed multiplicand and multiplier:

$$L_{Acc} = L_{multiplicand} + L_{multiplier} - 1 \tag{6.13}$$

Example 6.5 Determine the number of bits in an accumulator for multiplying two 8-bit signed numbers.

Solution In this case

$$L_{multiplicand} = 8 \text{ bits}$$
$$L_{multiplier} = 8 \text{ bits}$$
$$L_{Acc} = 8 + 8 - 1 = 15 \text{ bits}$$

Binary product sign is determined by MSB testing. If both MSBs have the

Figure 6.24
Product sign bit. (*a*) Truth table; (*b*) logic circuit.

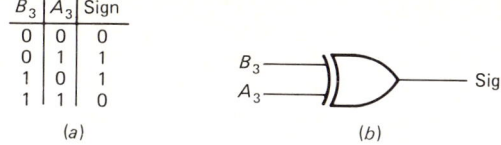

same sign, the product is positive; if the MSBs are opposite in sign, the product is negative. Figure 6.24 shows the truth table and circuit for determining product sign when two 4-bit numbers are multiplied.

Adapting a full-adder to perform multiplication is the remaining step. Multiplication is possible with either serial or parallel full-adders. A parallel multiplier shifts right after each partial product. This corresponds to left shifts in pencil-and-paper multiplication. Figure 6.25 shows the simplified schematic of a 4-bit parallel multiplier. Prior to use, the input register and accumulator are cleared. Then the sign bits are tested, and the product sign is stored in the EOR gate.

The multiplicand is now placed in the input register. In this case the product accumulator is 7 bits long and the 4 lowest-order accumulator bits contain the multiplier. This section is called the *MQ register*. The accumulator and the MQ register may physically be part of the same circuit, but separate chips are also possible. In either case, the accumulator and MQ register function as a single 7-bit shift-right register.

Multiplication begins by testing the LSB of the MQ register. When the LSB is 1, the input register is added to the accumulator; this is equivalent to multiplying by 1. When the LSB is 0, nothing is added to the accumulator;

Figure 6.25
Parallel multiplier circuit.

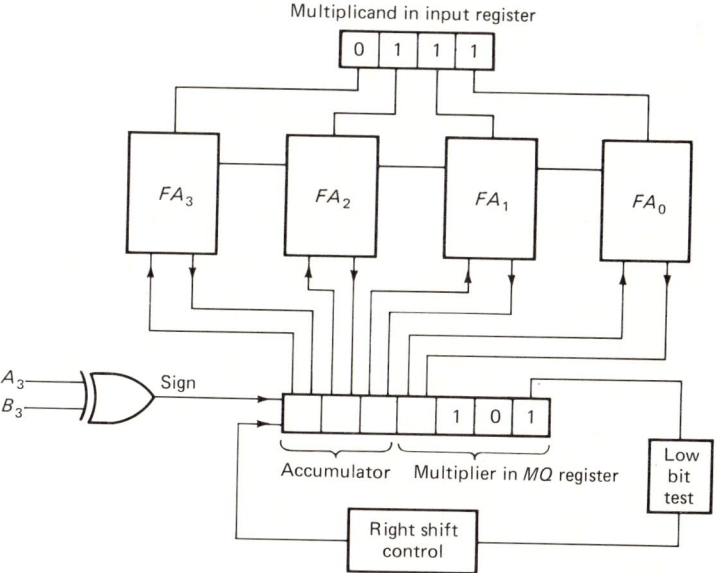

TABLE 6.4 Parallel Multiplication Procedure

Register	Description
Input Register `0 1 1 1`	Test sign bits and store result in EOR. Enter multiplicand magnitude in input register. Enter multiplier magnitude in MQ register.
Accumulator MQ `         1 0 1` LSB	
`0 1 1 1 1 1 0 1`	Test LSB. Since it is 1, add input register to accumulator.
`0 0 1 1 1 1 1 0`	Shift entire contents of accumulator/MQ register right 1 bit.
`0 0 1 1 1 1 1 0`	Test new LSB. Since it is 0, do not add input register to accumulator.
`0 0 0 1 1 1 1 1`	Shift entire contents of accumulator/MQ register right 1 bit.
`1 0 0 0 1 1 1 1`	Test new LSB. Since it is 1, add input register to accumulator (0111 + 0001 = 1000).
`0 1 0 0 0 1 1 1`	Shift entire contents of accumulator/MQ register 1 bit right. Since the last multiplier bit has been shifted out, the accumulator/MQ register contains the product magnitude.
`0 1 0 0 0 1 1 1`	Transfer the sign bit from the EOR into the MSB. 0111 × 0101 = 0100011 (+7) × (+5) = +35

this is equivalent to multiplying by 0. In either case the entire accumulator/MQ register is shifted right by 1 bit. This shifts a bit out of the MQ register, and the next higher-order bit becomes the new LSB. The new LSB is then tested, and addition or no addition of the input register to the accumulator results. Again, the entire contents of the accumulator/MQ register are shifted right by 1 bit. Testing, adding, and shifting continue until the last multiplier bit is shifted out of the MQ register. At this time the accumulator/MQ register contains the product magnitude. Placing the sign bit in the MSB position is the final step. Table 6.4 illustrates parallel multiplication using 0111 as the multiplicand and 0101 as the multiplier.

Each multiplier magnitude bit requires two clock pulses, one for adding the input register to the accumulator and one for shifting. Thus, obtaining the product magnitude of an n-bit multiplier requires $2n - 2$ clock pulses. Clearing the registers for multiplication and entering the sign bit each require a clock pulse. Therefore the total number of clock pulses $C_{p(\text{mult})}$ required for an n-bit multiplier is $2n - 2 + 2$ or

$$C_{p(\text{mult})} = 2n \tag{6.14}$$

Example 6.6 An 8-bit signed number is multiplied by a 4-bit signed number. If the clock frequency is 1 MHz, how long does the multiplication take?

Solution Since $n = 4$,

$$C_{p(\text{mult})} = 2 \times 4 = 8 \text{ clock pulses}$$

For $f = 1$ MHz

$$T = 1/1 \times 10^6 = 1 \times 10^{-6} \text{ s}$$

Therefore the total time is

$$\text{Time} = 8 \text{ pulses} \times 1 \times \frac{10^{-6} \text{ s}}{\text{pulse}} = 8 \text{ } \mu\text{s}$$

This description of parallel multiplication used signed magnitude rather than 2s complement numbers. Some computers still use signed magnitude to perform multiplication. However, ICs are available which perform 2s complement multiplication. The technique, known as Booth's algorithm, is an extension of the signed magnitude method.

Except for shifting in the opposite direction, computer multiplication is the same as the pencil-and-paper version. A similar situation applies to computer division. Since division is the most involved of the four basic arithmetic operations, a brief review is in order.

The result of dividing a dividend by a divisor is a quotient

$$\frac{\text{Dividend}}{\text{Divisor}} = \text{quotient}$$

and the following example demonstrates binary pencil-and-paper division:

```
                    10   quotient
      divisor 11 ) 110   dividend
                   11
                   ──
                   00   first remainder
                   00
                   ──
                   00   second remainder
```

The first step is aligning the divisor bits with the appropriate dividend bits. Next a trial divisor is selected. If the remainder is negative, the trial divisor is too large and a smaller divisor must be tried. An acceptable divisor is obtained when the remainder is positive and smaller than the divisor. This method continues until all dividend bits are processed. Division consists of:

- Divisor and dividend alignment
- Position-by-position divisor testing
- Subtraction

Machine division mimics each of these steps. The dividend is placed in the accumulator and the divisor in the input register. The quotient will appear in the MQ register. Additional control signals are needed, but the same accumulator/MQ register, full-adder, and input register are the circuits for performing division.

When performing pencil-and-paper division, aligning the divisor with the dividend is performed "by eyeball," by using our experience with division. In machine division, the input register is shifted left. Shifting continues until the most significant divisor 1 aligns with the most significant dividend 1. The number of shifts determines the minimum quantity of remainders needed to complete division.

$$\text{Remainders}_{min} = \text{shifts} + 1 \tag{6.15}$$

For example, if two shifts are required to align the divisor, then at least three remainders are required to complete division. Since bit alignment determines how many subtractions are involved, accumulator and input register must have the same length: a 4-bit accumulator requires a 4-bit input register, etc. Additional operations may be required to obtain quotients.

After alignment, the next step is subtracting the data in the input register from the accumulator. If the remainder is positive, the divisor fits, and a 1 is placed in the LSB of the MQ register. If the remainder is negative, the divisor is too large, and a 0 is placed in the MQ register. If the divisor is too large, the dividend must be restored by adding the divisor to the negative remainder. Remainder sign determines if the divisor is correct or too large—a negative remainder indicates an incorrect divisor. Regardless of the remainder sign, the next step is to left-shift the accumulator/MQ by 1 bit. Restoring the dividend is the machine version of trial-and-error divisor testing.

After the required number of subtractions have been performed, the quotient magnitude is in the MQ register. Then the sign bit is placed in the accumulator MSB. Sign conventions are the same for multiplication or division; thus, sign testing is performed with the same EOR used for multiplication. Table 6.5 illustrates division using restoration of the dividend. The problem is to divide 0110 by 0011.

Division requires more steps and more time. Multiplication requires one clock pulse for adding and another for shifting. Each divisor bit requires three clock pulses for comparing dividend with divisor, subtracting, and then shifting, so that

$$C_{p(div)} = 3 \cdot n \tag{6.16}$$

This equation shows that division is 50 percent slower than multiplication.

Talented programmers are aware of the time difference between multiplication and division and adjust their programs accordingly. A program such as finding the average score for a group of students requires dividing each student's point total by the number of exams. If four exams were given, then the sum of each student's grades is divided by 4, etc. However, dividing by 4 is

TABLE 6.5 Signed Division Procedure

Register	Description

Input Register
`0 0 1 1`
Accumulator   MQ
`0 1 1 0` `□ □ □ □`
	Test sign bits and store result in EOR. Enter dividend into accumulator and divisor into input register.

`0 1 1 0` `0 1 1 0` `□ □ □ □`	Shift input register 1 bit left to align MSB of divisor with MSB of dividend. With 1 shift two remainders are required.
`0 0 0 0` `□ □ □ 1`	Subtract input register from accumulator. Since result is positive, place a 1 in LSB of MQ register.
`0 0 0 0` `□ □ 1 □`	Shift accumulator/MQ 1 bit to the left.
`0 0 0 0` `□ □ 1 □`	Subtract input register from accumulator. Since result is negative, add input register to accumulator in order to restore remainder.
`0 0 0 0 0 1 0`	Since subtraction result was negative, place 0 in LSB of MQ and shift left. Two subtractions are completed. The quotient magnitude is in the MQ.
`0 0 0 0 0 1 0`	Transfer sign bit into MSB. $\underline{0}110 \div 00\underline{1}1 = \underline{0}010$ $(+6) \div (+3) = (+2)$

the same as multiplying by the reciprocal of 4, which is 0.25. The result is the same, but multiplying is preferable because time is saved.

Restoring the remainder is the basic method for division. Refined versions are faster.

6.7 FLOATING-POINT ARITHMETIC

Preceding arithmetic circuits perform integer arithmetic. The binary point, while not mentioned, was assumed to be at the end of each number. Integer numbers are also called *fixed-point*. While fixed-point arithmetic is important, a general-purpose computer must process fractions and mixed numbers.

Fixed-point arithmetic circuits can be adapted for noninteger arithmetic. Consider the following examples of binary addition:

```
   1010      1.010
 + 0010      0.010
   ----      -----
   1100      1.100
```

In both cases the bit pattern in the sum is the same, but the sums are different. Positioning a binary point determines the magnitude of the result. When performing pencil-and-paper arithmetic, we solve a problem as if the numbers were fixed-point; only after a result is obtained do we position the decimal point by "counting decimal places." For example, to multiply 3.5 by 7.2 we multiply 35 by 72 and then move the decimal point two places to the left.

One technique for locating the binary point in a computer is to permanently assign part of each register to the integer component and the remainder to the fractional component. In a 4-bit register, 2 bits could be used for the integer portion and 2 bits for the fraction. This approach severely restricts the range of numbers which can be added or subtracted. Multiplying and dividing are even more difficult because data and results can differ by orders of magnitude. Early computers used fixed-point arithmetic and located the decimal point by pencil-and-paper methods. The program had to include rules for positioning the decimal point.

Even before computers, a fixed-point calculator called the slide rule existed. The only numbers on a slide rule are 0 through 10. After a slide rule result is obtained, the "programmer" adjusts the decimal point location. A slide rule user keeps track of the decimal point location by using scientific notation. Suppose we want to determine how many inches in a mile by using a slide rule. Since 12 and 5280 are both out of slide rule range, the numbers are written in scientific notation

$$1.2 \times 10^1 \times 5.28 \times 10^3$$

and the slide rule result is read as approximately

$$6.34 \times 10^4$$

The general formula for multiplying in scientific notation is

$$A \cdot 10^x \cdot B \cdot 10^y = A \cdot B \cdot 10^{x+y} \tag{6.17a}$$

Multiplication in scientific notation requires two separate operations, multiplying the magnitudes, and adding the exponents. Similarly, division in scientific notation

$$\frac{A \cdot 10^x}{B \cdot 10^y} = \frac{A}{B} \cdot 10^{x-y} \tag{6.17b}$$

requires two operations, dividing the magnitudes and subtracting exponents.

Modern computers use a variation of scientific notation called *floating-point*. The magnitude of a floating-point number is a fraction which is at least 0.1. In floating-point form the determination of inches in a mile is

$$0.12 \times 10^2 \times 0.528 \times 10^4$$

The results of floating-point computations are the same as those of computations with standard scientific notation, but the floating-point form uses register bits more efficiently. Of course computers perform calculations in binary. The floating-point forms for binary multiplication and division are:

$$A \cdot 2^x \cdot B \cdot 2^y = A \cdot B \cdot 2^{x+y} \tag{6.18a}$$

and

$$\frac{A \cdot 2^x}{B \cdot 2^y} = \frac{A}{B} \cdot 2^{x-y} \tag{6.18b}$$

where A and B are binary fractions which are at least 0.1_2.

Figure 6.26 Eight-bit floating-point number.

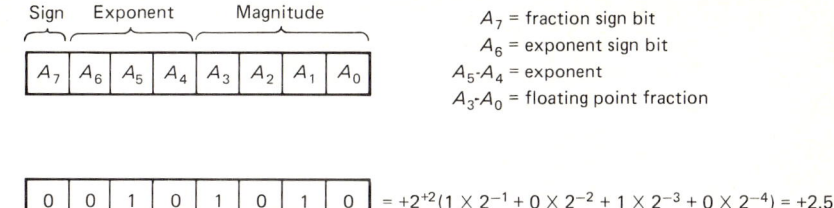

A_7 = fraction sign bit
A_6 = exponent sign bit
A_5-A_4 = exponent
A_3-A_0 = floating point fraction

In a floating-point register some bits represent magnitude, and the remaining bits contain the exponent. Magnitude bits are processed separately from the exponent bits according to Eqs. (6.18). Figure 6.26 shows a 1-byte floating-point number. Number sign conventions are the same as for fixed-point numbers.

Example 6.7 Find the largest 1-byte floating-point number.

Solution A floating-point number is largest when:

> The exponent is positive and as large as possible
> The fraction is positive and consists entirely of 1s

For a 1-byte number the result is:

$$\underline{0}011\ 1111 = +2^{+3}(1 \times 2^{-1} + 1 \times 2^{-2} + 1 \times 2^{-3} + 1 \times 2^{-4})$$
$$= +8(0.5 + 0.25 + 0.125 + 0.0625)$$
$$= +7.5$$

Similarly the most negative number is 1011 1111, or -7.5_{10}. If signed-magnitude floating-point numbers were used, the smallest nonzero positive number would be $\underline{0}111\ 1000$, which is 0.0625. Most newer machines use 2s complement floating-point numbers, and the smallest number is different.

Regardless of whether signed-magnitude or 2s complement floating-point numbers are used, the number range of a single byte is too limited for practical applications. Realistic numbers are obtained when 12 or more bits represent a floating-point number. Figure 6.27 shows a 16-bit floating-point number.

Figure 6.27 Sixteen-bit floating-point number.

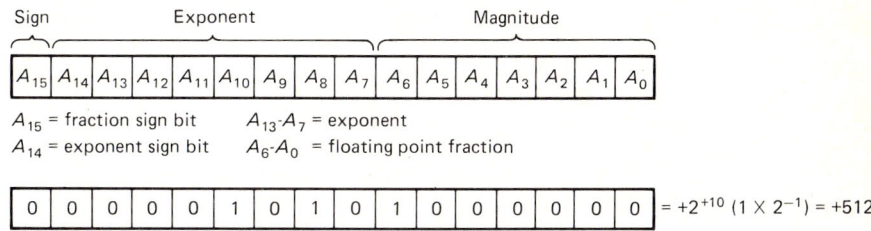

A_{15} = fraction sign bit
A_{14} = exponent sign bit
A_{13}-A_7 = exponent
A_6-A_0 = floating point fraction

Example 6.8 Determine the largest 16-bit floating-point number.

Solution The principle is the same as in the previous example. Since 7 bits exist for the exponent value:

$$2^{127} = 10^x$$

$$127 \log 2 = x \log 10$$

$$x = 127 \frac{\log 2}{\log 10} \simeq 127 \times \frac{0.301}{1} \simeq 38.2$$

This example shows a highest number of about 10^{38} and a lowest number of about 10^{-38}. Such a range is sufficient for many applications. In physics, for example, the charge on a single free electron is in the 10^{-19} coulomb range, and there are more than 10^{24} free electrons in a cubic centimeter of copper. In the field of economics the U.S. gross national product is about 10^{13} dollars. Although the range 10^{38} to 10^{-38} is quite large, many computers and some hand calculators have a range of 10^{99} through 10^{-99}.

A 16-bit floating-point number is called a *single-precision* number. Greater accuracy can be achieved by joining two 16-bit words together and treating the combination as a single 32-bit, or *double-precision*, word. As shown in Fig. 6.28, the MSB of the first word is the fraction sign. The next 8 bits are the exponent, and the remaining 23 bits are the fraction.

In both single- and double-precision numbers, the exponent is 7 bits. Thus, the range of double- and single-precision numbers are the same. Double precision is more accurate because the additional bits represent magnitude. Using single-precision numbers, the floating-point value of π is 3.125, which is more than 0.5 percent lower than the real value. On the other hand, with double precision π is 3.14159265. The penalty for improved accuracy is the additional 16 bits required to implement double precision.

Exponents greater than 10^{38} are obtained by combining more than two 16-bit words. This allows more than 8 bits for the exponent and maintains double-precision accuracy.

Offset is an alternate method of working with the exponent of floating-point numbers. With offset the largest possible exponent value is added to each exponent. For example, 128 is the largest exponent in both single- and

Figure 6.28 Double-precision number.

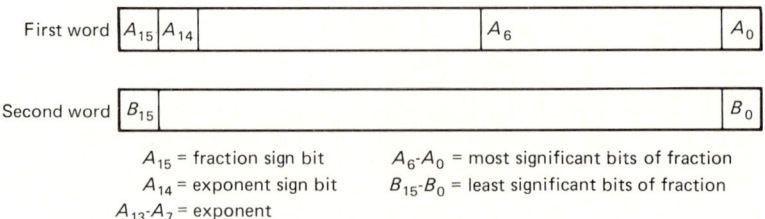

A_{15} = fraction sign bit
A_{14} = exponent sign bit
$A_{13}\text{-}A_7$ = exponent
$A_6\text{-}A_0$ = most significant bits of fraction
$B_{15}\text{-}B_0$ = least significant bits of fraction

double-precision numbers. If the actual exponent is 127, the offset exponent is $128 + 127$, or 255. Similarly, an exponent of 0 has an offset exponent of 128, and an exponent of -128 has an offset value of 0. For a 7-bit exponent the offset value O_v is 128 plus the actual value A_v:

$$O_v = A_v + 128 \tag{6.19}$$

Example 6.9 An offset floating-point number has an exponent of 122. Find the actual exponent.

Solution In this case $O_v = 122$. Therefore

$$122 = A_v + 128$$

$$A_v = -6$$

and 2^{-6} is 0.015625.

Using offset does not change the range of exponents. However, offset eliminates negative exponents and removes the need for sign testing of an offset exponent. The disadvantage of offset is the need to subtract 128 from each exponent.

Floating-point arithmetic requires converting input data to floating-point form. Since the rules for combining magnitudes and exponents of floating-point numbers are different, each segment must be processed separately. Different methods for working in floating point exist. One method uses ICs to convert data into floating point and relies on logic circuits to separate exponent from magnitude. Another method uses a self-contained program to perform conversion to floating point and also to separate exponent from magnitude. Circuits are faster than programs. While the program approach is slower, fewer components are required. Generally, large computers use circuits while desk-top computers and hand calculators use a built-in program.

The same combination of input register, full adder, and accumulator used in mathematical operations can also perform Boolean operations such as AND and OR. As with floating-point arithmetic, logical operations can be implemented either with circuits or built-in programs. Boolean operations are performed by combining bits in the input register and accumulator according to Boolean rather than arithmetic rules. The ability to perform Boolean operations is, of course, useful in its own right.

Boolean algebra is also used in arithmetic procedures. When working with floating-point numbers, Boolean techniques are used to separate exponent and magnitude. The result of ANDing with 0 is always 0, but the result of ANDing with 1 depends on what 1 is ANDed with. Exponent and magnitude are separated by ANDing the magnitude bits with 0 and the exponent bits

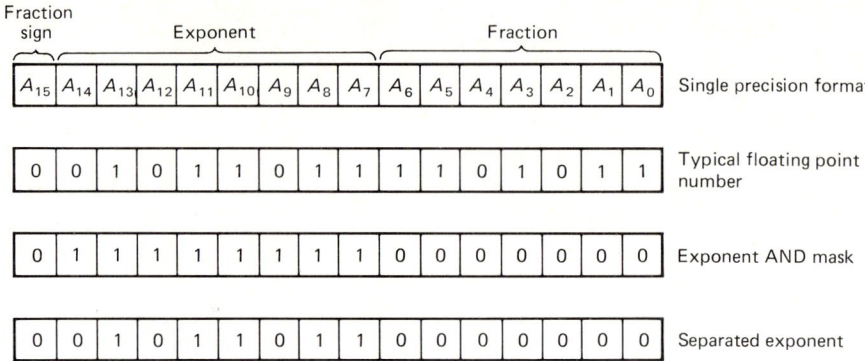

Figure 6.29 AND mask separation.

with 1. This process, called *masking*, is illustrated in Fig. 6.29 for a single-precision number. The same components perform mathematical and logic operations, and the device is called an *arithmetic logic unit* (ALU). TTL and MOS ALU chips are available with a variety of options.

6.8 BINARY-CODED DECIMAL

Because semiconductors switch on and off rapidly, computers operate most efficiently in binary. On the other hand, we are most comfortable with decimal. Therefore, computers need circuits to convert decimal to binary at the input and binary to decimal at the output. Many early computers and most modern pocket calculators use a number system which straddles the fence between decimal and binary. In *binary-coded decimal* (BCD) each of the 10 digits is represented by its 4-bit binary equivalent. As shown in Table 6.6, BCD requires 4 bits per digit even when leading bits are 0.

Each bit of a BCD number has the same positional value as an ordinary

TABLE 6.6 Decimal/BCD Equivalents

Decimal	BCD
0	0000
1	0001
2	0010
3	0011
4	0100
5	0101
6	0110
7	0111
8	1000
9	1001

binary number. The MSB has a weight of $2^3 = 8$, the next lower bit has a weight of $2^2 = 4$, etc. Since there are always 4 bits per digit, BCD is also called an 8421 or *natural* code. Other 4-bit codes are not sequentially weighted. In BCD:

- A 2-digit number requires two 4-bit groups
- A 3-digit number requires three 4-bit groups, etc.

For example

$$29_{10} = 0010\ 1001_{BCD}$$

and $\quad 834_{10} = 1000\ 0011\ 0100_{BCD}$

Arithmetic operations can be performed in BCD

Decimal		BCD
12	=	0001 0'010
+ 46	=	0100 0110
58		0101 1000

In this example, a carry results within a BCD group and the result is correct. When a carry requires a transfer from one BCD group to the next, results are incorrect. For example

Decimal		BCD
19	=	000'1 1001
+ 07	=	0000 0111
26	≠	0010 0000

Another situation with BCD is the possibility of invalid digits. The numbers 10 through 15 have 4-bit binary equivalents but are not BCD. For example

Decimal		BCD
4	=	0100
+ 7	=	0111
11	≠	1011

This sum, while correct in binary, should be $0001\ 0001_{BCD}$. These two sources of error, invalid 4-bit groups and carrys from one BCD digit to the next are correctable. A circuit detects the error, and Karnaugh maps are, as always, convenient. Figure 6.30a shows 1s for the invalid numbers 10 through 15. The best Karnaugh reduction for invalid-number error I_E is

$$I_E = D \cdot C + D \cdot B \tag{6.20a}$$

Figure 6.30
BCD error detection. (*a*) Karnaugh map; (*b*) invalid number circuit; (*c*) complete error detector.

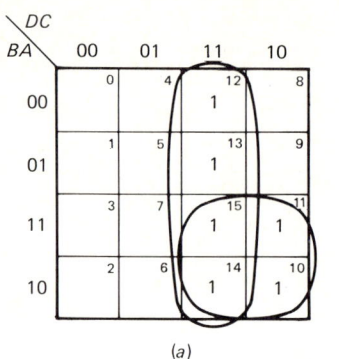

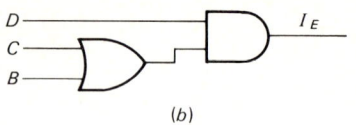

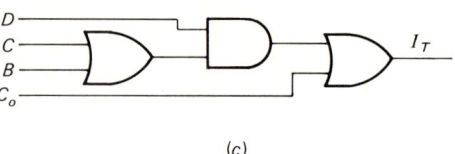

Ordinary Boolean algebra can further simplify this expression to:

$$I_E = D(C + B) \tag{6.20b}$$

This circuit is shown in Fig. 6.30*b*. The other source of error, a carry between BCD groups, is detected by including C_0 of the MSB in the error-detecting circuit. Thus the total invalid error I_T is:

$$I_T = C_0 + D(C + B) \tag{6.21}$$

and this circuit is shown in Fig. 6.30*c*. Each 4-bit BCD digit requires an error detector consisting of 2 OR gates and 1 AND gate.

The arithmetic to correct both BCD errors is the same. Adding 6_{10}, which is 0110_{BCD}, adjusts the sum. The previous invalid group illustrates correction by adding 6_{10}

Decimal		BCD	
4	=	0100	
+ 7	=	0111	
11	≠	1011	invalid group
		0110	add 0110
		0001 0001	to obtain correct sum

Similarly, for the previous carry between BCD groups

Decimal		BCD	
19	=	0001 1001	
+ 07	=	0000 0111	
26	≠	0010 0000	incorrect
		0110	add 0110
		0010 0110	to obtain correct sum

Figure 6.31
Single-digit BCD adder.

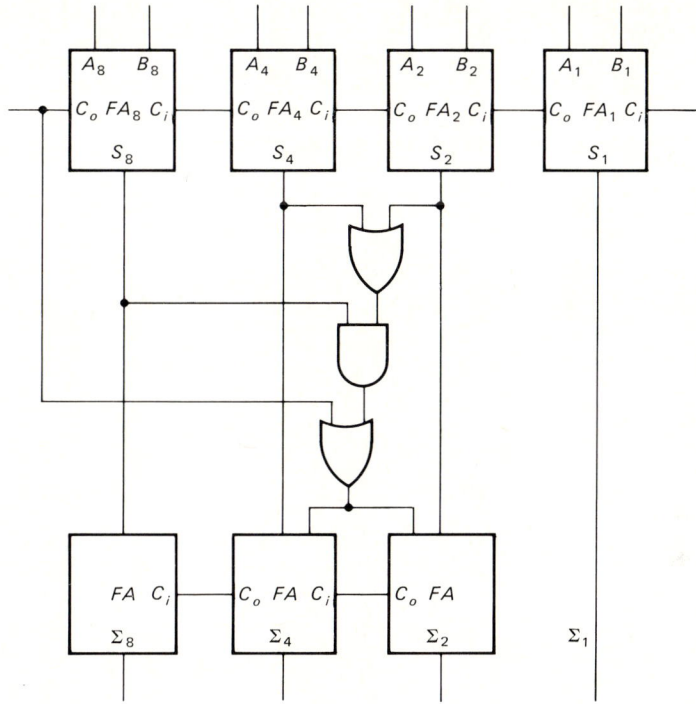

A single-digit BCD adder sums the weighted bits of two numbers, A_8 with B_8, A_4 with B_4, etc. Also, 1s must be added to the middle sum bits, S_4 and S_2, to add 0110 whenever an error occurs. Connecting the output of the error-detecting circuit to the inputs of S_4 and S_2 accomplishes this. The completed outputs are labeled Σ_8, Σ_4, Σ_2, and Σ_1. A single-digit BCD adder is shown in Fig. 6.31.

The adders Σ_8 and Σ_2 each have two inputs and can be replaced with less complex half-adders. This circuit also contains C_i and C_o leads to permit cascading single-digit BCD adders. For example a 4-digit BCD adder requires four identical single-digit BCD adders.

Example 6.10 Obtain the sum of 569 and 473 using BCD.

Solution Convert the numbers into BCD. Add 0110 when invalid numbers or carry between numbers occur.

```
Decimal              BCD
    569    =    0101 0110 1001
  + 473    =    0100 0111 0011
  ─────         ─────────────────
   1042    ≠    1001 1101 1100    incorrect
                     0110 0110    add 0110
                1010 0100 0010    incorrect
                0110              add 0110
                ─────────────────
                0001 0000 0100 0010    to obtain correct sum
```

Signs are the same for BCD numbers as for ordinary binary: a leading $\underline{0}$ in front of a BCD number indicates positive and a leading $\underline{1}$ indicates negative. For example, $\underline{0}\ 1000_{BCD}$ is $+8$ while $\underline{1}\ 1000_{BCD}$ is -8 in signed magnitude BCD. The same three methods—signed magnitude, 1s complement, and 2s complement—are used in BCD. As already discussed, the decimal equivalent of 1s complement is 9s complement. Similarly, the decimal equivalent of 2s complement is 10s complement.

It is a relatively simple matter to obtain the 1s complement of a number, which is accomplished by inverting each bit. But BCD is not a true binary-based system, and it is more difficult to obtain the 9s complement. For example, the 9s complement of 0111_{BCD} is not 1000_{BCD} but is 0010_{BCD}. Table 6.7 compares 9s complements in decimal and BCD. BCD numbers are indicated as $A_8\ A_4\ A_2\ A_1$ and the complements as $C_8\ C_4\ C_2\ C_1$.

Conditions shown in Table 6.7 are the basis of designing a 9s complement generator. Comparing each BCD number with its complement shows that:

1. C_1 is always the opposite of A_1 and is formed by inverting A_1.

2. C_2 is always the same as A_2. Only a jumper wire is required.

TABLE 6.7
9s Complement Numbers

Decimal Number	Decimal Complement	A_8	A_4	A_2	A_1	C_8	C_4	C_2	C_1
0	9	0	0	0	0	1	0	0	1
1	8	0	0	0	1	1	0	0	0
2	7	0	0	1	0	0	1	1	1
3	6	0	0	1	1	0	1	1	0
4	5	0	1	0	0	0	1	0	1
5	4	0	1	0	1	0	1	0	0
6	3	0	1	1	0	0	0	1	1
7	2	0	1	1	1	0	0	1	0
8	1	1	0	0	0	0	0	0	1
9	0	1	0	0	1	0	0	0	0

Figure 6.32
9s complement operations. (a) 9s complementer; (b) 9s complement subtraction.

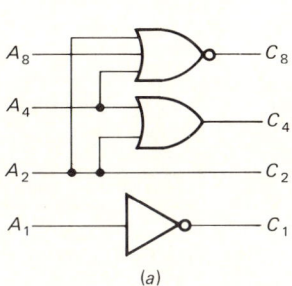
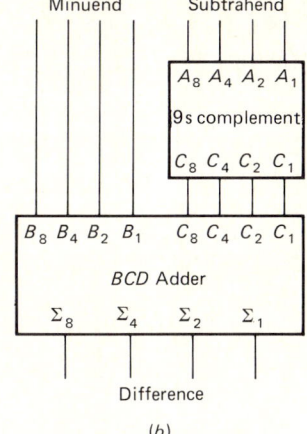

3. C_4 is 1 when A_4 and A_2 are opposite but 0 when A_4 and A_2 are the same. These conditions are satisfied with an EOR gate.

4. C_8 is 1 when A_8 AND A_4 AND A_2 equal 0. Then
$$C_8 = \overline{A}_8 \cdot \overline{A}_4 \cdot \overline{A}_2$$

This condition can be satisfied with three inverters and an AND gate. However, De Morganizing yields a single-gate circuit for obtaining C_8.

$$C_8 = \overline{A_8 + A_4 + A_2}$$

Figure 6.32a shows a BCD 9s complement circuit. This complementer can be combined with a BCD adder to perform subtraction as shown in Fig. 6.32b. This circuit is the BCD version of a binary add/subtract circuit. Control signals determine whether addition or subtraction is performed.

Example 6.11a Obtain the sum of $+7$ and -4 using 9s complement BCD.

Solution In BCD $+7$ is $\underline{0}\,0111$ and in 9s complement -4 is $\underline{1}\,0101$. As in 1s complement, all bits, including the sign bit, are added.

Decimal		BCD	
$+7$	$=$	$\underline{0}\,0111$	
-4	$=$	$\underline{1}\,0101$	
$+3$	$\neq$	$1\,1100$	Incorrect
		0110	Add 0110
		①$\,0\,0010$	
		↳1	Perform end-around carry
		$\underline{0}\,0011$	To obtain correct result

Negative 2s complement numbers are obtained by adding 1 to the negative 1s complement version. The same method is used for negative BCD numbers. Adding 1 to a negative 9s complement number produces the 10s complement; the advantage of 10s complement in BCD is the same as that of 2s complement in ordinary binary. Carry-outs are ignored, and the result is a simple circuit.

Example 6.11b Obtain the sum of $+7$ and -4 using 10s complement BCD.

Solution In 10s complement BCD -4 is $\underline{1}\,0110$.

Decimal		BCD	
$+7$	$=$	$\underline{0}\,0111$	
-4	$=$	$\underline{1}\,0110$	
$+3$	$\neq$	$\underline{1}\,1101$	incorrect
		0110	add 0110
	①	$\underline{0}\,0011$	ignore carry-out; result is correct

BCD adders use the same circuits as binary adders, namely, input registers, accumulators, clocks, etc. Serial BCD adders process one digit at a time, and parallel BCD adders process all digits simultaneously. Unfortunately, 4-bit representations of digits makes it difficult to apply conventional binary multiplication and division techniques to BCD numbers. There are no standard approaches for BCD multiplication and division. Some pocket calculators store multiplication and division tables and "look up" the required partial products and quotients, which are then combined by using conventional binary methods.

When pocket calculators are designed, less hardware is more important than high-speed operation. Whether a pocket calculator performs arithmetic in microseconds or milliseconds is not as important as reduced physical size. With BCD each 4-bit group converts directly into a single digit. This condition does not apply to ordinary binary. For example, a result of 1011_2 requires additional decoding before 11 can be displayed. The relative ease of encoding and decoding reduces the size of BCD circuits. Besides calculators, instruments such as voltmeters, data plotters, and printers have BCD outputs. BCD instruments can be connected to computers.

SUMMARY

1. A half-adder circuit adds 2 bits and outputs a sum and a carry-out. The Exclusive-OR gate simplifies half-adder construction and is useful in other circuits.

2. A full-adder adds 3 bits, addend, augend, and carry-in. The carry-in bit is the carry-out bit from the bit of next lower order.

3. A serial adder sums numbers by computing each bit sum separately. The accumulator is a sum register to store partial sums and final results.

4. A parallel adder requires a full-adder for each bit but adds all bits simultaneously. Carry anticipation circuits are needed to complete an addition during a single clock pulse.

5. Signed add/subtract circuits are practical in both 1s and 2s complement. 2s complement arithmetic uses few circuits and is more popular.

6. Multiplication and division are performed with the same input register and an accumulator/MQ. Division is more involved and takes longer than multiplication.

7. Computers work with floating-point binary numbers. The arithmetic logic unit performs mathematical and Boolean operations.

8. Binary-coded decimal, which uses 4-bit groups for each decimal number, simplifies encoding and decoding to decimal.

PROBLEMS

1. Using an EOR half-adder, determine the outputs when the inputs are: (a) $A = 1$, $B = 0$; (b) $A = 1$, $B = 1$.

   Answer: (a) $S = 1$, $C_0 = 0$; (b) $S = 0$, $C_0 = 1$.

2. The word 1101 is entered into a true/complement generator. Determine the output when the control line is: (a) low; (b) high.

3. For the digital comparator shown in Fig. 6.8, the word in register A is 110 and register B contains 101. Determine the comparator output.

   Answer: The output is low.

4. The data 1011 is entered into a parity checker. Owing to transmission line noise the word appears as 1010. Determine the original and final parity.

5. Follow the full-adder inputs $C_i = 1$, $B = 0$, $A = 0$ through the circuit shown in Fig. 6.10.

   Answer: $C_i \cdot \bar{B} \cdot \bar{A}$ high. Therefore $S = 1$. Since B and A are both low, $C_0 = 0$.

6. Follow the inputs $C_i = 1$, $B = 0$, $A = 1$ through the circuit shown in Fig. 6.11.

7. Follow the inputs $C_i = 0$, $B = 1$, $A = 1$ through the circuit shown in Fig. 6.12.

   Answer: $S_\mathrm{I} = 0$, $C_{0\mathrm{I}} = 1$; $S_\mathrm{II} = 0$, $C_{0\mathrm{II}} = 0$. Therefore $S = 0$, $C_0 = 1$.

8. Using the block diagram of a full-adder, show a circuit which adds two 3-bit numbers together. Describe how an overflow can be detected.

9. Find the largest number which does not result in an overflow in a 16-bit accumulator.

   Answer: 65,535.

10. The binary numbers 1001 and 0111 are added in a 4-bit accumulator. Determine the result.

11. If T_p of a serial adder is 2.5 μs, how long does it take to add two 10-bit numbers?

    Answer: 25 μs.

12. How many bits are required in an accumulator which adds to 1 million?

13. How many clock cycles are required to add (*a*) 1010 to 0101; and (*b*) 0111 to 0110?

    Answer: (*a*) 1; (*b*) 3.

14. How many clock cycles are required to add the numbers in the previous problem by using a parallel fast adder?

15. If T_p, the clock pulse for a parallel fast adder, is 125 μs, how long does it take to add 1000 + 0001 + 0100 + 0010?

    Answer: 1 ms.

16. Design a carry anticipation circuit for the bit which is adjacent to the LSB.

17. Using 8 bits find (*a*) the most positive and (*b*) the most negative numbers which can be represented in 2s complement.

    Answer: (*a*) 0111 1111 = +127; (*b*) 1000 0000 = −128.

18. Represent +19 and −19 with 8-bit numbers in: (*a*) signed magnitude; (*b*) 1s complement; (*c*) 2s complement.

19. Determine the results when (*a*) 0101 and (*b*) 1011 pass through the circuit shown in Fig. 6.22.

    Answer: (*a*) 1011; (*b*) 0101.

20. Use a Karnaugh map to plot the 2s complement 4-bit overflow detection equation and construct a minimum sum-of-products logic circuit.

21. An accumulator is designed to multiply a signed 8-bit number by a signed 4-bit number. (*a*) How long is the accumulator? (*b*) At a 750 kHz clock rate, how long does multiplication take?

    Answer: (*a*) 11 bits; (*b*) 10.64 μs.

22. Show the status of the registers when 0100 is multiplied by 1110.

23. The multiplier and multiplicand of Problem 21 are interchanged. How long does multiplication take?

    Answer: 21.28 μs.

24. Signed 16-bit and 8-bit numbers are multiplied at a 1-MHz clock rate. How long does multiplication take if: (*a*) the multiplier is 8 bits; (*b*) the multiplier is 16 bits?

25. Compare the time required to (*a*) multiply by and (*b*) divide by a 4-bit number in a computer which uses a 1-MHz clock.

    Answer: (*a*) 8 μs; (*b*) 12 μs.

26. Show the original and final register status for performing $\underline{1}110/\underline{0}010$.

27. A class of 30 students has taken two exams. Assume that both test scores have been added and compare the time required to divide by 2 with the time required to multiply by 0.5.

    Answer: Division takes 180 clock pulses and multiplication takes 123 clock pulses.

28. Assume a 1-byte signed-magnitude floating-point machine. Find the decimal values of (*a*) $\underline{1}010\ 1100$ and (*b*) $\underline{0}110\ 1100$.

29. A computer operates in the single-precision floating-point signed-magnitude mode. Find the decimal values of: (*a*) $\underline{0}010\ 0000\ 0100\ 0000$; (*b*) $\underline{1}001\ 0000\ 0101\ 0000$.

    Answer: (*a*) $\simeq 9.22 \times 10^{18}$; (*b*) $\simeq -2.68 \times 10^9$.

30. The double-precision values of the numbers in the previous problem are: (*a*) $\underline{0}010\ 0000\ 0100\ 0000\ 0000\ 0000\ 0000\ 0000$; (*b*) $\underline{1}001\ 0000\ 0101\ 0000\ 0000\ 0000\ 0000\ 0000\ 0000$. Find the decimal values.

31. The offset values of a double-precision exponent are (*a*) 140 and (*b*) 100. Find the decimal equivalents of these exponents.

    Answer: (*a*) 4096; (*b*) $\simeq 3.73 \times 10^{-9}$.

32. In BCD perform; (*a*) 23 + 45; (*b*) 703 + 258.

33. Compare the number of BCD adders with the number of binary adders needed for sums which are less than 10 million.

    Answer: 7 BCD adders, 24 binary adders.

34. Perform 45 − 23 using 9s-complement BCD.

35. Perform 45 − 23 using 10s-complement BCD.

    Answer: $\underline{0}\ 0010\ 0010$.

7
MEMORY CIRCUITS

7.0 INTRODUCTION

Computers require internal memory to store instructions and data for the program being processed. An internal memory must retain information at least until results are obtained.

The first computers stored programs on segments of a rotating magnetic drum. Drum memory is permanent and programs are retained even after power is shut down. However, a program can be replaced with a newer program whenever necessary. Speed is a major limitation of drum memory. Mechanical rotation and sequential storage cause time delays in obtaining data from a drum.

Core memory is another magnetic technique. Small toroidal cores are the magnetic equivalent of flip-flops. Each core stores a single bit. Cores do not move, and core storage is faster than drum. Since no specific selection sequence is needed, core memory is an example of a *random-access memory* (RAM).

IC RAMs are now less expensive and considerably smaller than magnetic memories. Addressing, reading, and writing techniques developed for magnetic storage were adapted to semiconductor RAMs. Semiconductor RAMs use an *RS* flip-flop as the basic storage element. Bipolar RAMs are faster, but MOS RAMs dissipate less power and are less expensive. A newer memory technique uses charge stored on internal MOSFET capacitance. Periodic charge refreshing is required. Despite this added circuit complexity, MOSFET charge storage is smaller, dissipates less power, and is less expensive than any other IC RAM.

Besides a temporary memory for programs, a computer also needs permanent memory. Permanent memory stores internal computer instructions. Internal instructions are used but not altered. A permanent memory is called a *read-only memory* (ROM). ROMs also store mathematical tables. ROMs store data in either matrix or combinational logic gate form. A matrix ROM is a sequence of locations, each containing a single word. The combinational logic ROM is an assembly of gates to generate Boolean expressions.

7.1 DRUM MEMORY

Binary is an efficient computer number system because electronic devices have excellent bistable characteristics. Other bistable devices are compact enough to store large quantities of data in a relatively small volume. In particular, magnetic materials can have north-to-south or south-to-north polarity. Magnetization of randomly oriented iron or Alnico is accomplished by electric induction.

Whenever current flows through a conductor, a magnetic field exists around the conductor. Magnetic field intensity is proportional to current intensity, and polarity is determined by current direction. The right-hand rule is a convenient method of determining magnetic polarity. A conductor is grasped, at least conceptually, in the right hand. The thumb should point in the conventional current direction. Magnetic field polarity is indicated by the fingers of the right hand. If the thumb points left, the magnetic field is clockwise. If the thumb points right, the magnetic field is counterclockwise. Induced magnetic polarity is essentially permanent but can be reversed with a current in the opposite direction.

Figure 7.1 shows a typical magnetic hysteresis loop. Magnetic flux intensity H in ampere-turns per meter is plotted against magnetic flux density B in webers per square meter. Actual values of H and B vary from one material to another. However, the general shape is similar for all ferromagnetic materials.

When flux density is sufficiently large, saturation at B_{max} results. I_{max} is the current necessary to cause B_{max}. The same value of current flowing in the opposite direction I_{min} causes saturation B_{min}. Flux remains at or near saturation after current is removed. Typically, B_{max} represents a binary 1 and B_{min} a binary 0. Early computers used induced magnetization to store data on a *magnetic drum* by coating a cylinder with precise dimensional tolerances with

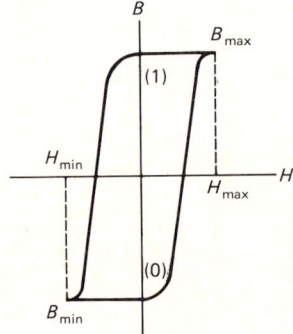

Figure 7.1 Magnetic hysteresis.

magnetic material and placing data on the drum by magnetizing separate drum segments in the appropriate direction.

Entering data on a drum is called *writing*. When written data is required, determining induced magnetization direction is called *reading*. Both functions are performed by a *read/write head*. Energizing the write coil with a logic 1 current pulse induces an N-S magnet. Similarly, energizing with a logic 0 current pulse induces an S-N magnet. Reading is performed by reversing the writing process. As the drum rotates past a head, the induced magnet generates a small voltage across the read coil. Amplification of read coil voltage yields logic level signals. Once a magnet has been written onto the drum, magnet orientation is retained even after power is turned off. However, data can be altered at some later time by entering new information on the write head.

There is an air gap between drum and read/write heads. Narrow air gaps reduce power required for both reading and writing. Also, a narrow gap reduces the size of the induced magnet, and more data can be stored on a drum. Noise reduction is another advantage of narrow air gaps. Typical gaps between head and drum are 0.001 in. With such a small clearance, thermal expansion of the drum could cause physical contact between drum and heads, which destroys the heads and the drum surface. Clever mechanical design maintains a constant air gap while allowing for thermal expansion.

A drum is divided into circumferential bands called *tracks*. Each track in Fig. 7.2 contains individual bits. Data can be read or written in serial or parallel. In the serial mode, bits are entered sequentially around the circumference, and a single head processes the entire word. However, reading or writing a serial word requires as many clock pulses as bits per word. A 1-byte word, for example, requires eight clock pulses. In the parallel mode an entire word, regardless of length, is processed during a single clock pulse. But parallel read and write requires as many heads as bits per word. Drum storage capacity is the same in either serial or parallel. Usually there are fewer heads than tracks, and heads must be mechanically shifted. A drum spins at

Figure 7.2 Magnetic drum. (*a*) Configuration; (*b*) head detail.

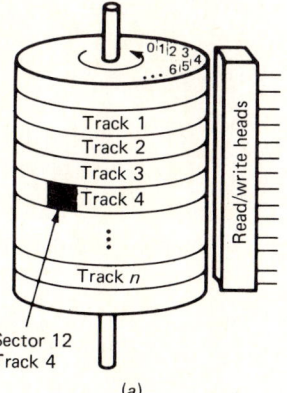
(a)

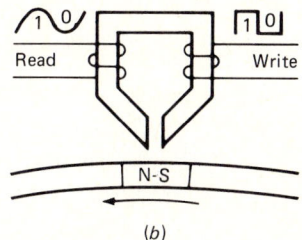
(b)

constant speed. Each portion of the drum can be read or written only when it is beneath the read/write heads. Some drums use separate heads for reading and writing.

Each word on a drum has a unique location called an *address*. On a cylinder two coordinates locate an address. One coordinate is a track number and the other is a *sector* number. A reference position on the circumference is sector 0, and sectors are numbered in sequence 0, 1, 2, 3, etc. Each sector number corresponds to a number on a counter which is driven by the clock and reset at the completion of each drum revolution. Sector numbers locate a point along the drum circumference and track the numbers along the drum length. Taken together, the track number and sector number are the address of each particular word on a drum. A desired sector and track number are entered into a word comparator. Data are read or written at a particular address when the word comparator agrees with sector and track number.

Drum speed is limited by mechanical considerations. A cylinder is about 1 ft in diameter and $\frac{1}{2}$ ft long. Fast drum speeds are about 8000 r/min. Selecting a drum speed is a compromise between locating a given word rapidly and storing many words: high speed reduces the time required for any word to reach a read/write head, and low speed permits more data to be stored. Typical drum speeds are about 4000 r/min.

The time needed to obtain a specific word is called *access time* and is related to the drum period T. Ideally, a desired word is under the read/write heads at the precise time it is needed, in which case access time is zero. Worst-case access time occurs when the desired word has just passed the head, in which case access time is the entire period of revolution. Drums are compared on the basis of average access time t_a, which is one-half of a drum period:

$$t_a = 0.5 \cdot T \tag{7.1}$$

Example 7.1 Determine the average access time for a drum turning at 4000 r/min.

Solution Drum period is the reciprocal of speed. Thus

$$T = \frac{1 \text{ min}}{4000 \text{ r}} \cdot \frac{60 \text{ s}}{\text{min}} = 15 \text{ ms/r}$$

Since

$$t_a = 0.5 \cdot T$$
$$t_a = 0.5 \cdot 15 \text{ ms} = 7.5 \text{ ms}$$

Drum access time depends on mechanical rotation. Moving devices are inherently slower than electronic switching speeds. Physical limitation restricts drum access times to the 10-ms range. Other parts of a computer are much faster than drum memory. MOS operates in microseconds and TTL in nanoseconds. Long access time is the principal drawback of drum storage. It is inefficient to process data thousands of times faster than the time required to obtain and store data in memory. However, some improvement in drum access time is possible.

Just as a clever programmer chooses multiplication because it is faster than division, programming can reduce access time. During a program, each word in memory is required at a specific time. By knowing the time required for each step in a program, data can be placed in the most appropriate sector number. Clever programming does not reduce average access time. This is a mechanical consideration. However, programming can reduce the delay caused by access time.

Nevertheless, access time places a limitation on overall computer speed. On the other hand a good deal of data is stored on a drum surface. The number of stored bits N is determined by:

- Track density d
- Drum height h
- Bit packing density b
- Drum circumference c

Thus

$$N = d \cdot h \cdot b \cdot c \tag{7.2}$$

Example 7.2 Determine the capacity of a drum which is 1 ft in diameter and 6 in high. There are 50 tracks per inch and packing density is 1000 bits per inch.

Solution The circumference of a cylinder is $\pi \cdot D$. Thus

$$c = \pi \cdot D = \pi \cdot 12 \text{ in}$$

and since

$$N = d \cdot h \cdot b \cdot c$$

$$N = 50 \frac{\text{tracks}}{\text{in}} \cdot 6 \text{ in} \cdot 1000 \frac{\text{bits}}{\text{in}} \cdot \pi \cdot 12 \text{ in}$$

$$N = 1.13 \times 10^7 \text{ bits}$$

If this drum is byte-organized, its capacity is approximately 1.4 million bytes. Actual capacity is slightly less because one track is reserved for the sector number.

7.2 CORE MEMORY

Magnetic materials actually swtich orientation rapidly. Drum storage is slow because cylinder rotation is slow. Eliminating physical motion results in a magnetic memory with speeds comparable with those of MOS switching.

A *core* memory contains no moving parts. Magnetic cores are made of powdered iron compounds and binders fired together in a furnace. Cores are small and doughnut-shaped. Typical core dimensions are height 0.015 in, outside diameter 0.05 in, and inside diameter 0.025 in. With such small dimensions, 30,000 cores occupy less than 1 in³. A fine wire through the core center carries the magnetizing current. Polarity is found by applying the right hand-rule. In Fig. 7.3 the magnetization is clockwise when current flows towards the left and counterclockwise when it flows to the right. Current direction determines whether 1 or 0 is written into a core.

Magnetic force is the product of current and the number of turns of wire. Because cores are small, only one write wire passes through a core. With only one turn, high current is needed to obtain sufficient ampere-turns to cause saturation. About 0.5 A is required for writing into a core. However, current is only required until saturation is achieved, and 1 μs is sufficient to cause saturation. Thus the writing current is large, but the saturation power is small.

Each core stores 1 bit and retains data after the computer is turned off. A write wire is called an *input line*. The input line serves the same purpose as the write head of a drum memory; however, each core requires a separate input line. Another wire through the core, called the *sense* line, serves the same function as the read head.

Relative motion is required to obtain an indication of magnetic state. Thus, reading core data is more complicated than reading drum data. Two methods exist for determining magnetic state, one physical and the other electrical. The physical method requires motion between the magnet and sense coil. A rotating drum and fixed read head use physical motion.

A completely stationary system, such as core memory, uses electrical sensing. Reading a stationary magnet requires forcing the core into a known state with an external current pulse. Specifically, a negative current sufficient to saturate the core at B_{min} is injected on the input line. If the core stores a 0, it is very close to B_{min}. Therefore, the flux change is negligible and very little voltage is induced in the sense line. Similarly, the core is very close to B_{max}

Figure 7.3
Core magnetization. (*a*) Current flows left; (*b*) current flows right.

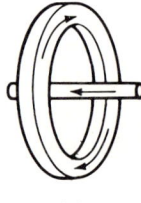
(a)

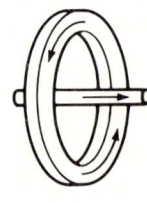
(b)

Figure 7.4
Reading a core. (a)
Construction; (b) typical
waveforms.

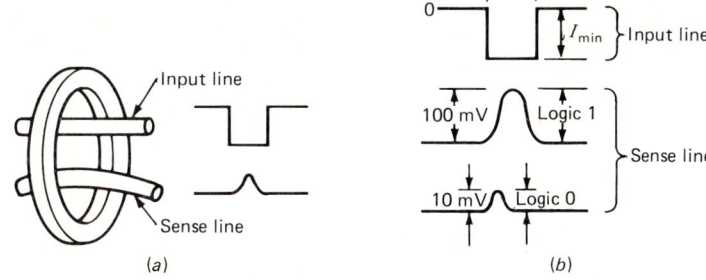

when storing a 1. In this case, the negative current through the input line switches the core to B_{min}. A large flux change occurs, and a large voltage is induced in the sense line. The amplitude of the sense line voltage indicates whether a 1 or a 0 is stored. Figure 7.4 shows typical voltage conditions. Sense voltage for a 1 is larger and lasts longer.

Electrical sensing creates a problem which does not exist for physical sensing. If a core stores a 1, reading reverses the core polarity. Regardless of initial core state, applying I_{min} to read a core results in a core which is at B_{min}. Drum memory is unaffected by reading, but data stored in core memory is destroyed. However core data can be restored by storing sense line status in a flip-flop. After sense line data is read, the flip-flop resets the input line.

Operation sequence is controlled by a clock cycle which contains read and write periods. Reading is first. During the read period, data is taken from core, and the core goes to B_{min} regardless of initial state. During the write cycle the core is either reset or else new data is entered.

Core memories of more than 1 million bytes exist. In this case, separate input and sense lines for each core requires complicated cables and control circuits. Wiring cores in *matrix* form is more practical. Each core matrix consists of a single plane containing horizontal and vertical input lines. A core is placed at the intersection of each horizontal and vertical input line. Saturation current is divided between horizontal and vertical lines. If I_{max} is 0.5 A, then 0.25 A flowing through a horizontal and a vertical line results in one saturated core. The core at the intersection of the two current-carrying lines is the only core to saturate. Other cores on the selected lines see only $0.5\,I_{max}$, which is not enough to cause saturation. Cores on neither of the selected lines do not experience any current. Typically, core planes are square. There are an equal number of horizontal and vertical lines.

The vertical input lines in Fig. 7.5 are labeled X_n, and the horizontal input lines are Y_n. No current flows through the vertical lines X_0, X_1, X_3 or the horizontal lines Y_0, Y_1, Y_3. Half-saturation currents flow through input lines X_2 and Y_2. Under these conditions, the address of the only saturated core is $X_2 Y_2$. This core remains saturated after the write currents are removed. Since saturation only occurs at a core where currents coincide, a core matrix is

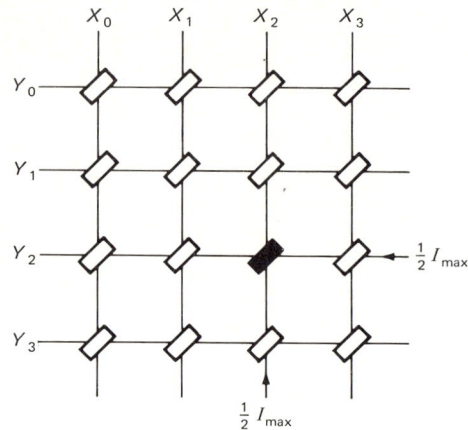

Figure 7.5
16-Bit coincident core matrix.

called a *coincident* current memory. Coincident core currents are the magnetic equivalent of a 2-input AND gate.

The 16-bit core shown in Fig. 7.5 requires 8 input lines, 4 horizontal and 4 vertical. This is less than the 16 lines needed when individual input lines for each core are used. The difference between individual and coincident core input lines becomes more significant as the number of cores increases. For example, a coincident memory with 100 cores requires 10 horizontal and 10 vertical lines for a total of 20 input lines. This is significantly better than the 100 lines required when individual core lines are used. The number of lines N in a coincident matrix increases as the square root of the number of cores n:

$$N = 2\sqrt{n} \tag{7.3}$$

Example 7.3 How many input lines are needed to construct a 4096-bit coincident core plane?

Solution Since the number of bits is equal to the number of cores

$$N = 2\sqrt{n}$$
$$= 2\sqrt{4096}$$
$$N = 128 \text{ lines}$$

of which 64 are horizontal and 64 are vertical. The improvement is dramatic compared with 4096 lines for individual core input lines.

Reading cores may also be accomplished by the coincident method. This requires $0.5\,I_{min}$ to be applied to one horizontal and one vertical line. For a 0.5-A saturation current, reading requires -0.25 A through both of the selected lines. Continued storage requires flip-flops for rewriting.

Figure 7.6
Core matrix with read and write lines.

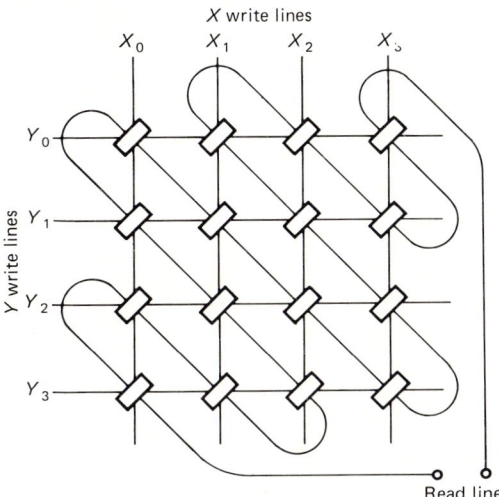

The magnetic drum is a *nonvolatile* and *sequential* memory system, nonvolatile because memory is not destroyed by reading and sequential because each drum sector can only be read in turn as it passes under the read heads. On the other hand, core memory is *volatile* and *nonsequential*, volatile because memory is destroyed by reading and must be restored when data is needed for later use and nonsequential because any core can be accessed at any time without waiting for a specific timing pulse. A storage system in which all addresses are equally accessible is called a *random-access memory* (RAM).

Core access time is in the 1-µs range. Any core can be read in this time without regard to specific core address. Core memory is about 10,000 times faster than drum memory. An access time of 1 µs makes core storage compatible with MOS and somewhat slower than TTL.

At any given time only one core in a plane can be addressed. Therefore fewer lines are needed for reading than for writing. A single sense line threaded through each core in a plane is sufficient for reading an entire plane. If the particular core being addressed shows a significant flux change during read, the core is at logic 1; if the flux change is negligible, the core is at logic 0.

The 16-bit coincident core plane shown in Fig. 7.6 contains a total of 9 input lines:

- Writing requires 4 horizontal and 4 vertical lines.

- Reading the entire matrix requires 1 line.

Example 7.4 How many lines are needed to perform reading and writing for a 64-bit-coincident core plane?

Solution Writing requires

$$N = 2\sqrt{n}$$
$$= 2\sqrt{64}$$
$$N = 16 \text{ lines}$$

and reading requires 1 line. Therefore reading and writing 64 cores require a total of 17 lines.

Thus three wires pass through each core of a coincident core memory plane, two wires to write and one wire to read. Some core designs use four wires through each core. The fourth wire is an *inhibit* line which goes through each core. An inhibit line allows independent 1s or 0s to be written in individual cores when core planes are connected together. Other core memories maintain three wires. In this case the read line becomes an inhibit line during the write portion of the read/write cycle.

7.3 CORE RAMS

Core memory is a practical method for storing data. However, constructing an entire memory on a single XY plane is impractical. Real core memories contain many planes which are connected together in a three-dimensional array. There are as many planes as bits in a computer word. If the computer is byte-organized, then 8 core planes are needed. Planes are numbered P_0, P_1, P_2,... to correspond with bit number.

In a three-dimensional core, corresponding half-current lines are connected in series. The X_0 line of P_0 is connected to the X_0 line of P_1, which in turn is connected to all other X_0 lines. Corresponding Y half-current lines are also connected in series. However, each plane requires a separate sense line.

The memory shown in Fig. 7.7 contains four core planes. Since there are four X lines and four Y lines, this core stores sixteen 4-bit words. In a three-dimensional array the total number of cores n_T is the product of the number of planes and the numbers of X and Y lines:

$$n_T = P \cdot X \cdot Y \tag{7.4}$$

Example 7.5 Determine the configuration of a byte-organized core memory containing 16,384 words.

Solution An 8-bit word requires eight core planes. Assuming square planes

$$\sqrt{16{,}384} = 128 \ X \text{ lines} \quad \text{and} \quad 128 \ Y \text{ lines}$$

Therefore the total number of cores is

$$n_T = 8 \cdot 128 \cdot 128 = 131{,}072 \text{ cores}$$

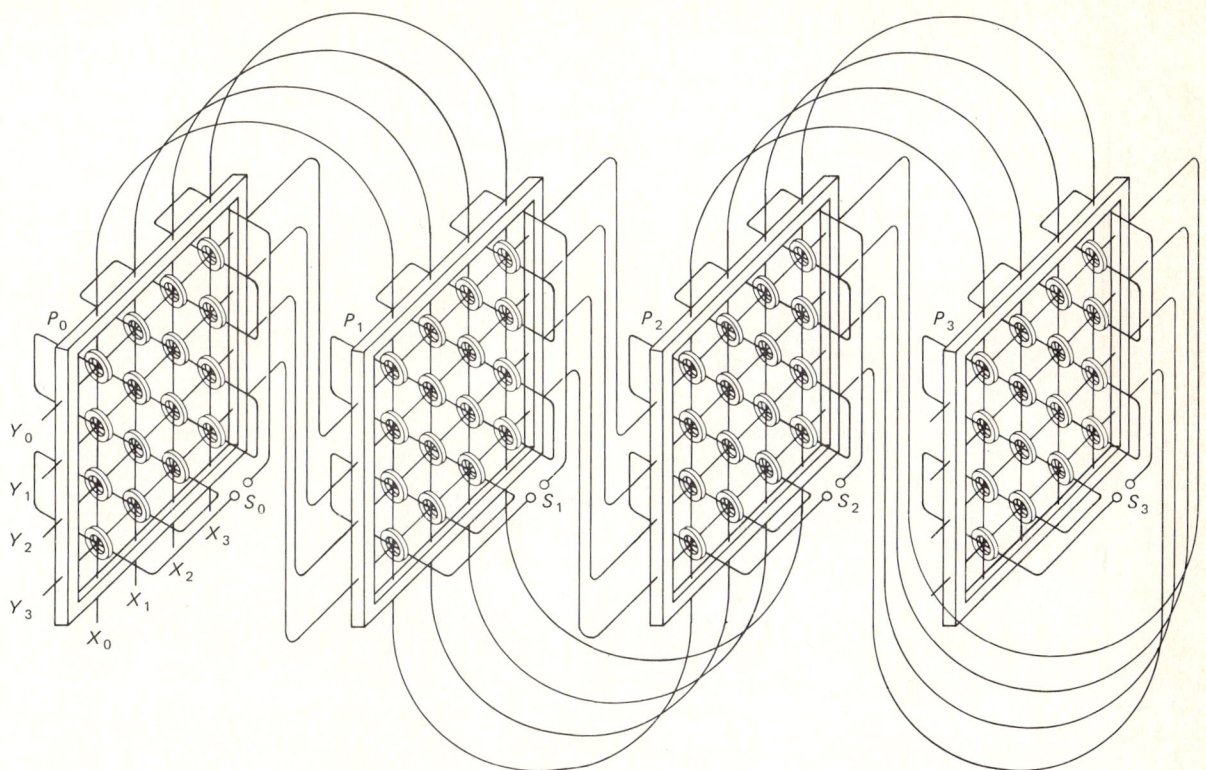

Figure 7.7 Three-dimensional core planes.

Core memories usually store integer powers of 2: $2^8 = 256$, $2^{10} = 1024$, etc. When specifying memory, it is common practice to round off to the nearest thousand because $2^{10} \simeq 1$ K. In Example 7.5 the memory is called a 16 K by 8, or 16 K × 8, core memory.

Three-dimensional core memories also use coincident currents. Assume that 0.5 I_{max} flows through the X_3 and Y_3 lines. The X_3Y_3 core in each plane will store a 1 because corresponding lines are wired in series. Since computer words are almost always combinations of 1s and 0s, there must be a technique for entering 0s. Inhibit lines enter 0s by sending 0.5 I_{min} to each plane which must contain a 0. The total coincident current I_T is the sum of X, Y, and the inhibit line I_i:

$$I_T = I_X + I_Y + I_i \tag{7.5a}$$

When an inhibit line is activated

$$I_T = 0.5\, I_{max} + 0.5\, I_{max} - 0.5\, I_{max}$$

or

$$I_T = 0.5\, I_{max} \tag{7.5b}$$

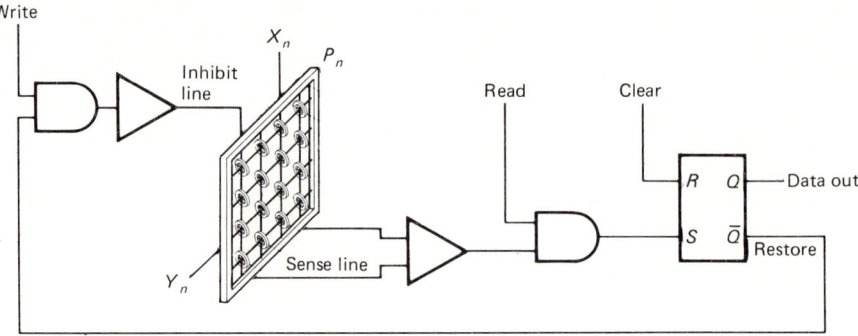

Figure 7.8 Data-restoring circuit.

Since a total of I_{max} is required to write a 1, this particular core cannot saturate. The core in each plane with an activated inhibit line stores a 0. If the word to be stored in $X_3 Y_3$ of Fig. 7.7 is 1010, the X_3 and Y_3 lines will both be 0.5 I_{max} while the inhibit lines of P_2 and P_0 will be 0.5 I_{min}.

Reading a three-dimensional core is also accomplished with coincident currents. To read the word $X_3 Y_3$, 0.5 I_{min} is sent through the X_3 and Y_3 lines. In this case, outputs from the P_3 and P_1 sense lines will be much greater than from P_2 and P_0. Thus the sense line outputs represent 1010.

As described, a core is always at B_{min} after being read. If the original data is still required, they must be restored immediately after reading. Data restoration is controlled by sense line status. Sense line output is connected to an amplifier, which drives a flip-flop through an AND gate. This flip-flop serves two functions. Data which has been read is temporarily stored at the Q output. The $\bar{Q}$ output controls the inhibit line through a gate and amplifier combination. Figure 7.8 shows a single-plane data restoration circuit. Each plane of a three-dimensional array requires a separate circuit.

Operation begins with a clear pulse to reset the flip-flop. After reset, the read AND gate is enabled by driving the read line high. Amplified data from the sense line provide the other AND gate input. For example, if the particular line $X_n Y_n$ being read is at logic 1, Q goes high while $\bar{Q}$ goes low. When reading is completed, the read enable signal goes low and the write enable goes high. In this case, $\bar{Q}$ is low and the AND gate output is low. Thus the inhibit line amplifier has a low output and there is insufficient drive to generate 0.5 I_{min}. The X_n and Y_n lines restore the original 1 using coincident currents.

On the other hand, data read from the sense line can also be a logic 0. In this case, Q goes low while $\bar{Q}$ goes high. During the write time, the high at $\bar{Q}$ drives the inhibit line amplifier with sufficient input to cause 0.5 I_{min} through the inhibit line. This inhibits core $X_n Y_n$ and a logic 0 is stored. The circuit shown in Fig. 7.8 restores data after reading and is called a *nondestructive readout* (NDRO). This particular NDRO contains separate sense and inhibit lines. However, one line can perform both the sense and inhibit functions. A

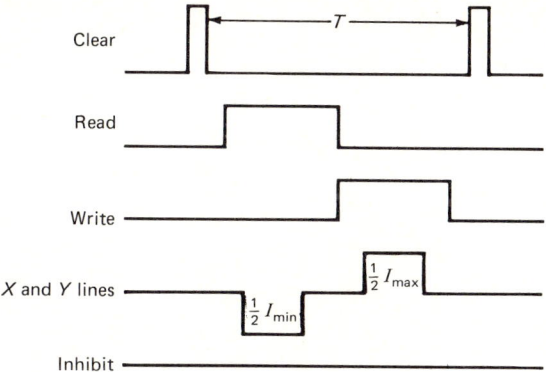

Figure 7.9
Core memory timing cycle.

single line is possible because sense occurs only during reading and inhibit only during writing.

Core memories operate on a *memory cycle* which consists of the clear-read-write sequence just described. Reading is destructive and must precede writing. During the write portion of the memory cycle, data are restored or else new data are entered. Figure 7.9 shows a memory timing cycle. One memory cycle takes about 1 μs.

The X and Y line conditions shown in Fig. 7.9 write a 1 into core. Since a 1 is written, the inhibit line to this plane must be off. But when a 0 is written, both X and Y lines must have 0.5 I_{min} during the write portion of the memory cycle. In either case 0.5 I_{min} is required through the X and Y lines during reading. X and Y drive currents do not last as long as read and write enable pulses. Using shorter drive than enable pulses reduces noise. This is important because noise occurs when cores switch states.

The basic NDRO circuit requires modification to write new data. Only two additional gates per plane are needed to convert an NDRO circuit into a complete read/write circuit. Figure 7.10a shows one more AND and one more OR gate. New data enter during the write portion of the memory cycle. This is accomplished by driving the data enable line high, while new data are entered at the other input to the same AND gate. Outputs from the read gate and new data gate are connected to an OR gate. The OR gate sets the flip-flop. Regardless of whether the new data gate or the read gate is enabled, the flip-flop stores data temporarily and also activates the particular core location. Core activation from a complete read/write circuit is the same as in a basic NDRO circuit.

The flip-flop shown in Fig. 7.10a stores a single bit from a particular plane. As described in Chap. 6, flip-flops can be combined in a variety of ways to form registers. When flip-flops from each individual plane are combined, the circuit is called a *memory buffer register*. This register provides temporary storage of an entire word. Figure 7.10b shows a 4-plane, or 4-bit, memory buffer register. For clarity, only the data input and restore lines are shown.

Figure 7.10 Processing core data. (*a*) Single-core read/write circuit; (*b*) 4-plane memory buffer register.

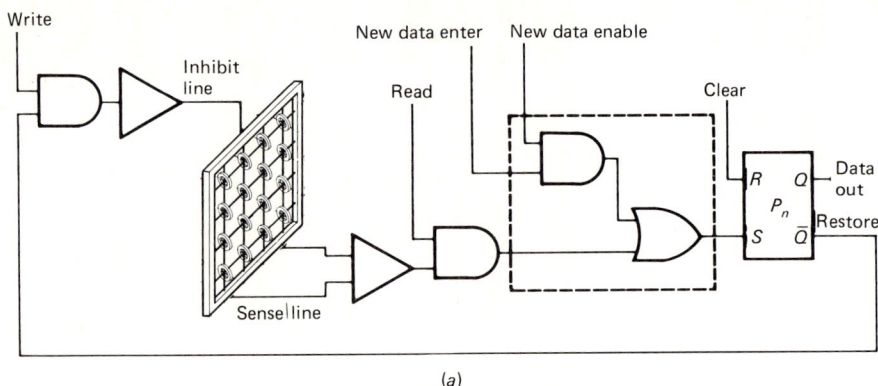

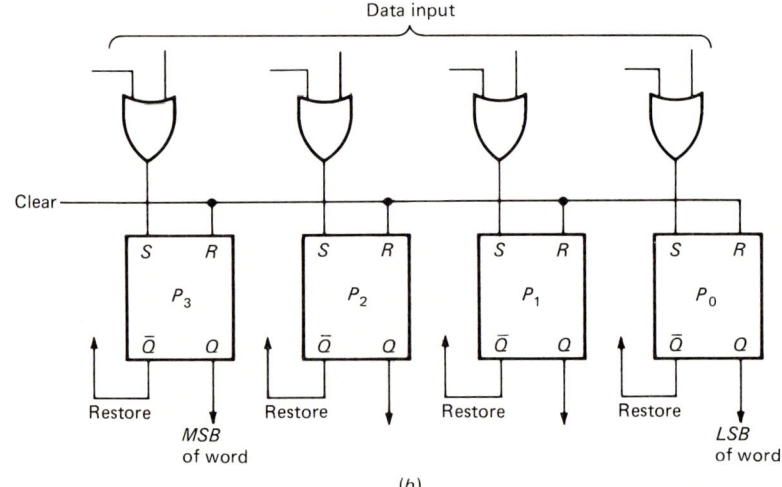

The remainder of the circuit for each flip-flop is as shown in Fig. 7.10a. Memory buffer registers store the word until transfer to or from the arithmetic logic unit or some other section of the computer is required.

Appropriate X and Y lines are activated to select a particular address during the read and write portions of a memory cycle. The desired address is entered into a *memory address register*. This register is a set of flip-flops containing both the X and Y line addresses. Typically, the most significant flip-flops contain the X line address and the least significant flip-flops contain the Y line address. These flip-flops store both Q and $\bar{Q}$. The number of flip-flops required to store a complete address is given by

$$n = 2^{(x+y)} \tag{7.6}$$

Example 7.6 How many flip-flops are needed to select a specific core in a 64-core plane?

Solution   In this case $n = 64$. Therefore

$$64 = 2^{(x+y)}$$

$$\log 64 = (x + y) \log 2$$

$$x + y = \frac{\log 64}{\log 2} = \frac{1.80618}{0.30103}$$

$$x + y = 6 \text{ flip-flops}$$

This condition is satisfied when

$$x = 3 \text{ flip-flops} \quad \text{and} \quad y = 3 \text{ flip-flops}$$

Figure 7.11 shows the details of a 64-core select circuit. $FF_5 - FF_4 - FF_3$ contains the X line address and $FF_2 - FF_1 - FF_0$ contains the Y line address. The memory address register might contain 101110. In this case the X_5 and Y_6 lines are selected to address core $X_5 Y_6$. Core selection occurs during read and write portions of a memory cycle.

Actually, the memory address register activates the X and Y line address decoders. These decoders are similar to the 1-of-n decoder described in Chap. 3. A 1-of-n decoder requires inverters to present both the inverted and noninverted signals. Since flip-flops contain both Q and $\bar{Q}$ outputs, X and Y line decoders can be operated directly from the memory address register without inverters. Each AND gate output shown in Fig. 7.11b is activated by a unique set of input conditions. The Y line decoder is connected in like manner by using inputs from $FF_2 - FF_1 - FF_0$.

While square core planes are typical, rectangular planes are possible. Rectangular planes use an odd power of 2. For example, 2^5 results in a 32-core plane. This core could be decoded by using a 5-flip-flop memory address register. An arrangement of three X line flip-flops and two Y line flip-flops or some similar arrangement uniquely locates each core.

During the write phase of a memory cycle, cores are driven high or low depending on whether 1s or 0s are written. Cores are always driven low during the read phase. Core driver amplifiers must process negative as well as positive current pulses. Thus a core driver will be either a source or a sink for current, and current must be steered in the correct direction. Two diodes connected at opposite ends direct the current.

7.4 BIPOLAR RAMS

Sequential locations on a magnetic drum were the memory storage units in the 1950s. At that time computers used vacuum tubes, and drum access time was compatible with computer circuits. By the early 1960s computers were using discrete transistors, which are less expensive, smaller, and faster than vacuum tubes. Core RAMs were used with discrete transistor circuits.

The bipolar transistor RS flip-flop, described in Chap. 5, is another

294 Computer Circuit Concepts

Figure 7.11
Address selection. (*a*) *x–y* core selector; (*b*) *x* line decoder detail.

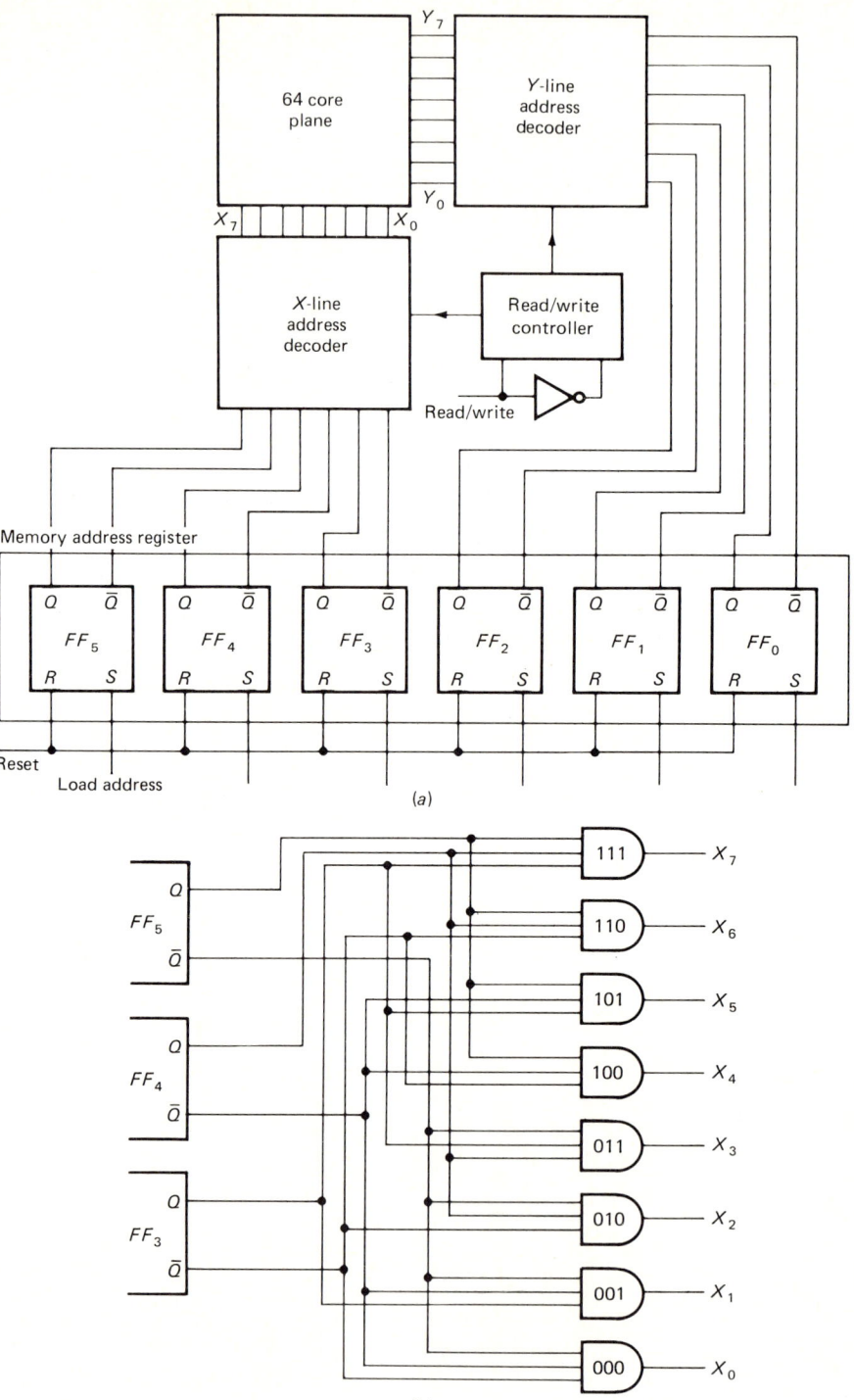

memory device. Triggering drives one transistor into saturation and the other into cutoff. Both transistors are stable as long as power is maintained, and either transistor can be read by measuring its collector voltage. At any time transistor states can be altered by triggering the appropriate base. Although flip-flops are faster than cores, the cores were initially less expensive and smaller.

By the early 1970s significant improvements had been made in IC fabrication. In particular, the multi-emitter transistor simplified circuit manufacture. Multi-emitter transistors have no discrete transistor counterpart, but bias requirements are the same. A bipolar NPN device conducts if the base is more positive than the emitter and the collector is more positive than the base. Conventional transistors operate by connecting the collector and emitter to fixed voltages, the base drive then being adjusted as required. Multi-emitter transistors operate by connecting collector and base to fixed voltages. Then emitter drive is adjusted as required.

IC flip-flops and TTL gates use multi-emitter transistors. These circuits operate at compatible voltages and can be connected to each other without voltage translation circuits. The development of multi-emitter ICs continued. Multi-emitter IC RAMs are now less expensive, faster, and smaller than magnetic core RAMs. Thus, it is unlikely that core memories will be considered in new computer designs.

A multi-emitter transistor is shown in Fig. 7.12a. Base and load resistors are connected to V_{CC}, which is usually $+5$ V. Whether or not this transistor conducts depends on the emitters. The transistor will saturate if at least one emitter is at 0 V. This is a logic low. But if the emitters are at $+5$ V or else open, the transistor is cut off. This is a logic high.

Details of a multi-emitter RS flip-flop are shown in Fig. 7.12b, where Q_1 and Q_2 are cross-coupled to form the RS flip-flop. A discrete transistor RS flip-flop requires external base leads for triggering and collector leads for sensing. Thus, the normal state of a memory cell is similar to a discrete

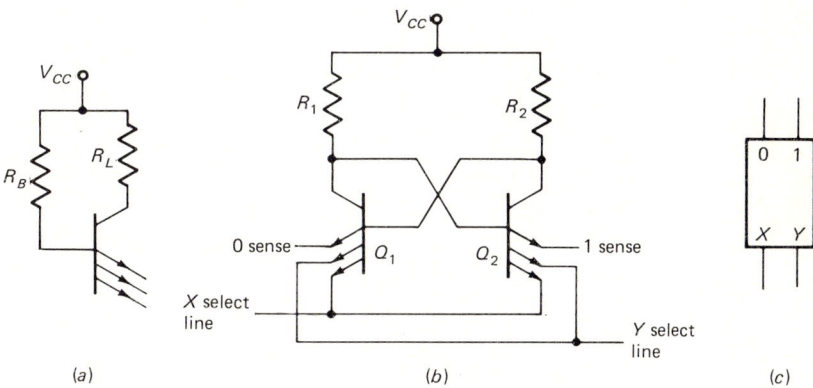

Figure 7.12 Bipolar memory cell. (a) Transistor; (b) flip-flop; (c) block diagram.

transistor *RS* flip-flop. If Q_1 is conducting, Q_2 is cut off and vice versa. The multi-emitter version is simpler because all external leads are connected to multi-emitters. Since no signal leads are connected to the base or collector, interconnection and printed circuit board layout are simplified. One flip-flop stores 1 bit. Each flip-flop in an IC RAM is called a *memory cell*.

XY address techniques developed for magnetic RAMs were transferred to IC RAMs. Figure 7.12b shows one emitter from Q_1 and one emitter from Q_2 connected together to form the *X* select line and another emitter from each transistor tied together to form the *Y* select line. This leaves one uncommitted emitter in each transistor, and these are the sense lines. An emitter of Q_1 can sense 0 while the emitter of Q_2 senses 1. Sense lines perform reading and writing.

When memory cells are not sensed, *X* and *Y* select lines are at logic 0. A specific memory cell is selected by driving the appropriate *X* and *Y* lines high, which disables the *X* and *Y* emitters of both transistors. With the *X* and *Y* lines disabled, control shits to the sense line emitters. Assume Q_2 was conducting and Q_1 was cut off prior to driving the select lines high: with the select lines high, the sense line emitter of Q_2 conducts and the sense line emitter of Q_1 remains cut off. A conducting sense line represents a logic high, and a nonconducting sense line represents a logic low.

Amplifiers are connected to the sense lines. Current from the conducting emitter drives the amplifier output high. At the same time, the amplifier connected to the nonconducting emitter is low. The amplified outputs, driven by the sense line emitters, are the memory cell read signals. If power is maintained, reading does not alter the state of either transistor.

Writing into a memory cell also requires driving the select lines high. As with reading, this shifts control to the sense line emitters. Writing requires amplification to the opposite direction. When reading, the sense emitter is an input whose amplified output is the logic state, but when writing, the sense emitter receives an input. Turning a multi-emitter transistor on to write a logic high on the sense line requires a low output from the write amplifier. Similarly, turning a multi-emitter transistor off, to write a logic low, requires a high output from the write amplifier. Thus, read and write sense line amplifiers perform opposite functions: write amplifiers drive the sense line emitters, whereas read amplifiers are driven by the sense line emitters. After writing is completed, the memory cell is disconnected from the select lines. This returns the *X* and *Y* lines to logic low. Both transistors retain the written states.

The same memory cycle developed for core RAMs is used for IC RAMs. Reading occurs before writing. Selecting a specific address in an IC RAM is also accomplished with a memory address register, and decoding uses the same AND gate technique developed for core storage. Figure 7.13 shows a 16-bit IC RAM.

Similarly, selecting 1 out of 16 possible memory cells requires eight address lines, four *X* and four *Y* lines. A single line connects all 0 sense terminals and

Figure 7.13
16 × 1 bipolar RAM.

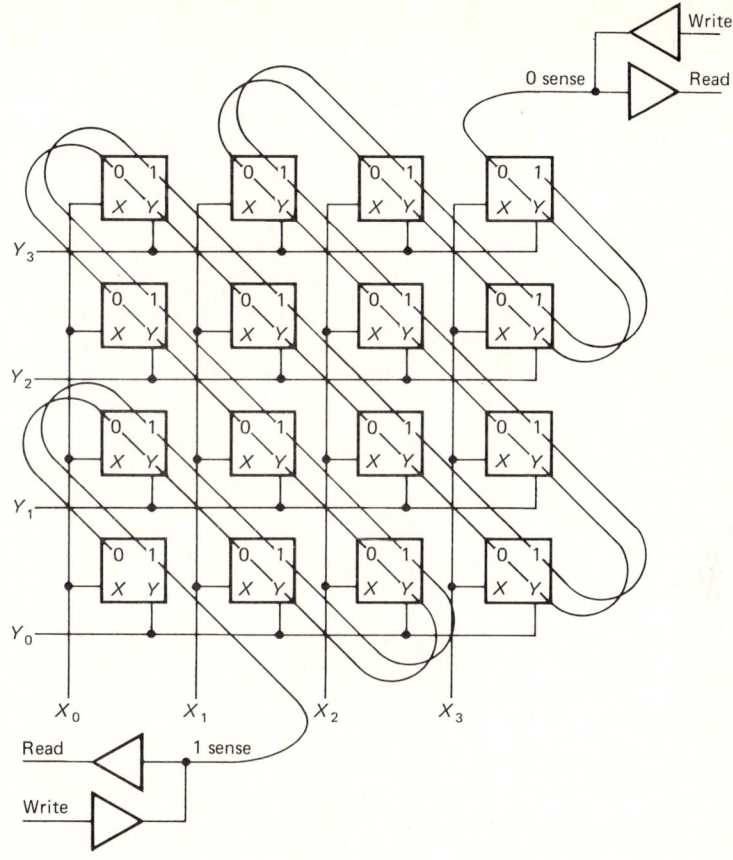

another line connects all 1 sense terminals. Reading and writing require separate amplifiers. Thus a 16 × 1 IC RAM requires

$$4 \ X \text{ select}$$
$$4 \ Y \text{ select}$$
$$2 \text{ read/write } 0$$
$$2 \text{ read/write } 1$$

for a total of 12 signal leads

Two additional leads are required for power supply connections. Therefore a 16 × 1 RAM just fits into a standard 14-pin DIP.

Example 7.7 Assuming external X and Y select lines, how many leads are needed for a 256 × 1 IC RAM?

Solution The number of flip-flops for the memory address register is given by

$$n = 2^{(x+y)}$$

In this case $n = 256$. Therefore

$$256 = 2^{(x+y)}$$

$$\log 256 = (x+y) \log 2$$

$$(x+y) = \frac{\log 256}{\log 2} = \frac{2.40824}{0.30103} = 8$$

Since 8 flip-flops contain 16 outputs, 16 address lines are needed. The total IC pin count is therefore

8 X select

8 Y select

2 read/write 0

2 read/write 1

2 power supply

---

22 leads

Therefore a 22-pin IC is needed.

External X and Y select lines are feasible for small RAMs. With larger RAMs it is practical to include address decoding circuits on the same chip as the memory cells and sense amplifiers. Combining functions on a single chip reduces external address lead count to the number required by the address decoder. This cuts the address line count in half. Thus the 256 × 1 RAM discussed in Example 7.7 fits into a 14-pin DIP because 8 rather than 16 leads are needed for address decoding.

Eliminating either the read/write 0 or read/write 1 sense line eliminates two more leads. This is possible because a low at Q_1 is the same as a high at Q_2 and vice versa. Removing two additional leads results in a 1024 × 1 RAM, which fits into a 14-pin package:

10 address lines

2 read/write 0 or read/write 1

2 power supply

---

14 leads

Bipolar IC RAMs are extremely fast, typical access times being in the 20 to 40 ns range. Depending on design details, per cell power dissipation varies between 0.25 and 2 mW per bit. Some manufacturers label the chip diagrams

Figure 7.14 RAM chip control.

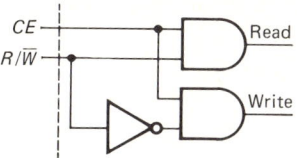

data I or data input, which corresponds to write. Similarly, data O or data output corresponds to read.

Memory can be expanded by using more than one RAM chip. With several RAMs it is not always necessary to activate each chip. RAM ICs usually contain chip enable (CE), and real/write ($R/\overline{W}$) leads, which are used to simultaneously select the appropriate function and activate a chip. Figure 7.14 shows the internal control logic. Two external leads and three internal gates perform CE and $R/\overline{W}$. A RAM chip is disabled when CE is low and enabled when CE is high. A combination of a high on CE and a high at $R/\overline{W}$ enables the read AND gate. Because of the inverter, a high on CE and a low at $R/\overline{W}$ enables the write AND gate.

7.5 MOS RAMs

The MOSFET, described in Chap. 3, is an alternate device for constructing semiconductor RAMs. MOSFETs contain an insulating layer of silicon dioxide, which is glass between the gate and transistor body. Thus no current flows into the gate terminal. A gate regulates source-to-drain current by controlling the electric field through the glass insulator.

For a *p*-channel enhancement mode MOSFET, the drain load resistor is connected to a negative voltage supply. The source is connected to ground, and conduction between drain and source occurs when the gate voltage is negative. For an *n*-channel enhancement mode MOSFET, the load resistor is connected to a positive supply, and the source is grounded. An *n*-channel MOSFET conducts when the gate voltage is positive. As described, substrate arrows should be used to distinguish between *n*-channel and *p*-channel MOSFETs, but they are omitted in most MOSFET schematics. With no arrows, power supply polarity is the only method of determining MOSFET type. For example, the MOSFET shown in Fig. 7.15a is understood to be *n*-channel because the drain is connected to $+V_{DD}$.

Although MOSFETs and bipolar transistors are based on different principles, voltage polarities are similar. An *n*-channel MOSFET and an *NPN* transistor use the same bias polarity. Both are cut off when the input voltage is zero and both saturate if the input voltage is sufficiently positive. Similarly, a *PNP* transistor and *p*-channel MOSFET are both cut off when the input voltage is zero and both saturate at sufficiently negative input voltages.

Originally, because of manufacturing ease *p*-channel MOSFET RAMs were dominant. However *p*-channel requires two power supplies and is

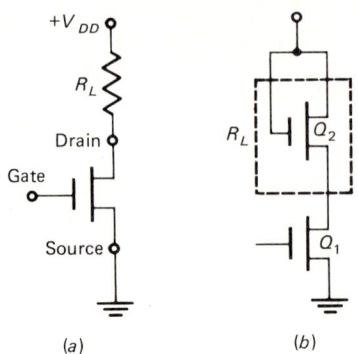

Figure 7.15
MOS stage. (*a*) *n*-Channel MOSFET; (*b*) with MOSFET as R_L.

slower than *n*-channel. Also, since *p*-channel MOSFETs operate at negative voltage, interfacing circuits are needed to connect with TTL. Improvements in MOS manufacturing have resulted in practical *n*-channel MOSFETs.

MOS transistors are inherently smaller and require fewer manufacturing steps than bipolar transistors. When ICs are manufactured, resistors require more area than transistors. However, MOSFETs can be used as load resistors for other MOSFETs. Since gate area determines drain current, adjusting gate area and connecting the gate lead directly to V_{DD} results in a MOSFET with fixed current and voltage drop; this is equivalent to a fixed resistor. Thus, a physically small MOSFET takes the place of a physically large resistor. In Fig. 7.15*b* Q_2 is the load resistor for Q_1.

The basic MOS memory cell is also an *RS* flip-flop. Figure 7.16*a* shows drains which are cross-coupled to gates of the opposite MOSFETs. Load resistors R_1 and R_2 are MOSFETs which are used as fixed resistors. However, a schematic is easier to follow if functions rather than components are shown. When Q_2 is saturated, the drain of Q_2 is almost 0 V. This same voltage exists at the gate of Q_1, and Q_1 is cut off. Therefore Q_1 is cut off when Q_2 is saturated and vice versa. MOS RAMs use the same convention as bipolar RAMs. The conducting MOSFET is a logic high and the cutoff MOSFET is a logic low. The addition of select lines and read/write lines converts a basic *RS* flip-flop into a memory cell.

MOSFETs are smaller and easier to manufacture than bipolar transistors. However, no MOSFET equivalent of the multi-emitter transistor exists—there is no "multi-source" MOSFET. Therefore, a basic MOS memory cell contains more components, but the reduced size of MOS components still results in a MOS memory cell which is considerably smaller than the bipolar version. Figure 7.16*b* shows a complete MOS cell. Outputs from Q_3 and Q_4 are sense lines; Q_3 senses 0 and Q_4 senses 1. The gates of Q_3 and Q_4 are connected to the *Y* select line, and similarly the gates of Q_5 and Q_6 are connected to the *X* select line.

MOS cells are also addressed by driving the appropriate *X* and *Y* lines high. Thus in Fig. 7.16*b*, if Q_2 is assumed to be on and Q_1 off, when the select

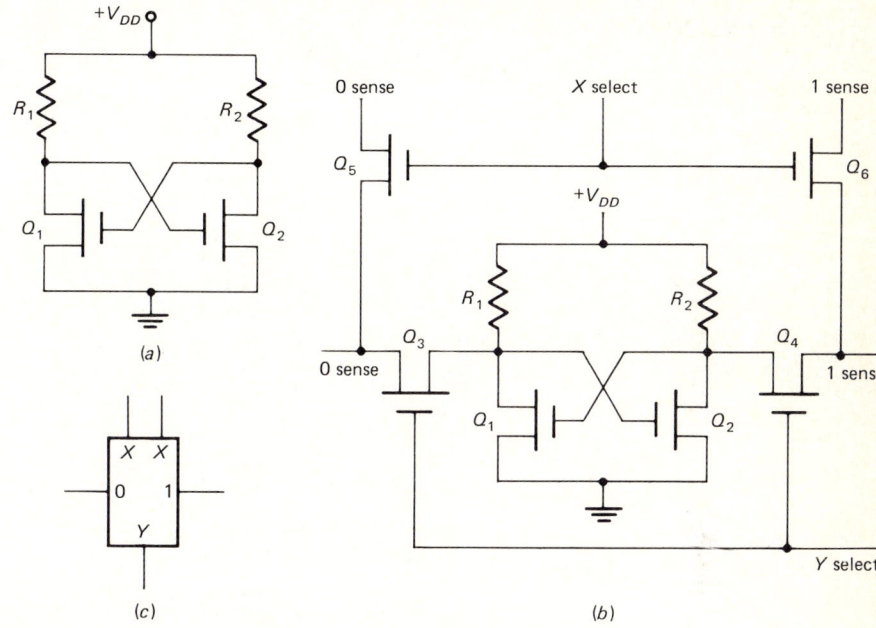

Figure 7.16 MOS memory cell. (*a*) Basic flip-flop; (*b*) complete schematic; (*c*) block diagram.

lines of this cell are driven high, Q_4 turns on because the drain of Q_2 is approximately 0 V. Since Q_1 is off at this time, driving the Y select line high cannot turn Q_3 on. Similarly, a positive signal at the gates of Q_5 and Q_6 will only turn Q_6 on, but Q_5 remains cut off because the source of Q_4 is on while Q_3 is off.

Figure 7.17 shows the structure of a 16 × 1 RAM using MOSFET cells. Q_5 and Q_6 in the X_0 select line are the same as in Fig. 7.16. These two MOSFETs are the select circuit for the entire X_0 column. Corresponding MOSFET pairs perform select for other X lines. MOS RAMs use differential amplifiers for reading and writing. The same memory cycle, read before write, is used with MOS memories.

MOS RAMs dissipate less power than bipolar RAMs with the same cell count, lower speed being their only disadvantage. Bipolar RAMs have access times in the 20 to 40 ns range, but MOS RAMs require 400 to 600 ns.

Memories described thus far contain 1-bit words: 256 × 1, 512 × 1, 1024 × 1, etc. Word length is increased with a separate read/write amplifier in each RAM column. For example, the common amplifier lines shown in Fig. 7.17 can be broken, and individual read as well as write amplifiers can be attached to each X select column. This converts a 16 × 1 RAM into the 4 × 4 RAM shown in Fig. 7.18. Total bit capacity is the same, but word length is increased.

Cascading is another technique for increasing word length. Figure 7.19 shows four RAMs, each of which has a 16 × 1 capacity combined into a

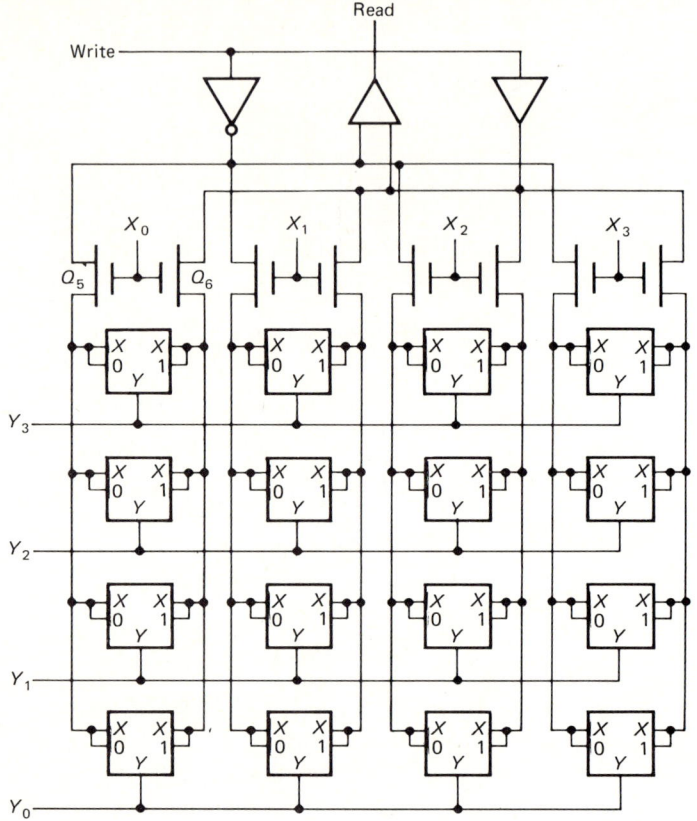

Figure 7.17
16 × 1 MOS RAM.

16 × 4 memory. CE and $R/\overline{W}$ lines are operated in parallel. Address lines are also connected in parallel from the memory address register. The data lead of each RAM is individually connected to the memory buffer register. If the memory address register contains $X_2 Y_3$, then RAM_0 contains the LSB of word $X_2 Y_3$ and RAM_3 contains the MSB.

Example 7.8 Describe the construction of a 1K × 1 byte memory using 1024 × 1 RAMs.

Solution A byte contains 8 bits. In this case eight RAMs which are 1024 × 1 should be connected by using the circuit in Fig. 7.19.

An alternate cascading method permits increased address count while word length remains constant. In Fig. 7.20 the address lines and $R/\overline{W}$ lines are still operated in parallel, but the CE lines are operated individually and all RAM data terminals are connected to a 1-bit memory buffer register. Corresponding addresses of each RAM are connected in parallel, but only

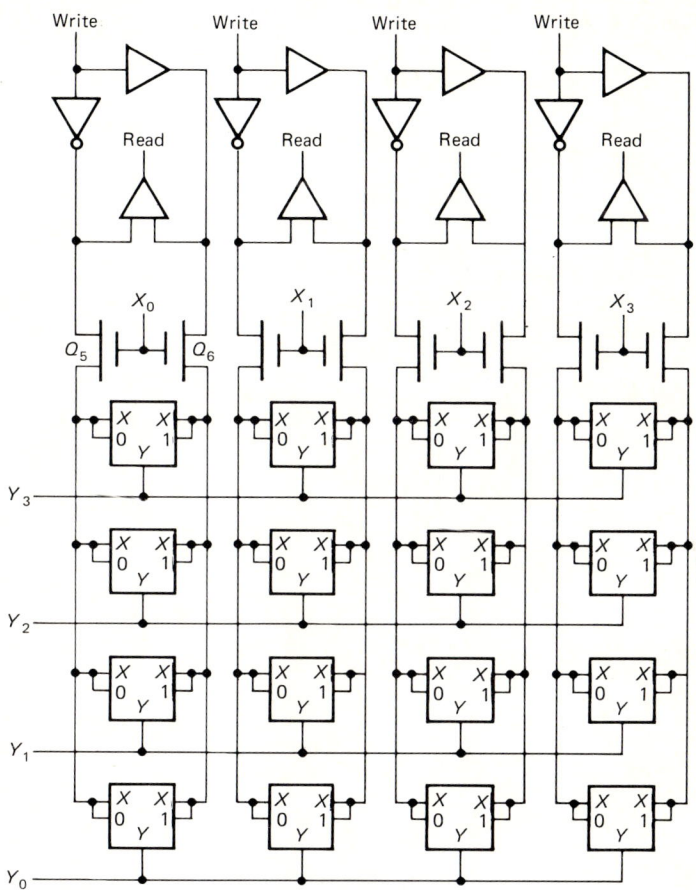

Figure 7.18
4 × 4 MOS RAM.

one CE line is enabled at any given time. If $X_2 Y_3$ is addressed and CE_1 is enabled, RAM_1 is the only RAM which can be read or written. In effect the CE lines are part of the memory address register. Figure 7.20 shows the construction of a 64 × 1 memory using four 16 × 1 RAMs.

Example 7.9 Describe the construction of a 1024 × 1 memory using 128 × 1 RAMs.

Solution Since 1024/128 = 8, eight RAMs are required. Connections are shown in Fig. 7.20. Address lines, $R/\overline{W}$ lines, and data lines are connected in parallel, but CE lines are enabled separately.

It is also possible to combine both cascading techniques. This increases address count and word length. If two 512 × 1 RAMs are connected as

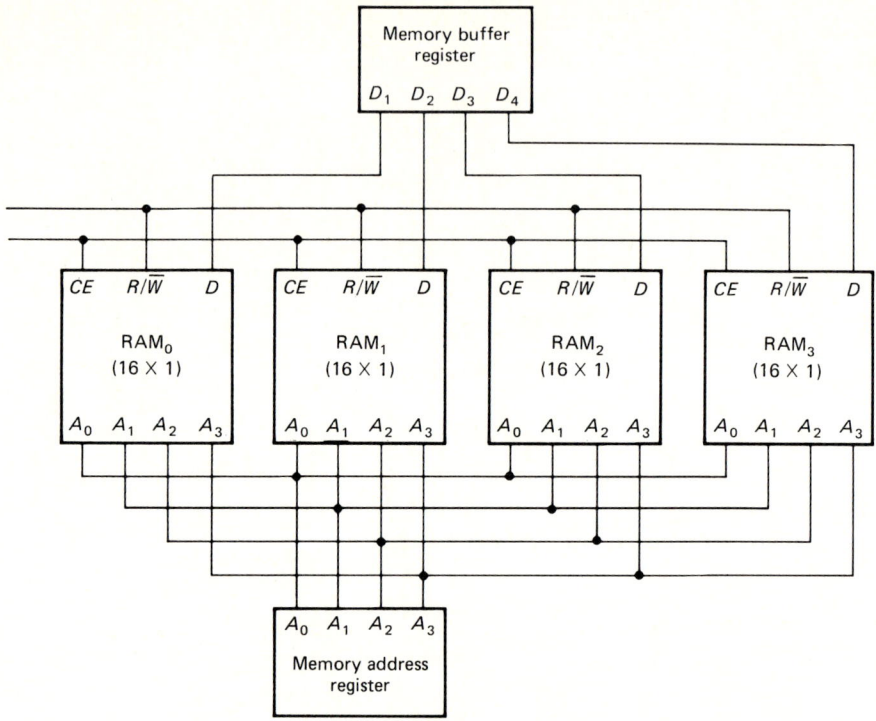

Figure 7.19
16 × 4 RAM.

shown in Fig. 7.20, the result is a 1024 × 1 RAM. Then, if eight of these 1024 × 1 RAMs are connected as shown in Fig. 7.19, the result is a 1024 × 8 memory. Thus, 16 RAMs which are 512 × 1 can be combined to form a 1K × 1 byte memory. The same methods for expanding MOS RAM capacity apply to bipolar RAMs.

7.6 DYNAMIC RAMs

RAMs described in previous sections use similar memory cells. Whether bipolar or MOS, two inverter stages are cross-coupled from input to output and form an *RS* flip-flop. If power supplies remain on, the flip-flops retain data. Flip-flop-type memory cells are classified as *static*.

Static memory cells are practical but another MOSFET storage technique exists. A quality capacitor stores charge for a long time and charge storage has binary capability. A charged capacitor can represent a logic 1; a discharged capacitor can represent a logic 0. MOSFET input resistance is about 10^{12} Ω. High resistance makes the gate to substrate capacitor C_{GS} a quality component. Figure 7.21a shows data storage using MOSFET C_{GS}. A logic 0 is written onto C_{GS} of Q_1 by charging C_{GS} to 0 V. Not charging C_{GS} keeps Q_1 cut off, and the resultant drain voltage is $+V_{DD}$. An uncharged C_{GS}

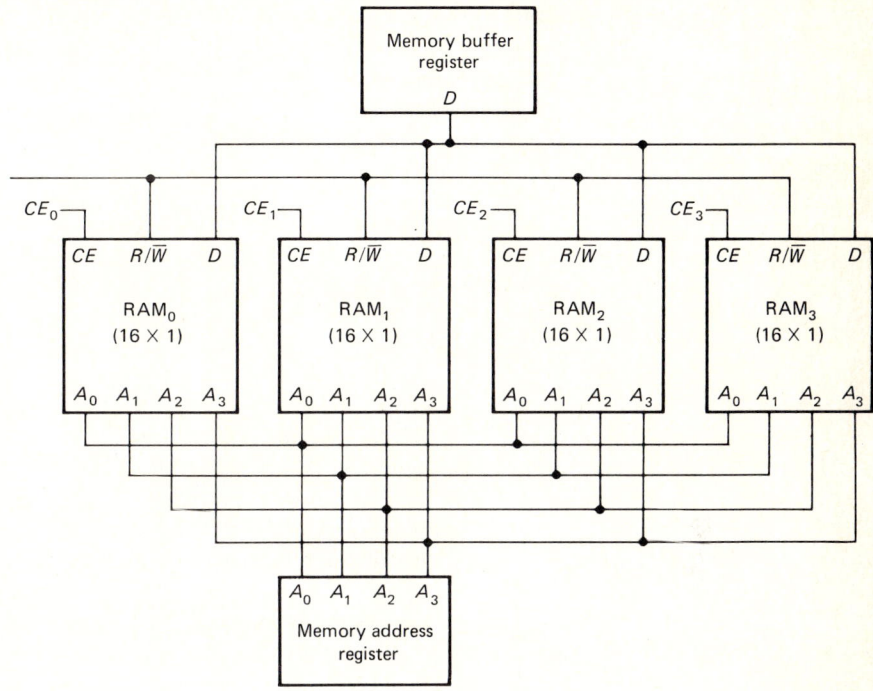

Figure 7.20
64 × 1 RAM.

stores a logic 0, which is read as a positive voltage at the MOSFET drain. A logic 1 is written by charging C_{GS} to a positive voltage. The charging voltage must be high enough to drive Q_1 into saturation. Maintaining charge on C_{GS} stores a logic 1 and is read as a zero drain voltage.

A cutoff MOSFET is the normal state. Storing a logic 1 is more involved because C_{GS} discharges through the MOSFET input resistance. While C_{GS} is a quality capacitor, it is so small that measurement is difficult. Typical values of gate-to-substrate capacitance are in the 0.04 to 0.07 pF range.

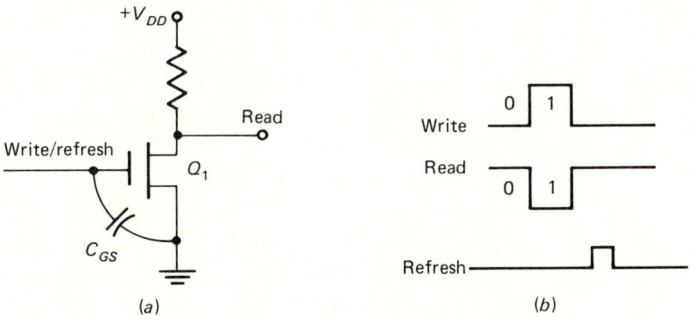

Figure 7.21
Capacitive data storage.
(*a*) *n*-Channel MOSFET;
(*b*) waveforms.

Example 7.10 Assume that a saturated MOSFET cuts off in one input time constant. How long can C_{GS} store a logic 1?

Solution Input resistance is

$$R \simeq 10^{12}\ \Omega$$

and $\quad C_{GS} \simeq 0.04$ pF

Since $\quad t = R \cdot C$
$\qquad\quad = 1 \times 10^{12} \times 0.04 \times 10^{-12} = 0.04$ s

$\quad t = 40$ ms

This is an extremely short time for storing data and must be increased. Extending data storage time requires periodically *refreshing* or recharging C_{GS}. Refresh pulses must be of sufficient amplitude and duration to prevent C_{GS} from discharging when a logic 1 is stored. However, refreshing must not drive a cutoff MOSFET into conduction. Manufacturers of charge storage RAMs recommend one refresh pulse every 2 ms. As opposed to static memory cells, which automatically retain data, memory cells which require periodic refreshing are called *dynamic*.

Three different dynamic memory cell circuits exist. The cell shown in Fig. 7.22a contains a single component. Because 1 MOSFET constitutes the entire cell, many cells fit into a small chip area. When a particular cell is enabled, writing into C_{GS} occurs: charging C_{GS} to a positive voltage writes a logic 1, and not charging C_{GS} writes a logic 0. Reading is accomplished by sensing drain voltage. The read amplifier is on the same chip, as shown in Fig. 7.22b.

The high packing density resulting from a single MOSFET cell is a mixed blessing. Small gate-to-substate capacitance requires a high-gain read amplifier. In fact, C_{GS} is many times smaller than the distributed capacitance of the common read line. Thus, separate high-gain read amplifiers are required for each column of the RAM. Also, read/write timing and refresh synchronization are critical. Single MOSFET memory cells are practical but additional support circuitry is required. As a result, the high packing density is partially offset by the need for auxiliary circuits.

Another dynamic cell design is shown in Fig. 7.23a. The cell also uses C_{GS} of Q_1 as the data storage element, but there are two additional MOSFETs. A 3-MOSFET cell requires extra access lines but is about the same size as a single MOSFET cell. Q_1 in the 3-MOSFET cell amplifies as well as stores data; since Q_1 performs both functions, capacitance and size can be reduced. Size reduction results in comparable areas for 1- and 3-MOSFET cells.

A memory cycle begins by charging read and write line capacitance to $+V_{DD}$. Next, the appropriate read select line is enabled by driving the gate of Q_2 high. Consider what happens if Q_1 stores a 1: the drain of Q_2 goes low,

Figure 7.22
Dynamic MOSFET
memory. (*a*) Single cell;
(*b*) 16 × 1 RAM.

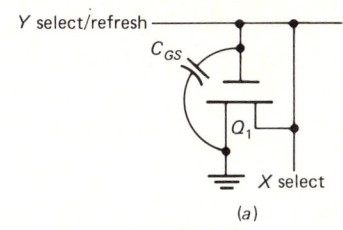

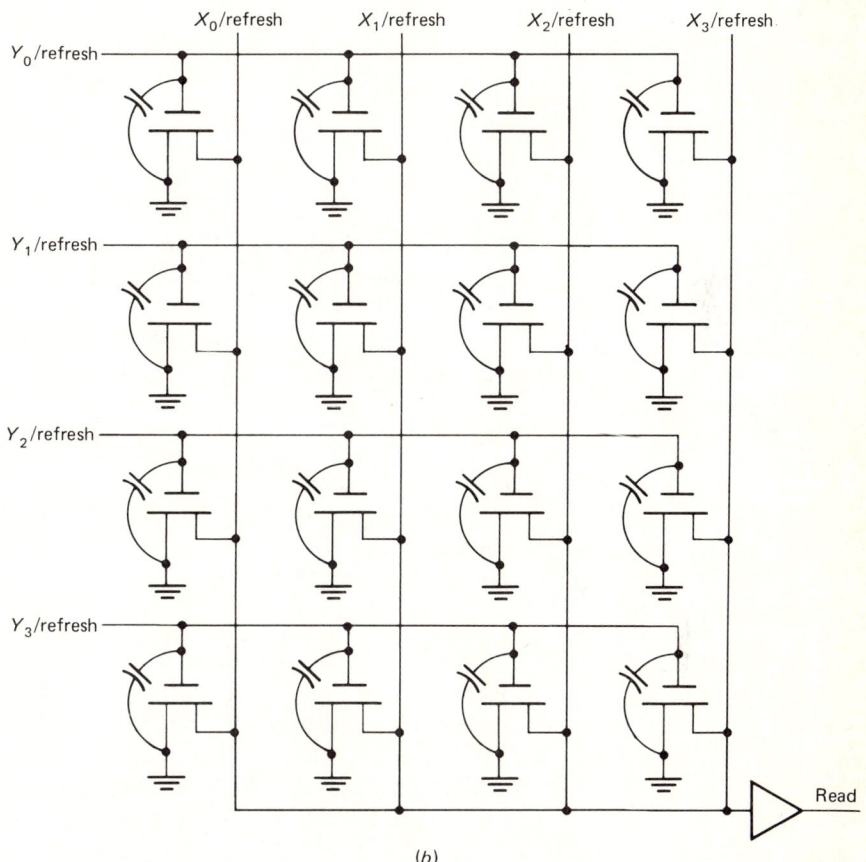

because storing a 1 on Q_1 makes the drain of Q_1 low, and since the source of Q_2 is connected to the drain of Q_1, Q_2 also saturates. On the other hand, Q_1 may store a logic 0, in which case C_{GS} is uncharged and Q_1 is cut off. With Q_1 cut off, Q_2 cannot conduct when the gate of Q_2 is enabled. Reading a logic 0, as in the single MOSFET cell, results in a high voltage on the read line. Data are restored by recharging the read and write lines. After recharging, new data can be written into the same cell by enabling the write select line. This activates the gate of Q_3.

Figure 7.23
Multitransistor dynamic cells. (*a*) 3 MOSFETs; (*b*) 4 MOSFETs.

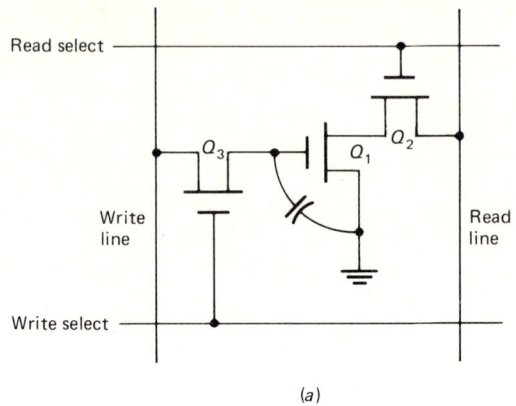

(a)

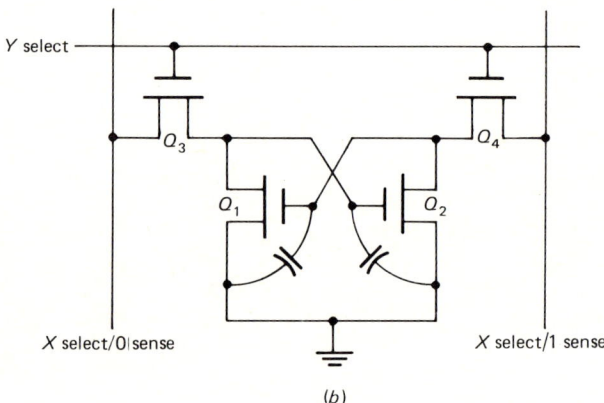

(b)

Figure 7.23*b* shows a 4-MOSFET dynamic cell, which seems similar to a static MOSFET cell in that:

- There are flip-flops in both versions.
- Data and complemented data are available.
- The X sense line and data line are common.

However, in a dynamic flip-flop, data is actually stored on the gate-to-substrate capacitances of Q_1 and Q_2. The dynamic flip-flop does not contain load resistors.

Operation of the 4-MOSFET cell also begins by charging the data lines. In this case, it is the 0 sense and 1 sense lines which are initially charged to $+V_{DD}$. When a cell is enabled, Q_1 drives Q_3 into conduction or else Q_2 drives Q_4 into conduction. This is the read portion of the memory cycle.

The gate-to-substrate capacitance of Q_2 may store a logic 1. In this case Q_2

conducts. When Q_2 conducts, Q_4 goes low because the source of Q_4 is connected to the drain of Q_2. This action drives the 1 sense line low and simultaneously causes Q_3 to refresh Q_1. Original data is restored.

The dynamic flip-flop requires one more MOSFET than the three-transistor version. However, refreshing and sense amplifier are simpler. As a result, 3- and 4-MOSFET dynamic cells have about the same area.

Regardless of MOSFET count, periodic refreshing is required to maintain stored charge. Several methods of refreshing exist. In the *distributed* refresh mode, each row is sequentially recharged during a refresh period.

Example 7.11 A 16,384 × 1 dynamic RAM uses distributed refresh. How often does a refresh occur?

Solution Refresh is required at least every 2 ms. A square 16,384 RAM contains

$$\sqrt{16,384} = 128 \text{ rows}$$

Thus an individual row is refreshed every

$$\frac{2 \times 10^{-3} \text{ s}}{128} = 15.6 \text{ }\mu\text{s}$$

Another method of refreshing is the *burst* mode. When burst refresh is used, all RAM cells are recharged during the single pulse which occurs every 2 ms. In either case about 0.5 μs is required to replenish an individual row. If burst refreshing is used with the RAM of Example 7.11, the refresh pulse is 64 μs wide.

Still another technique is the *hidden* refresh mode, which combines features of distributed and burst refresh. Several rows at a time are recharged while some other row is read. With hidden refresh, the computer is "unaware" that recharging occurs. Most dynamic RAMs can be refreshed by any of these methods. Distributed, burst, and hidden refresh each have advantages and disadvantages. Selection is based on specific design details.

On chip address selection using flip-flops to reduce RAM leads has been described. This method is called *full address decoding*. A fully decoded 16,384 × 1 RAM requires 7 bits for row address and 7 bits for column address. *Multiplexed address decoding* reduces the leads by half; thus a multiplexed 16,384 × 1 RAM only needs 7 address leads. Thus, an entire 16K RAM fits into a 16-pin DIP. Multiplexing requires separate transmission of row and column address. Typically, the row address is sent first and latched into the internal address decoder. Then the column address is transmitted to the RAM along the same leads. After the column address is latched, the on chip decoder locates the specific cell.

Expansion techniques described for static RAMs also apply to dynamic RAMs. Thus, it is possible to increase dynamic RAM word length as well as

TABLE 7.1 Typical RAM characteristics

Model	Type	Organization	P_{diss} (mW)	$t_{acc.}$ (ns)	Cost/Bit
74S206	Bipolar	256 × 1	450	40	1.00
93415	Bipolar	1,024 × 1	800	50	0.26
1101	Static	256 × 1	100	650	0.40
2102	Static	1,024 × 1	150	350	0.05
4027	Dynamic	4,096 × 1	470	250	0.04
5298	Dynamic	8,192 × 1	465	200	0.01
4164	Dynamic	65,536 × 1	465	200	0.006

total address count by cascading RAMs. Usually RAM chips in an expanded RAM are refreshed by a *refresh controller*, which is a separate chip and can be programmed for any refresh mode.

Originally, static and dynamic MOSFET RAMs required three power supplies. Two supplies, usually −5 and +12 V, were needed for MOSFET opeation and a +5-V power supply was needed for the self-contained interface to TTL levels. Single-power-supply MOSFET RAMs operating at +5 V are now available.

Although refreshing requires additional circuits, dynamic RAMs with refresh controllers are smaller, dissipate less power, and cost less than static RAMs with the same cell count. Table 7.1 compares features of several bipolar, static MOS, and dynamic MOS RAMs. The 64K dynamic RAMs are readily available, and 256K versions can be obtained. Dynamic MOS RAMs, while slower than bipolar, are somewhat faster than static MOS. Dynamic RAMs have a distinct power dissipation advantage. An entire 64K dynamic RAM dissipates about the same power as a 256 × 1 bipolar RAM. If actually built, a 64K bipolar RAM would dissipate about 115 W. The last column in Table 7.1 compares per bit cost relative to that of a 256 × 1 bipolar RAM. For example, if the 74S206 costs 2 cents per bit, then a 64K bipolar RAM would cost more than $1300. By contrast the 4164, a 64K dynamic RAM, costs less than $8.

An appreciation of the pace of computer progress is realized by tracing RAM development. In the middle 1950s computers used magnetic drum storage, and a 2K RAM was typical. Less than 30 years later some hand-held computers have 16K RAMs and the typical desk-top computer contains a 64K RAM.

7.7 ROM PRINCIPLES

The computer uses RAM to store a user program and data. During operation, the RAM also stores intermediate and then final results. Once results are obtained, the program is removed from the RAM by storing the program for later use on some external device such as a magnetic tape or disk. Permanent destruction of the program stored in a RAM is a more drastic removal technique. Destruction is accomplished by writing a new program into the same RAM locations or else shutting the computer off.

Besides temporary storage provided by RAM, a computer also needs permanent memory, called a *read only memory* (ROM). A ROM stores internal instructions for operating the computer and may also contain mathematical tables such as logarithms and trigonometric functions. Some ROMs also store constants such as π and the base of natural logarithms.

As with RAM, a ROM stores data which are read by selecting the appropriate address. Writing is a different matter—a RAM can and should be written into, but data enters a ROM during manufacture and is permanent. A ROM is truly nonvolatile. Actually, semiconductor devices, whether bipolar or MOS, lose data when power is shut off, but semiconductor ROMs automatically restore data when power is reapplied. There is no need to reload ROM data each time a computer is turned on.

Two methods of designing semiconductor ROMs exist, based on either combinational gate logic or else a two-dimensional matrix. In effect, ROMs based on combinational gate logic have been discussed. Half- and full-adders use gates to store the rules for binary addition. During manufacture, necessary gates are permanently connected to perform a specific function. Therefore the general procedure for designing a logic-gate ROM has been presented. First the problem is defined and the required truth table is generated. Then reduction techniques are applied in an attempt to minimize component count. The resultant circuit is a ROM which accepts data and produces results according to a set of rules. A logic gate half-adder, for example, is a ROM which adds two independent input bits and outputs the correct sum and carry-out bits.

The initial steps in designing a matrix-type ROM are the same. A truth table follows from the problem definition. However when a matrix is designed, no attempt is made to reduce component count. The matrix ROM is a grid of mutually perpendicular lines, with conductive devices at some grid intersections but not at others. Conducting intersections store a logic 1, and nonconducting intersections store a logic 0. Unfortunately, simple connections at grid intersections do not provide isolation between devices connected to the input and output of a ROM. Diodes or other unilateral components are required as conductive devices; the presence of a diode at a junction indicates a logic 1, and the absence of a diode indicates a logic 0.

Matrix ROMs have rows and columns. Typically, addresses are the rows, while the stored data are taken from the columns. A matrix ROM contains input address lines, an address decoder, the actual stored word, and data output lines. Consider the ROM shown in Fig. 7.24a, where A_1 and A_0 are the input address lines and D_1 and D_0 are the data output lines.

The two address lines control the 1-of-4 address decoder, which, as previously described, consists of 2 inverters and 4 AND gates. Each AND gate in the decoder is labeled with the $A_1 A_0$ address line sequence needed to drive that particular AND gate high. AND-gate outputs are connected to the

Figure 7.24
Diode matrix ROM. (a) Schematic; (b) data table; (c) block diagram.

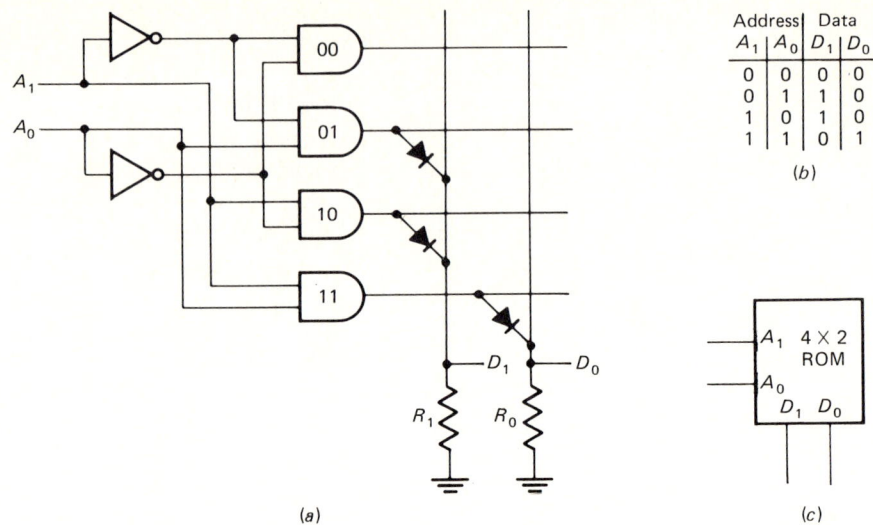

data storage section of the ROM. The data outputs are taken across the load resistors R_1 and R_0 and are high or low depending on the selected address. If the selected address is 00, then AND gate 00 is enabled. No diodes are connected at the intersections of the 00 and the D_1 or D_0 columns. Therefore $D_1 D_0$ data are 00 when the address is 00.

Next consider the results when the address is 01. AND gate 01 is connected to the diode matrix in a different way, so that there is a diode at the intersection of the 01 address line and the D_1 line. There is no diode at the intersection of the 01 address line with the D_0 line. If the address lines are driven by +5-V signals, the D_1 output will be about +4.4 V while the D_0 output remains 0 V. Thus, in terms of TTL logic levels, the ROM data output is 10 when the input address is 01. Output for the other two addresses can also be traced through the ROM, and the results are tabulated in Fig. 7.24b.

This is a simple 4 × 2 ROM, with only three diodes required to store data, and yet the results are rather impressive. For example, D_1 can be considered as a 2-input EOR gate in which case D_0 performs as a 2-input AND gate. Another way of interpreting the same results exists: D_1 is the sum output of a half-adder, and D_0 the carry-out.

Example 7.12 Design a full-adder diode matrix ROM.

Solution The full-adder truth table shown in Fig. 7.25a is the same as shown in Fig. 6.9b for the logic gate version. In this case the address lines A_2, A_1, and A_0 replace carry-in, input B, and input A. In particular, A_2 is C_i, A_1 is B, and A_0 is A. Similarly, S and C_0 are D_1 and D_0, respectively.

For simplicity, the circuit shown in Fig. 7.25b contains only the eight decoded address lines and two-column diode matrix portion of the ROM. A diode is connected whenever a 1 in the data portion of the truth table occurs.

Figure 7.25
Matrix full-adder. (*a*) Data table; (*b*) 8 × 2 ROM.

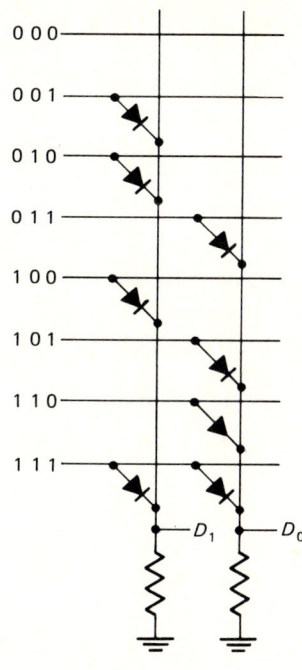

Address			Data	
A_2	A_1	A_0	D_1	D_0
0	0	0	0	0
0	0	1	1	0
0	1	0	1	0
0	1	1	0	1
1	0	0	1	0
1	0	1	0	1
1	1	0	0	1
1	1	1	1	1

(*a*)   (*b*)

A complete schematic would show the address decoder, which in this case requires 3 inverters and eight 3-input AND gates. Clearly, any function or any data word can be stored permanently using either a matrix ROM or a combinational gate logic ROM. The diode matrix half-adder and full-adder ROMs can be compared with logic gate versions discussed in previous chapters.

A truth table is needed to design both the combinational gate logic and the matrix ROM version. The logic gate ROM uses input data to present output data while the matrix ROM uses input addresses to present output data. A matrix ROM is an addressable sequence of completely independent single data words with permanent memory. On the other hand, a gate ROM uses data rather than addresses to perform computations in order to present a result. It is easier to manufacture a large matrix than a sequence of unrelated functions using logic gates. Large IC ROMs are usually matrices.

As shown in Fig. 7.26, Boolean expressions can also be stored in a matrix ROM. This table shows that:

- D_3 is a 3-input AND gate.

- D_2 is a 3-input NAND gate.

- D_1 is a 3-input OR gate.

- D_0 is a 3-input NOR gate.

Figure 7.26
Three-input logic gate ROM. (a) Data table; (b) 8 × 4 ROM.

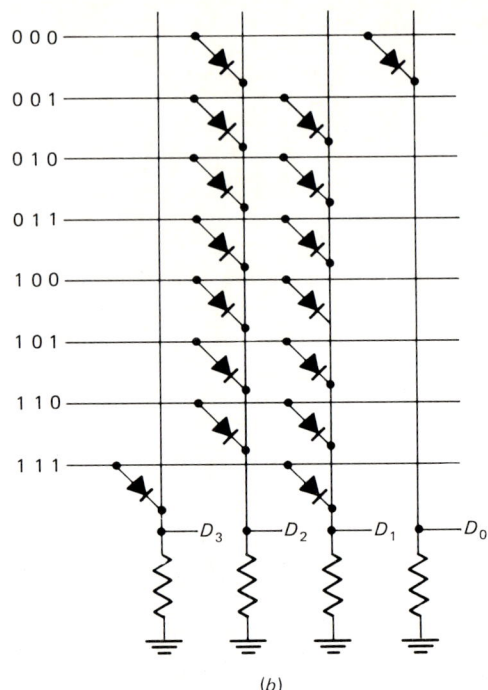

Address			Data			
A_2	A_1	A_0	D_3	D_2	D_1	D_0
0	0	0	0	1	0	1
0	0	1	0	1	1	0
0	1	0	0	1	1	0
0	1	1	0	1	1	0
1	0	0	0	1	1	0
1	0	1	0	1	1	0
1	1	0	0	1	1	0
1	1	1	1	0	1	0

(a)   (b)

7.8 ROM METHODS

ROM design is based on fixed connections between input and output lines. Once a ROM has been constructed, the data stored in it are permanent. The information contained in a ROM can be read as often as necessary but cannot be altered.

A diode matrix ROM demonstrates a basic approach but has the same loading problem as that associated with diode logic gates. Improving drive capability requires adding transistors to the output of a diode matrix or else replacing the diodes with transistors. In either case, using transistors necessitates a power supply.

Figure 7.27a is the transistor equivalent of the diode half-adder shown in Fig. 7.24a. Transistors are located at the same intersections as the diodes. Collectors are connected to the load resistors and emitters are grounded. The transistors are enabled from the base leads, which are connected to the address lines. With no voltage at the bases, the transistors are cut off. Since the data lines are directly connected to the load resistors, data outputs are normally high. When an address line is enabled, that particular transistor saturates and the data output drops nearly to 0 V. Therefore, the circuit in Fig. 7.27a has an output which is the invert of an actual half-adder. Inversion occurs because the transistors are connected as common emitters. In Fig. 7.27b, standard half-adder outputs are obtained by connecting inverters to

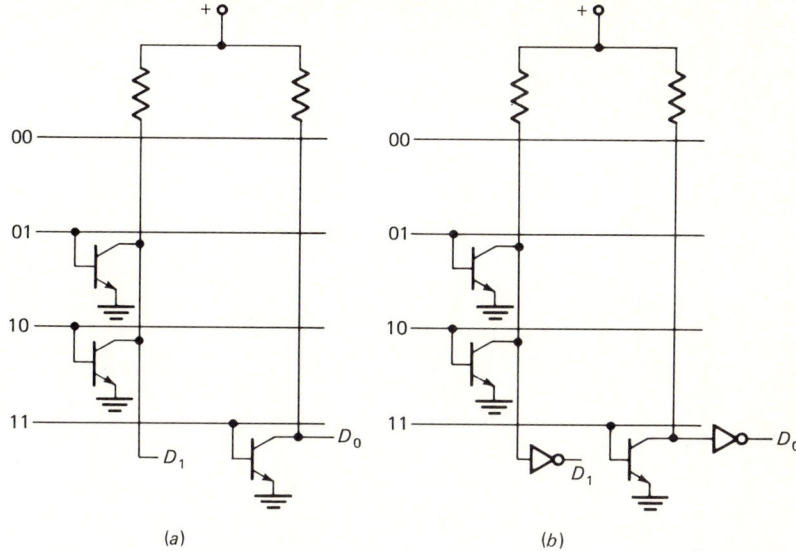

Figure 7.27 Transistor matrix ROM. (*a*) Transistors replace diodes; (*b*) inverters added.

each data line. The inverters are common emitter stages located on the same chip. Half-adder outputs could be obtained directly if common collectors replaced the common emitters. However, since it is difficult to manufacture IC common collectors, inverters are a practical IC solution.

Most chip manufacturers offer a family of ICs which function together. A particular ROM contains the instructions needed to operate a specific ALU, RAM, and other components. Compatibility facilitates design of a general-purpose computer. It is also possible that a specific application may not require all the features available from a general-purpose ROM.

Consider the ROM needed to control a self-service gasoline pump or a computing scale for groceries. There is no need for addition, subtraction, division, or trigonometric, or logarithmic tables. It is only necessary to multiply a variable price by a variable quantity, such as dollars per gallon times gallons, or dollars per pound times pounds. If many of the pumps or scales are needed, it is less expensive to use a custom-designed ROM, in which case a designer submits a prepared form, containing the binary address and binary data to be stored in each location, to the chip manufacturer.

Chip manufacturers produce general-purpose as well as custom-designed ROMs. It is less expensive if all ROMs have many manufacturing steps in common. Usually ROMs, regardless of final application, are manufactured with transistors at each matrix intersection. For ROMs of equal size, the only difference between various ROM types is which transistor bases, or gates, are connected to address lines and which are not. A different interconnection scheme results in a different ROM. During manufacture, connecting or not connecting control elements of transistors in a ROM is determined by a single step. This step is the layout on a photographic negative, which in

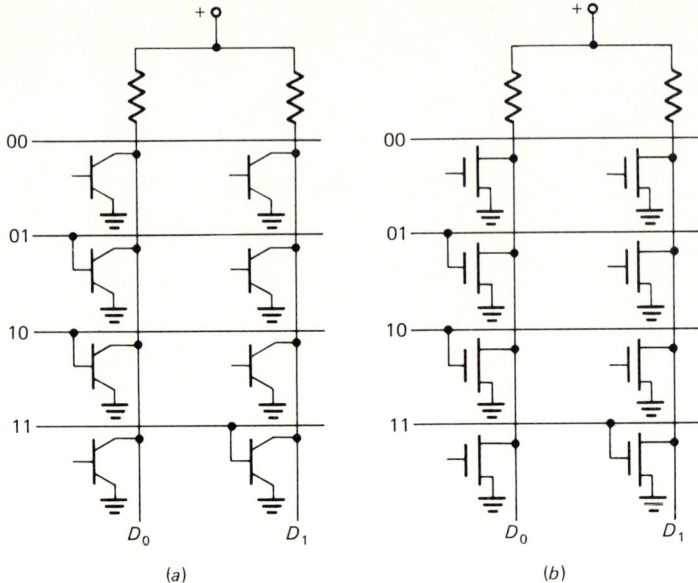

Figure 7.28 Half-adder MROM. (*a*) Bipolar; (*b*) MOS.

semiconductor parlance is called a *mask*. ROMs produced in this manner, whether general-purpose or custom, are called *maskable read-only memories* (MROMs).

Of course once an MROM is manufactured, data are permanent and unalterable. Bipolar as well as MOS MROMS exist. Figure 7.28 shows MROM half-adders. Necessary base leads are connected and the remaining base leads are left open for the bipolar version. The same technique applies to the gates in the MOS circuit. In the schematics of both versions, inverters are omitted for clarity.

The specifications of some typical MROMs are shown in Table 7.2. As with RAMs, bipolar MROMs are at least four times faster than MOS. However, if access times in hundreds of nanoseconds are acceptable, MOS MROMs are smaller, dissipate less power, and are less expensive. Originally, MROMs also required three power supplies, but newer models operate from a single $+5$ V supply.

TABLE 7.2 MROM Characteristics

Number	Type	Organization	t_{acc} (ns)	P_{diss} (mW)
93467	Bipolar	256 × 4	25	425
3604	Bipolar	512 × 8	60	650
93464	Bipolar	1K × 8	30	550
35151	MOS	512 × 8	600	300
2332	MOS	4K × 8	300	200
2364	MOS	8K × 8	300	200

MROMs can also store mathematical functions. There are two general methods for placing tables in an MROM. One technique stores the power series expansion of a particular function, in which case the ALU calculates a specific value. For example, the power series of the sine function, in radians, is given by

$$\sin(X) = X - \frac{X^3}{3!} + \frac{X^5}{5!} - \frac{X^7}{7!} + \ldots \tag{7.7}$$

The other technique stores selected values of a function in tabular form, and interpolates for intermediate values. When interpolation is required, the necessary computations are performed in the ALU. Values of the sine from 0 to 90° in 10° intervals can be stored in a 10 × 4 MROM. In this case, the address is the angle and the datum is the sine of the angle. With only 4 bits for the sine value, it is more accurate to normalize the sine and divide by the normalized value later in the computation. Since 15_{10} is the largest 4-bit number, the normalizing value N_V for a 4-bit data sequence is:

$$N_V = 15 \cdot \sin(X) \tag{7.8}$$

Table 7.3a shows the normalized as well as the rounded-off values, and Table 7.3b shows the binary equivalent of the rounded-off sine values.

This MROM is not particularly accurate. The sine of 15° is off by more than 15 percent, and the sine of 30° is almost 7 percent high. However, Table 7.3 is illustrative. A 10 × 8 MROM reduces these errors to the 0.5 percent range. Moreover, with practical-size MROMs such as 512 × 8 or larger, errors are negligible.

Whether an MROM from a compatible family or a custom memory is selected, the user is restricted to the original ROM design. Although it seems a contradiction in terms, there are applications for alterable ROMs. This is

TABLE 7.3 10° Interval Sin MROM

	(a) Data				(b) As 10 × 4 MROM				
Angle, °	Sin	N_V	N_V (rounded)		Angle, °	D_3	D_2	D_1	D_0
0	0.000	0.00	0		0	0	0	0	0
10	0.174	2.61	3		10	0	0	1	1
20	0.342	5.13	5		20	0	1	0	1
30	0.500	7.50	8		30	1	0	0	0
40	0.643	9.65	10		40	1	0	1	0
50	0.766	11.49	11		50	1	0	1	1
60	0.866	12.99	13		60	1	1	0	1
70	0.940	14.10	14		70	1	1	1	0
80	0.985	14.78	15		80	1	1	1	1
90	1.000	15.00	15		90	1	1	1	1

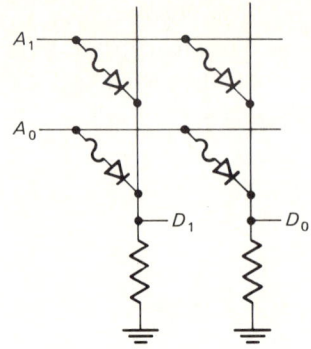

Figure 7.29
Fusible-link PROM.

particularly true during the development of a new computer. ROMs require changes as work progresses from initial design to final prototype. Also, depending on quantity it may be less expensive for designers to fabricate their own ROMs rather than to purchase custom MROMs. ROMs which can be altered after manufacture are called *programmable* ROMs (PROMs).

One type of PROM uses fusible links. During manufacture a metallic fuse is deposited at each junction of the matrix. Depending on manufacturing method, a fusible-link PROM originally contains either all 1s or all 0s.

The diode PROM shown in Fig. 7.29 initially contains 1s at each junction. When a specific location is addressed and a 25 to 35 mA current pulse is applied, the fuse blows. Each blown fuse breaks a junction, and each open circuit is a logic 0. Fuses which are not blown continue to store logic 1s. Normal read currents are much lower than the values needed to program a fusible-link PROM.

While a 2 × 2 fusible link PROM is small enough to be programmed manually, practical sizes range from 25 × 4 to 2K × 8. Manually blowing fuses in such large PROMs is not practical. The work is tedious, and the probability of blowing some incorrect fuses is high. Fusible-link programming circuits exist; they sequentially address each junction and either do or do not blow fuses as required. PROM program circuits contain a RAM, which is first loaded with the data to be programmed into the PROM. Most units can simultaneously blow fuses of six or more PROMs and are relatively inexpensive. Fusible-link PROMs are manufactured with bipolar transistors and have the same access times as bipolar MROMs.

While the remaining fuses of a PROM can be blown at a later time, it is impossible to restore blown fuses. Thus, for practical purposes a fusible-link PROM can only be programmed once. Other PROMs exist which can be programmed and reprogrammed as often as desired. Such memories are called *erasable programmable* ROMs, or EPROMs.

An EPROM is manufactured by using a special type of transistor called a *floating-gate* MOSFET. Besides the standard MOSFET components, a lead-

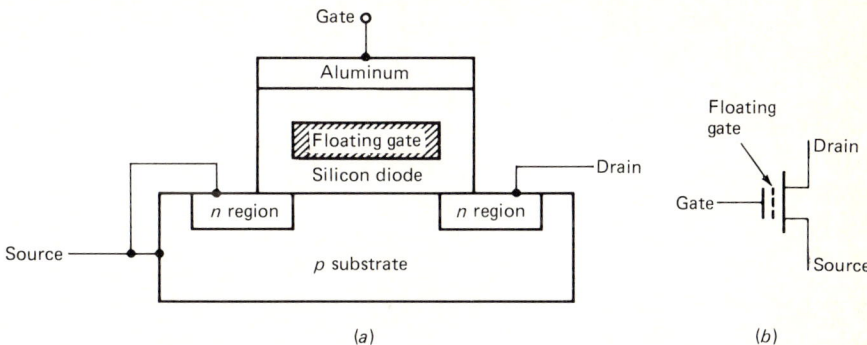

Figure 7.30
Floating-gate MOSFET.
(*a*) Structure; (*b*) symbol.

less gate is internally embedded in the silicon dioxide, as shown in Fig. 7.30*a*. The silicon dioxide completely isolates the floating gate from any other portion of the transistor. With an uncharged floating gate, the MOSFET performs in the conventional manner. Floating-gate MOSFETs are normally cut off and require positive gate voltage to conduct.

If sufficient voltage is applied to the external gate lead, charge is built up on the floating gate which opposes conduction. This induced charge can be made large enough to keep the MOSFET permanently cut off. A charged floating gate stores logic 0. Once a 0 is written into a floating gate, charge is retained after power is shut off. Floating-gate MOSFETs have retained charge for 10 years even under adverse conditions. Thus an EPROM is, for practical purposes, a permanent memory.

Before a new program can be written into an EPROM, charge stored on the floating gates must first be removed. Since leads are not connected to floating gates, standard discharge methods do not work. However external energy in the form of ultraviolet light discharges floating gates. EPROMs contain quartz windows which transmit light from ultraviolet lamps. When erasure is required, the lid covering the window is removed. Exposure to ultraviolet light for about 30 min is required. Complete erasure is tested by addressing each MOSFET and determining if logic 1s are present at each location. Sunlight and fluorescent lamps also contain significant percentages of ultraviolet. Exposure to sunlight will discharge floating gates in about 3 weeks, while fluorescent lamps require about 3 years. Accidental erasure of EPROMs is prevented by using opaque rather than translucent lids for the quartz windows.

When used in a computer, an EPROM is addressed and read in the conventional manner. If reprogramming is required, the EPROM is removed and erased again using ultraviolet light. The EPROM is then placed in a unit which is similar to the fusible-link programmer. Each address is sequenced and 0s are written where necessary. Writing 0 into a floating gate requires a 6-ms voltage pulse.

Example 7.13 How much time is needed to program a previously erased 2K × 8 EPROM?

Solution A 2K × 8 EPROM contains

$$2048 \times 8 = 16{,}384 \text{ bits}$$

At 6 ms per bit the total programming time is

$$6 \times 10^{-3} \frac{\text{s}}{\text{bit}} \times 16{,}384 \text{ bits} \simeq 98 \text{ s}$$

This example shows that the time required to reprogram an entire EPROM is negligible. This is especially true because six or more EPROMs are usually programmed at once. However, all gates are erased by ultraviolet exposure. Rewriting even a single bit requires the complete reprogramming procedure.

The *electronically erasable* PROM (EEPROM) is the newest programmable ROM. An EEPROM is actually a RAM with a permanent memory. It can be erased and reprogrammed without removal from the computer. If costs can be reduced, EEPROMs could replace external memory units such as tape and disk.

The MROM of a desk-top computer is at least 14K × 8. Most MROM manufacturers also provide pin-compatible PROMs, which allow a user to purchase a standard MROM and expand it for special requirements with the compatible PROM.

All MROMs and various PROMs are matrix-type memories. Combinational gate logic memories are also available in IC form and are called *programmable logic arrays* (PLAs). Input terminals are usually labeled I_0, I_1, I_2, etc. Inverters are included on the chip, and thus inputs, as well as their complements, are available. Each input along with its complement is connected to multiple-input AND gates, which are connected to OR gates that are the outputs of the PLA. For example, a 16 × 48 × 8 PLA contains 16 inputs, which, along with the inverts, are connected to 48 AND gates. These 48 AND-gate outputs are connected to 8 OR gates, and the OR gate outputs are the PLA read lines. Mask-programmable as well as fusible-link PLAs are available.

Because of the AND gate–to–OR gate PLA input–to–output sequence, all outputs will be in sum-of-products (SOP) form. For example, combinations such as $I_0 \cdot \bar{I}_1 \cdot I_2 + I_0 \cdot I_1 \cdot \bar{I}_2 + \ldots$ can be programmed. Thus PLAs can perform as half-adders, full-adders, or any other desired logic function. It is not necessary to use each input and each output line. Only the necessary functions are combined to produce SOP equations.

Originally, PLAs were less expensive than matrix ROMs of the same capability. However, ROM costs have decreased and PLAs now are competitive only for special applications.

SUMMARY

1. Magnetic drum storage is based on inducing magnets on a rotating cylinder. Megabytes can be stored in relatively small volumes, but data retrieval is slow compared with electronic switching speed.

2. Core storage is based on magnetizing mechanically fixed cores, which are wired in planes. Core storage is faster, but restoring is required to preserve data every time a core is read.

3. Core memories are arranged in planes with 1 bit of each word in each plane. The memory buffer register temporarily stores the specific word and the memory address register locates that word.

4. Bipolar IC RAMs are multi-emitter transistors connected as *RS* flip-flops to store data. Such RAMs are faster, smaller, and less expensive than magnetic core RAMs.

5. MOS RAMs are also based on the *RS* flip-flop. Although MOS cells require more components per cell, they are smaller and less expensive and dissipate less power than bipolar cells.

6. A dynamic MOS memory cell stores data in the small gate-to-substrate capacitance. Dynamic RAMs require refreshing but are less expensive and smaller and dissipate less power than other semiconductor RAMs.

7. A ROM stores data permanently and contains internal computer operating instructions. Data can be read by addressing the appropriate location, but new data cannot be entered.

8. In addition to permanent data MROMs, alterable ROMs also exist. Fusible-link PROMs, EPROMS, EEPROMs can be used to change ROM data.

PROBLEMS

7.1 A magnetic drum rotates at 3600 r/min. Find (*a*) the maximum and (*b*) the average access time.

Answer: (*a*) 16.7 ms; (*b*) 8.3 ms.

7.2 Find the speed of a drum which has an average access time of 5 ms.

7.3 A drum is 5 in high and has a 10-in diameter. There are 60 tracks per inch and the bit density is 800 bits per inch. Find the drum byte capacity.

Answer: 942,000 bytes.

7.4 For the previous problem determine the diameter required to store 1 million bytes.

7.5 How many lines are required to write into a coincident core memory which contains (*a*) 256 and (*b*) 1024 bits?

Answer: (*a*) 32 lines; (*b*) 64 lines.

7.6 How many lines are needed to perorm both reading and writing for the memories in the previous problem?

7.7 Cores which require 720 mA are used in a coincident core plane. Find the coincident currents for (*a*) writing and (*b*) reading.

Answer: (*a*) +360 mA; (*b*) −360 mA.

7.8 A coincident core memory is numbered from X_0 to X_{31} and Y_0 to Y_{31}. How many lines are needed to perform reading and writing if inhibit lines are used?

7.9 A three-dimensional core memory contains 65,536 single-precision words. Determine (*a*) the number of planes; (*b*) the number of lines per plane, and (*c*) the total number of cores, and (*d*) describe how this memory would be specified.

Answer: (*a*) 16; (*b*) 256 X and 256 Y lines; (*c*) 1,048,576 cores; (*d*) 64K by 16.

7.10 When the location of a byte-organized core memory is read, the sense line conditions are 10 mV for P_0, P_1, and P_6 and 100-mV signals for the other planes. Find the decimal equivalent of this word.

7.11 Assuming that sense and inhibit are performed on a single line, determine how many (*a*) flip-flops, (*b*) planes, and (*c*) cores are required for a 16K double-precision core memory.

Answer: (*a*) 32; (*b*) 32; (*c*) 524,288.

7.12 A single-precision three-dimensional core memory contains 16,384 cores. Determine: (*a*) the number of planes; (*b*) the number of words; (*c*) the length of the memory buffer register; and (*d*) the length of the memory address register.

7.13 A bipolar 512 × 1 RAM contains separate 0 and 1 sense lines. If address decoding is performed by a separate chip, how many pins does the RAM require?

Answer: 24 pins.

7.14 The chip of Prob. 13 is redesigned to include self-contained address decoding. How many pins does this RAM require?

7.15 CE and $\overline{R/W}$ lines are added to the chip of Prob. 14. At the same time the 0 sense lines are deleted. How many pins are needed?

Answer: 15 pins.

7.16 Each bit of a 512 × 1 bipolar RAM dissipates 0.5 mW. Find the total power dissipation.

7.17 How can a 64 × 1 MOS RAM be converted into a byte-size memory?

Answer: Separate read and write amplifiers for each column result in an 8 × 8 RAM.

7.18 Using 1024 × 1 RAMs, describe the construction of a 16K double precision memory.

7.19 Describe the construction of a 2K × 1 memory using 512 × 1 RAMs.

Answer: 4 RAMs connected as shown in Fig. 7.20.

7.20 How can a 2K × 1 byte memory be constructed with 512 × 1 RAMs?

7.21 The MOSFET of a dynamic cell has a gate-to-substrate capacitance of 0.07 pF, an input resistance of 1×10^{12} Ω, and discharges in one time constant. How long can this cell store (*a*) a logic 1; (*b*) a logic 0?

Answer: (*a*) 70 ms; (*b*) indefinitely.

7.22 If distributed refresh is used, how often is a row refreshed in a 65,536 × 1 RAM?

7.23 Assume that a bipolar RAM costs 1 cent per bit. Use Table 7.1 to determine the dollar cost of 65K RAMs using (*a*) bipolar and (*b*) dynamic MOS cells.

Answer: (*a*) $655.36; (*b*) $3.93.

7.24 With the aid of Table 7.1, determine the per bit power dissipation of: (*a*) a 256 × 1 bipolar RAM; (*b*) a 65,536 × 1 dynamic RAM.

7.25 A diode ROM is designed to convert octal numbers into binary. (*a*) What is the ROM organization? (*b*) How many diodes are required?

Answer: (*a*) 8 × 3; (*b*) 12.

7.26 Show a data table and schematic for a diode ROM which converts hex keypad into binary.

7.27 Determine (*a*) the organization and (*b*) the number of diodes in a ROM which is a 2-input OR gate.

Answer: (*a*) 4 × 1; (*b*) 3.

7.28 A diode ROM performs the Boolean expression $C \cdot (B + A)$. Show a data table and schematic.

7.29 Determine the per bit power dissipation of the largest (*a*) bipolar and (*b*) MOS MROMs listed in Table 7.2.

Answer: (*a*) 67 μW per bit; (*b*) 3.1 μW per bit.

7.30 Use Eq. (7.7) to determine the accuracy of sin $\pi/6$, which is 30°, to (*a*) one, (*b*) two, and (*c*) three terms of the power series.

7.31 Find the error in sin 30° if a 10 × 8 MROM table is used.

Answer: 0.39 percent.

7.32 After programming, a single-bit error is discovered in a 1K × 8 EPROM. After erasure, how long does writing the corrected program into the EPROM take?

8
PERIPHERAL DEVICES

8.0 INTRODUCTION

Regardless of application, computer input is obtained from and results presented to the outside world. Peripheral devices are the links between computer and outside world. As computers developed, a variety of peripheral devices also developed.

The first computers processed paper in the form of punched cards and tape. Paper is still popular. Large computers, numerically controlled machine tools, fast food chain cash registers, etc., use paper for input and output. Although not as old as paper, magnetic devices have been used for some time. Drum storage was developed as the memory element for vacuum tube computers. Core storage is faster and was used in discrete transistor computers. These techniques evolved into magnetic disks and magnetic tapes.

Cathode-ray tubes (CRTs) are another example of an older device that has been adapted to a computer. Originally, CRTs presented graphs of voltage versus time. In a computer CRTs display input information and output results. Data links such as teletypes and telephone modems are also used and will be discussed in this chapter.

Measurement and process control applications generally present data in analog form. Such analog signals must be converted to digital at the input. Conversely, computer-generated process control data must be converted to analog at the computer output. Computers can accept a large variety of inputs and present a choice of outputs. This flexibility is a principal reason for the dynamic growth of the computer industry.

8.1 PAPER INPUT/OUTPUT

Computers use many methods to interface with the outside world. Paper in various forms can be either an input or an output material. In fact, the first practical electric computer used paper input.

This computer was developed to satisfy more efficiently Article I, Sec. 2 of the U.S. Constitution, which requires a census every 10 years. Census data

Figure 8.1
Hollerith's punch card computer.

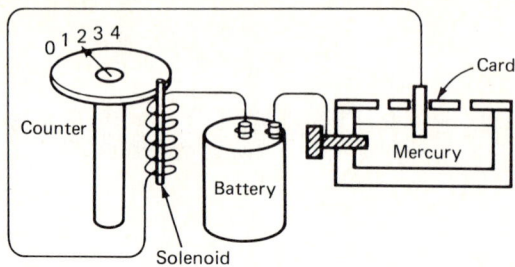

determine how many members each state sends to the House of Representatives. Herman Hollerith, a U.S. Census Bureau statistician, developed a method of entering data on paper cards and obtaining results electrically. Data were entered on cards by punching holes, and the cards were then placed between a set of metal pins and a mercury pool, as shown in Fig. 8.1. Whenever a hole occurs, a pin passes through the card into the mercury and completes a circuit, which contains a solenoid. Each time the solenoid is activated, a mechanical counter steps to the next position. Hollerith's machine, which he called a "recording tabulator," was first used for the 1890 census. Despite an increasing population, analysis of the census data took one-third of the time needed for the 1880 census. Work on the tabulator continued, and Hollerith founded the Tabulating Recording Company in 1911. This company later changed its name to International Business Machines Co., now IBM Corp.

The original method of coding data by punching holes on cards is still used and is named in Hollerith's honor. A punch card is shown in Fig. 8.2. It is $3\frac{1}{2}$ in wide and $7\frac{3}{4}$ in long. The most popular card configuration contains 12 rows numbered 12, 11, 0, 9, 8, ..., 3, 2, 1. Rows 12, 11, and 0 are called the *zone fields*, and rows 9 through 1 are called *digit* fields. Going from left to right, the columns are numbered 1 through 80. The Hollerith code contains letters of the alphabet and numbers; such a code is called *alphanumeric*.

The digit 0 is entered by punching a 0 in zone 0, and the digits 1 through 9

Figure 8.2
Typical punch card format. (Courtesy IBM Corp.)

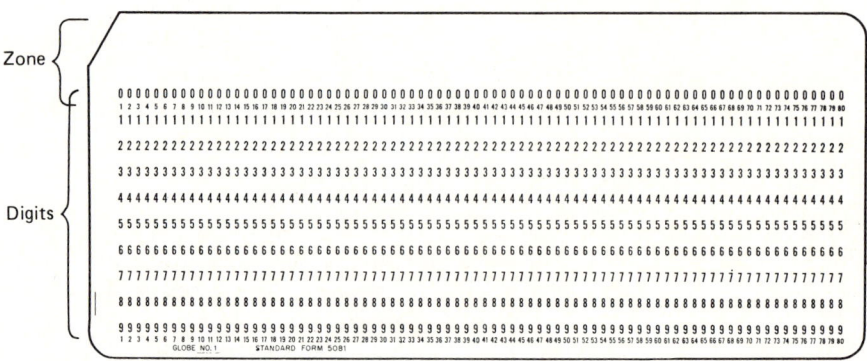

TABLE 8.1 Hollerith Alphanumeric Punches

	Fields			Fields			Fields			Fields			Fields	
Value	Zone	Digit	Value	Zone	Digit	Value	Zone	Digit	Value	Zone	Digit	Value	Zone	Digit(s)
A	12	1	J	11	1	S	0	2	0	0		blank		
B	12	2	K	11	2	T	0	3	1		1	=		3,8
C	12	3	L	11	3	U	0	4	2		2	,	0	3,8
D	12	4	M	11	4	V	0	5	3		3	&	11	3,8
E	12	5	N	11	5	W	0	6	4		4	.	12	3,8
F	12	6	O	11	6	X	0	7	5		5	—		4,8
G	12	7	P	11	7	Y	0	8	6		6	(	0	4,8
H	12	8	Q	11	8	Z	0	9	7		7	*	11	4,8
I	12	9	R	11	9				8		8	)	12	4,8
									9		9	/	0	1
												+	12	
												−	11	

are each entered by making a single punch in the appropriate digit field. For example, a rectangular hole in the 3 field represents the number 3_{10}. Letters of the alphabet are represented by two holes in the same column, one in the zone field and one in the digit field. Holes in the zone 12 and digit field of the same column represent the letters A through I. For example, the combination of a 12 punch and a 3 punch is the letter C. Similarly, an 11 punch together with a digit punch represents the letters J through R. The remaining letters, S through Z, are represented by a zone 0 punch with the digits 2 through 9. The Hollerith code also contains 12 punctuation marks and special symbols, and thus has a total of 47 characters. All letters in Hollerith code are uppercase. Table 8.1 shows the entire code.

Example 8.1   Begin at column 27 on a Hollerith card and show the coding for

101 BASE(2) = 5 BASE(10).

Solution   Including blank spaces, this message requires 25 columns. Appropriate punches are as indicated.

Statement	1	0	1		B	A	S	E	(	2	)		=		5		B	A	S	E	(	1	0	)	.
Column	27	28	29	30	31	32	33	34	35	36	37	38	39	40	41	42	43	44	45	46	47	48	49	50	51
Punches																									
Zone		0			12	12	0	12	0		12						12	12	0	12	0		0	12	12
Digit	1		1		2	1	2	5	4	2	4		3		5		2	1	2	5	4	1		4	3
Digit									8		8		8								8			8	8

The Hollerith code is a century old and is still popular. Businesses find many applications for it. Many large companies enter each employee's payroll data on a punch card and print the resultant paycheck on another punch card.

Each column of a punch card contains the information in a byte-sized word, and therefore a single punch card can store 80 data words. More efficient use is possible: for example, each row rather than each column could be arranged into bytes. In this case a single row would contain 80/8, or 10, bytes. Since there are 12 rows, a byte-organized card has a 12 × 10, or 120, byte capacity. This is a 50 percent improvement in punch card data storage.

Punching holes in a reel of tape is another method of storing data on paper. In fact, Hollerith experimented with paper tape before deciding on paper cards. The origins of paper tape for data storage go back to the early days of telegraphy. In 1858 Charles Wheatstone, of null-sensing bridge fame, developed an improved telegraph, in which the manually operated sending/receiving key was replaced by a punched tape. In Wheatstone's machine, holes were punched by hand with a hammer and steel punch, and the tape was then loaded onto a reel and fed past sensing pins. Punched-tape dots and dashes were faster than the hand-operated key.

Two years later Jean Baudot developed a 5-bit punched tape telegraph code. This code is still an accepted transmission standard. The term *baud*, used for data transmission in bits per second, was chosen in Baudot's honor. Each 5-bit column on a punched tape represents an alphanumeric character. With only 5 bits, it would seem only 2^5, or 32, different characters are possible and yet the Baudot code contains 26 letters, 10 digits, and a variety of other symbols. Increased character capability is achieved by special symbols called *letters* and *figures*. When the letter code is punched, the following 5-bit combinations represent the corresponding character in the letter column of the Baudot code. This situation continues until the code for figures is sent, at which time the symbols have the values indicated in the figure column. Each 5-bit combination has one of two meanings, the actual value being determined by whether it follows the letter code or the figure code. In this way 32 combinations represent upwards of 60 characters. Table 8.2 shows the Baudot code. As with the Hollerith code, only uppercase letters are possible.

Binary 1 is obtained by punching a hole and binary 0 by not punching a hole. Figure 8.3 shows the Baudot message "Hello."

A punched tape message begins with *line feed*, a control signal which turns the tape reel on. Next comes a space, which is followed by the letters symbol. After "Hello" is punched, the figure code must be punched before a period can be punched. The period is followed by a space and a *stop* signal. Stop turns the tape reel off.

8.2 PROCESSING PAPER DATA

Techniques for entering data onto punched tape and cards are similar; in both cases the operator uses a console with a typewriter-style keyboard. A

TABLE 8.2 Baudot Code Punches

Bit Position $2^5\ 2^4\ 2^3\ 2^2\ 2^1$	Letters	Figures
00000	Blank	Blank
00001	E	3
00010	Line feed	Line feed
00011	A	—
00100	Space	Space
00101	S	Bell
00110	I	8
00111	U	7
01000	Carriage return	Carriage return
01001	D	$
01010	R	4
01011	J	'
01100	N	,
01101	F	!
01110	C	:
01111	K	(
10000	T	5
10001	Z	"
10010	L	)
10011	W	2
10100	H	Stop
10101	Y	6
10110	P	0
10111	Q	1
11000	O	9
11001	B	?
11010	G	&
11011	Figures	Figures
11100	M	.
11101	X	/
11110	V	;
11111	Letters	Letters

paper tape console contains a reel with a blank roll of tape. Typewriter keys activate a column of steel punches, and striking a key activates the correct character punches. After punching, the feed mechanism advances the tape and another character is punched. The punch and advance sequence

Figure 8.3
Baudot code message.

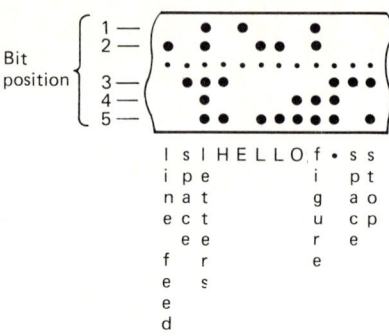

continues until an entire program has been transferred onto tape. The same time is needed to prepare a paper tape as to type information with an electric typewriter. Punched tapes contain 10 characters per inch.

Example 8.2 A program to evaluate the roots of a quadratic equation contains 420 alphanumeric characters. Determine the punched tape length.

Solution There are 10 characters per inch of punched tape.

$$\frac{420 \text{ characters}}{10 \text{ characters/in}} = 42 \text{ in}$$

This quadratic equation program is $3\frac{1}{2}$ ft long.

Once a program exists, it can be used as often as necessary. A tape reader transfers the punched tape program into a RAM. Tape readers contain sense pins to detect the presence or absence of holes. The sense pins mechanically activate switches which translate punched characters into 1s and 0s; the holes represent 1s and the blank spaces 0s. Punching and reading a tape are discontinuous operations. In both cases, the tape is in a fixed position while a character is actually processed. Then the tape is advanced.

Besides character punches, paper tape contains a row of smaller holes. These small holes engage a sprocket on the tape transport reel and are similar to movie film sprockets. The sprocket position also indicates when a character is in the read position. Reading punched tape is an electromechanical operation. Typical pin-sense tape readers operate at a speed between 10 and 50 *characters per second* (CPS). A 50-CPS tape reader enters the quadratic equation program of Example 8.2 into a RAM in about 8.5 s.

A photoelectric punched tape reader is faster because physical contact between tape and sense mechanism is eliminated. Photoelectric readers contain a light above each punch position and a light detector below. The lights and detectors are semiconductors. Tapes move continuously through photoelectric readers at a rate of 1000 CPS. A photoelectric reader enters the quadratic formula into the RAM in less than 0.5 s.

Punched cards are also prepared on typewriter-style consoles. A card is punched one column at a time, while the card is motionless. After each column is punched, the card mechanically advances to the next column. When a card is completed, it moves into a hopper, and the next card is punched. This process continues until the entire program has been punched onto cards. Many cards are required, and punching and reading cards takes longer than punching and reading tape. However, changing or adding information to a punch card program is easier. A single error on a punched tape invalidates the entire program, but a single punch card is easy to replace. However, a reel of tape is easier to handle than a stack of cards. Tape is more common for scientific programs and cards for business applications. Reading cards, by either mechanical or photoelectric readers, is slower than reading tape.

Hollerith cards can be coded with graphite pencils instead of holes. Since graphite is a conductor, reading is performed by completing a circuit through the graphite. Similarly, an unmarked space causes an open circuit. Complete circuits represent 1s and open circuits 0s. When Baudot is used, no data are transmitted for the letters and symbols codes. Since the letters and symbol characters are relatively frequent, data transmission is slow. In many applications, the 5-bit Baudot code has been replaced by the 7-bit ASCII, which stands for American Standard Code for Information Interchange and is pronounced "askee." Since ASCII is a 7-bit code, 128-code combinations exist. With so many combinations switching between letters and characters is not required, and thus ASCII is faster and more versatile than Baudot.

ASCII code is arranged in column-row format. Therefore coding, decoding, and storing ASCII in ROM is straightforward. The 3 MSBs of an ASCII character are the column number, and the 4 LSBs bits are the row number. Table 8.3 shows the complete code.

While Baudot and Hollerith codes are limited to uppercase letters, ASCII contains both upper- and lowercase letters. Because of the increase to 128 combinations, additional control signals exist, which makes ASCII even more versatile.

Example 8.3 Neglecting control signals, convert Hello! into ASCII.

Solution H is in column 100 and row 1000.

Thus H = 1001000
Similarly e = 1100101
l = 1101100
l = 1101100
o = 1101111
! = 0100001

TABLE 8.3 Seven-Bit ASCII Code

Row \ Column	000	001	010	011	100	101	110	111
0000	NUL	DLE	SP	0	@	P		p
0001	SOH	DC1	!	1	A	Q	a	q
0010	STX	DC2	"	2	B	R	b	r
0011	ETX	DCB	#	3	C	S	c	s
0100	EOT	DC4	$	4	D	T	d	t
0101	ENQ	NAK	%	5	E	U	e	u
0110	ACK	SYN	&	6	F	V	f	v
0111	BEL	ETB	'	7	G	W	g	w
1000	BS	CAN	(	8	H	X	h	x
1001	HT	EM	)	9	I	Y	i	y
1010	LF	SUB	*	:	J	Z	j	z
1011	VT	ESC	+	;	K		k	
1100	FF	FS	,		L		l	:
1101	CR	GS	−	=	M		m	
1110	SO	RS	.		N		n	
1111	SI	US	/	?	O		o	DEL

Control functions

NUL	Null	DLE	Data link escape
SOH	Start of heading	DC1	Device control 1
STX	Start of text	DC2	Device control 2
ETX	End of text	DC3	Device control 3
EOT	End of transmission	DC4	Device control 4
ENQ	Enquiry	NAK	Negative acknowledge
ACK	Acknowledge	SYN	Synchronous idle
BEL	Bell	ETB	End of transmission block
BS	Backspace	CAN	Cancel
HT	Horizontal tabulation (skip)	EM	End of medium
LF	Line feed	SUB	Substitute
VT	Vertical tabulation (skip)	ESC	Escape
FF	Form feed	FS	File separator
CR	Carriage return	GS	Group separator
SO	Shift out	RS	Record separator
SI	Shift in	US	Unit separator
		DEL	Delete

TABLE 8.4
Dog in ASCII

	Standard	Odd MSB Parity
D	100100	11000100
o	1101111	11101111
g	1100111	01100111

Transmission line noise can introduce errors. Unfortunately a change in a single bit alters the meaning of an entire program. Parity checking, as discussed in Chap. 6, improves the reliability of transmitted data. When parity is used, an additional bit is included with each character. Parity checking can be even or odd.

1. With even parity, if the number of bits, including the parity bit, is an even number, the parity bit is 1. If not, the parity bit is 0.

2. With odd parity, if the number of bits, including the parity bit, is odd, the parity bit is 1. If not, the parity bit is 0.

When parity is used, the parity bit is compared with its ASCII character. Incorrect parity indicates a transmission error. The parity bit can be either the MSB or the LSB, but in either case, including a parity bit converts ASCII from a 7-bit to a byte-size code. Table 8.4 compares the word Dog in ASCII with its odd MSB-parity version.

The *extended binary-coded decimal interchange code* (EBCDIC) is compatible with the zone-digit field format of the Hollerith code. IBM and manufacturers of associated equipment use EBCDIC, but most other computer companies use ASCII. If necessary, conversion between ASCII and EBCDIC can be performed in a ROM.

Computers can display results on either punched tape or cards. In many applications printed text is more convenient. Machines which present output in printed form are called *hard-copy printers. Formed-character* printers use electromechanical hammers to drive a character against an inked ribbon, which impinges on paper. This method is also used by electric typewriters. In fact, several typewriter manufacturers offer typewriters with cables to interface with computers. Typewriters with "golf ball" heads print output data at 10 to 15 CPS. Another impact printhead, shown in Fig. 8.4, has 92 separate ASCII characters, one on each arm. Because of the resemblance to the flower, this printhead is called a *daisy wheel.*

The golfball printer must rotate and tilt to obtain a character, but the daisy wheel, with only one character on each arm, need only rotate. Therefore daisy wheel printers are considerably faster and are available to 50 CPS. Type font of a formed character printer can be changed by changing golf balls or daisy wheels. Formed-character printers produce 80 or more characters per line.

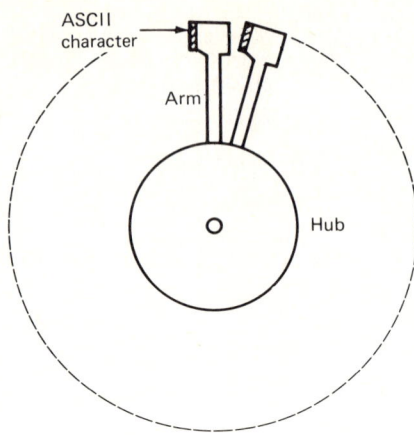

Figure 8.4 Daisy wheel print element.

Dot-matrix printers are different from formed character printers. Characters are generated from a matrix of horizontal and vertical dots. The characters of a 5 × 7 dot matrix consist of a rectangular array with 5 columns and 7 rows. As shown in Fig. 8.5a, this printhead has a single 7-pin column. As the printhead moves to each of the five column positions, the correct sequence of print pins is activated. Uppercase letter A is shown. More detail is available with larger dot matrices such as 7 × 7, 7 × 8, and 7 × 9. As shown in Fig. 8.5b, some matrix printers contain a double column of print pins. Double-column printers fill in most of the space between dots and approach formed-character legibility.

The font of dot-matrix printers can be changed during printing by including the desired formats in RAM or ROM. Thus it is possible to include standard ASCII characters, italic characters, subscripts, etc., on the same line. Special graphics can also be programmed into dot-matrix printers.

Dot-matrix printers are available with either 40 or 80 characters per line. A dot-matrix character is not as readable as a formed character, but the printer is less expensive and faster. Dot-matrix printers go up to 250 CPS.

Figure 8.5 Dot-matrix printer. (*a*) Single 5 × 7 element; (*b*) double 5 × 7 element.

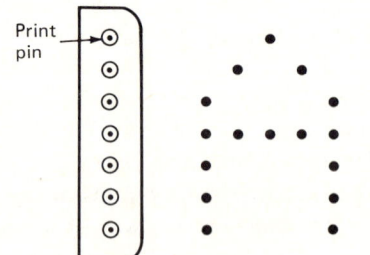
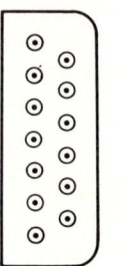

Example 8.4 A printed page contains about 2500 computer characters. How long does it take to print a page using (a) golf ball, (b) daisywheel, and (c) dot-matrix printers?

Solution (a) For a fast golf ball printer at 15 CPS

$$\frac{2500 \text{ characters}}{15 \text{ characters/s}} = 167 \text{ s}$$

(b) For a 50-CPS daisywheel printer

$$\frac{2500 \text{ characters}}{50 \text{ characters/s}} = 50 \text{ s}$$

(c) For a 250-CPS dot-matrix printer

$$\frac{2500 \text{ characters}}{250 \text{ characters/s}} = 10 \text{ s}$$

Printers for large computers are more sophisticated, each column having a separate printwheel. These printers have a self-contained RAM to store data and then print an entire line of up to 160 characters at one time. Printing speeds are 20 complete lines per second, which is equivalent to 3200 CPS. This is almost 13 times faster than dot-matrix printers.

8.3 MAGNETIC INPUT/OUTPUT

Paper is a practical and popular input/output (I/O) material. However, punched tape and cards are inconvenient for continuous updating of input data. Also, paper I/O information is truly permanent; once a hole is punched it cannot be relocated, so that writing new programs or revising old programs requires new tapes or new cards.

Paper is not the only I/O material. In fact, read/write methods developed for rotating drums have been applied to magnetic tape, which is like a "flattened" drum. As previously described, magnetic methods are slower than semiconductor devices; however, magnetic tape is faster than paper tape. Logic 1s and 0s are written by magnetizing small areas of tape in opposite directions. Similarly, reading tape requires sensing the direction of magnetization.

As with paper tape, each magnetic word is a vertical combination of bits. Magnetic tape data are more compact, normal density being 1600 magnetic characters per inch. This is 160 times better than paper tape. For example, the 42-in paper tape program for quadratic roots only requires 0.25 in of magnetic tape. A typical 10-in-diameter reel contains upwards of 2400 ft of magnetic tape.

Example 8.5 How many 300-page books can be stored on a reel of magnetic tape?

Solution As in the preceding example, use 2500 alphanumeric characters per page. There are

$$2500 \frac{\text{characters}}{\text{page}} \times 300 \frac{\text{pages}}{\text{book}} = 7.5 \times 10^5 \frac{\text{characters}}{\text{book}}$$

A 2400-ft magnetic tape can store

$$2400 \frac{\text{ft}}{\text{reel}} \times 12 \frac{\text{in}}{\text{ft}} \times 1600 \frac{\text{characters}}{\text{in}} = 4.61 \times 10^7 \frac{\text{characters}}{\text{reel}}$$

Therefore there are

$$\frac{4.61 \times 10^7 \text{ characters/reel}}{7.5 \times 10^5 \text{ characters/book}} = 61.4 \text{ books}$$

Recording this many books on tape requires continuous recording. A more practical approach is to tape each page as a separate record. Individual pages can then be located. Each record must be separated from the next with a blank gap, as shown in Fig. 8.6a. The time required for a stationary tape to reach operating speed determines gap length. Typically, the gap between each record is 0.75 in, and thus gap length is a significant portion of total tape length. In this case each gap is half as long as each page.

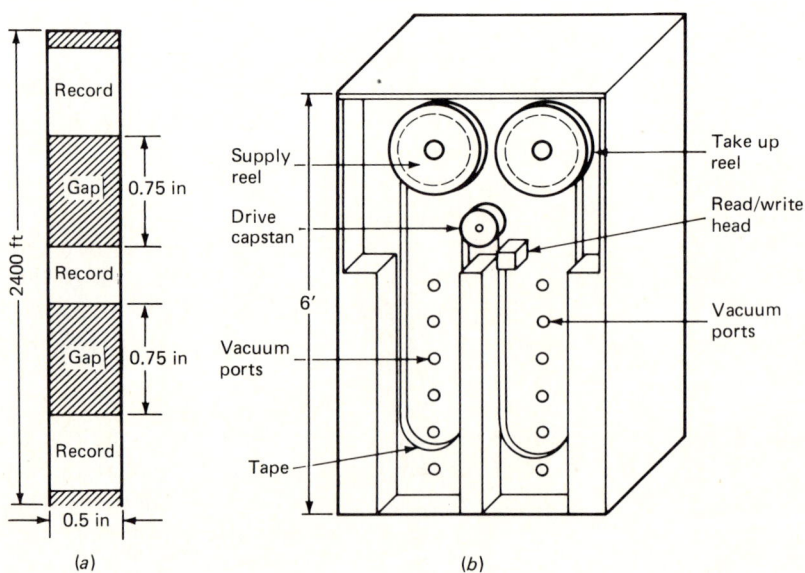

Figure 8.6 Magnetic tape I/O. (*a*) Tape with data; (*b*) tape transport unit.

Machines for reading and writing on magnetic tape are large. The unit shown in Fig. 8.6b is about 6 ft tall. A high-speed motor drives the capstan, near which read/write heads are located. The lower part contains a vacuum system, which mechanically isolates drive and take-up reels from the capstan. Mechanical isolation helps regulate tape speed. Although magnetic tape moves at 125 in/s, which is relatively fast, almost 4 min is required to search a 2400-ft reel for a record near the end. If rewinding is needed to locate a given record, the time is even longer. Such times are acceptable when data are stored externally. By contrast, locating data stored in the internal computer is a matter of microseconds.

Enclosed magnetic cassettes are smaller and less expensive than open reel tapes. The typical digital cassette contains 200 ft of tape. Transport speeds are between 30 and 50 in/s, and storage capacity is about 250K bytes per cassette. Digital cassettes are different from the ordinary audio cassettes used with inexpensive tabletop computers.

Reading and writing data on magnetic disks is another erasable input/output technique. Magnetic disks reduce the time required to locate a given record from minutes to tens of milliseconds. Large computers use a rigid plastic 11-in diameter disk coated with iron oxide. Read/write heads for disks are similar to magnetic tape heads. Magnetic disks look like phonograph records, but their operation is different. Information on a phonograph record is stored in a single mechanical groove which spirals towards the center hole, whereas a magnetic disk stores data on separate circular tracks without any grooves. Figure 8.7 shows the details of a magnetic disk system. Outer data tracks have larger circumferences than inner tracks; thus more data can be stored on outer tracks. However, it is more convenient to store an equal number of bytes on each track. The typical separation between tracks is 0.01 in, and there are about 200 tracks per disk. Track packing density is usually 4000 bits per inch.

Disks are mounted on a common spindle and all disks rotate at the same speed. Ten disks per spindle is typical. Each disk has a separate read/write head.

Figure 8.7 Magnetic disk I/O. (*a*) Single disk; (*b*) disk stack.

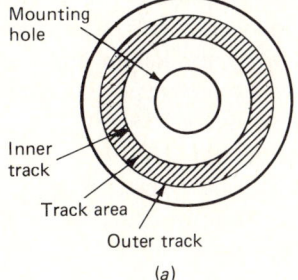
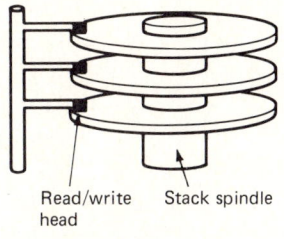

Example 8.6 Determine the storage capacity of a typical magnetic disk stack.

Solution Assume an average data track diameter of 10 in. The circumference is

$$C = \pi \times D = \pi \times 10 = 31.42 \text{ in per track}$$

At 4000 bits per inch, each data track contains

$$31.42 \frac{\text{in}}{\text{track}} \times 4000 \frac{\text{bits}}{\text{in}} = 1.26 \times 10^5 \frac{\text{bits}}{\text{track}}$$

With 200 tracks per disk

$$200 \frac{\text{tracks}}{\text{disk}} \times 1.26 \times 10^5 \frac{\text{bits}}{\text{track}} = 2.52 \times 10^7 \frac{\text{bits}}{\text{disk}}$$

For 10 disks the capacity of an entire stack is

$$2.52 \times 10^7 \frac{\text{bits}}{\text{disk}} \times 10 \text{ disks} = 2.52 \times 10^8 \text{ bits}$$

There are 8 bits per byte. Therefore

$$\frac{2.52 \times 10^8 \text{ bits}}{8 \text{ bits/byte}} = 3.15 \times 10^7 \text{ bytes}$$

is the information capacity of a typical stack.

This figure is in the same range as the 46 million characters calculated for a 2400-ft long magnetic tape. Double-sided disks or 20 disks per spindle are also available. Thus stack storage can exceed tape storage.

The entire disk stack spins at 3000 r/min, which is 50 r/s. This makes the average access time 10 ms if the desired record is on the track being read. If read/write heads must be moved to another track, additional time is required. Access time, including head shifting, is in the 40 to 60 ms range, which is about 2000 times faster than with magnetic tape. Bits on a disk track are read serially, and a SIPO register delivers parallel data to internal computer circuits.

Disk read/write heads do not touch the disk, and no disk wear occurs. The heads are suspended above the disks on cushions of air, and air pressure caused by disk rotation is usually sufficient to keep them "floating" there. A disk drive unit is about the same height and twice as wide as a tape drive.

Smaller, less expensive disk drives exist. The standard flexible, or floppy, disk has a 7.75-in diameter. Each floppy disk is permanently enclosed in an 8-in dustproof envelope. Floppy disks are not stacked. Figure 8.8 shows a floppy disk. When a floppy is inserted into a drive unit, it locks onto the spindle, which turns at 360 r/min. Read/write heads are positioned at the head slot hole in the envelope. At such slow speed the heads must contact the floppy and some wear is inevitable.

Figure 8.8 Floppy disk configuration.

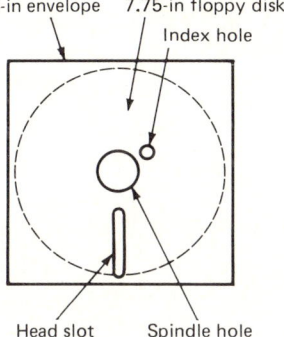

There are 78 tracks on a floppy disk. The average access time on any given track is 84 ms, but when time required to step read/write heads between tracks is included, the access time is about 250 ms, or about 500 times as fast as for magnetic tape.

One floppy disk track is for synchronizing the timing pulses. A light sensor is located below the floppy, and a counter is reset each time light is sensed from an LED above the floppy. Thus a reset occurs at the end of each revolution. All tracks except the timing track store I/O data. The typical storage capacity of a floppy disk is 250K bytes.

A *minifloppy* is a less expensive, smaller disk with a self-storing envelope combination. A 5.0-in-diameter disk is enclosed in a 5.25-in envelope. Read/write heads and disk drive mechanisms are scaled-down versions of the standard-size floppy. A minifloppy disk contains 40 tracks and can store up to 80K bytes. Minifloppies are used with desk-top computers.

Regardless of which magnetic I/O method is used, data storage is permanent but erasable. Care must be taken to prevent accidental erasure of magnetic data. If, for example, tapes or disks come near the magnetic field of a power transformer, all data are erased.

8.4 MAGNETIC RECORDING TECHNIQUES

Practical disk and tape storage techniques are extensions of methods originally developed for magnetic drums. Randomly oriented iron oxide moves past stationary read/write heads.

As shown in Fig. 8.9a, data is written by sending a current pulse through the sense coil. The current must be large enough to cause saturation of the magnetic flux. Current polarity determines flux direction according to the right-hand rule.

All magnetic paths are closed. In this case the closed path consists of an iron core and an air gap. At the air gap, flux interacts with the moving magnetic surface of the disk or tape. When the sense coil current is positive, the induced flux is clockwise. A clockwise flux writes an N-S magnetic spot

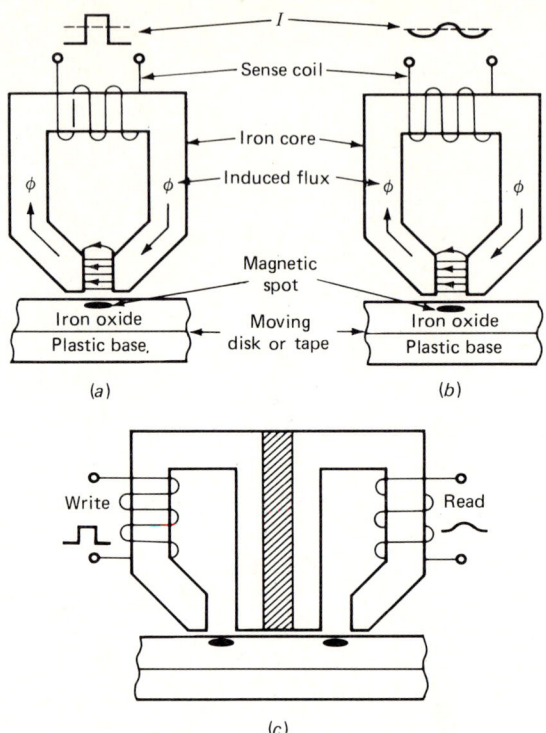

Figure 8.9 Magnetic data operations. (*a*) Writing; (*b*) reading; (*c*) combined head.

and represents binary 1. Similarly, a negative current pulse through the sense coil results in a binary 0.

Reading disk or tape is accomplished by reversing the procedure. As a recorded magnetic spot moves past the air gap, a magnetic flux is induced. The flux path is completed through the iron core. This induced flux generates a current in the sense coil winding. The polarity of the read current is the same as was used to originally write the magnetic spot. In Fig. 8.9*b* the same head is used for reading.

Separate read and write heads can be combined into a single unit, as shown in Fig. 8.9*c*. With two heads, the write head is positioned in front of the read head, an arrangement that permits checking each bit with the read head immediately after writing.

Reading and writing magnetic data require a separate head for each channel. Thus, serial magnetic bits are processed with a single read/write head, while parallel data bits require as many heads as bits per word. For example, the magnetic version of the Hollerith punch code contains 7 bits, of which 6 contain information while the seventh is for parity. Since there are 7 bits, 7 read/write heads are required to read or write.

Figure 8.10*a* shows the current waveforms to write 1011 in magnetic form. The time required to write a bit has two parts. First there is a current pulse,

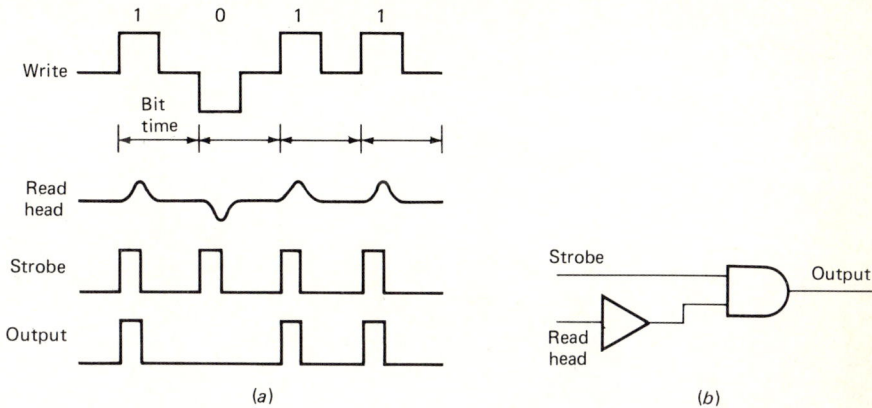

Figure 8.10
RZ recording. (*a*) Waveforms; (*b*) read circuit.

which is positive for writing 1 and negative for 0. Regardless of polarity each pulse is followed by a time of zero current. In other words, the current must return to zero after each bit is written. Writing and then returning to zero is called *RZ recording*.

Reading magnetic data requires a synchronizing pulse to enable the read circuit when data bits are present. Synchronizing pulses are derived from the computer clock and are called *strobe* pulses. Strobe pulses are one of the inputs to the read circuit shown in Fig. 8.10*b*. The other input is the read head voltage, which must be amplified to activate the AND gate. When a logic 1 read signal and strobe pulse coincide, the AND gate output is high. Similarly, when a strobe pulse and logic 0 coincide, the AND gate output is low. While the RZ write voltage is dual-polarity, the output of the read AND gate is single-polarity.

RZ recording contains blank spaces, and blank spaces cannot be erased. Problems arise when new data is recorded over old data; old data remain unless the new data bits are recorded in exactly the same position as the old data bits. Figure 8.11 shows the result of trying to record 1110 over 1011

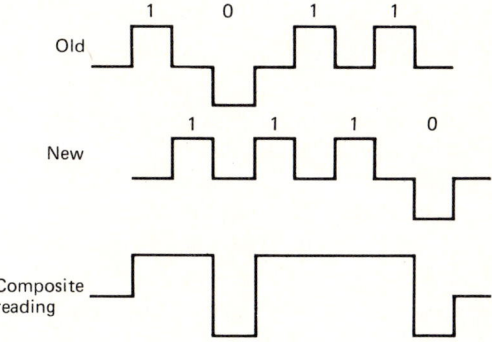

Figure 8.11
Possible RZ rewriting problem.

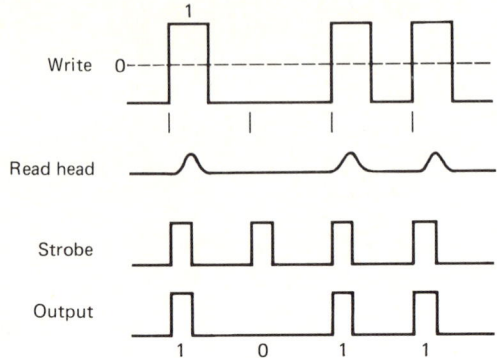

Figure 8.12 RB waveforms.

when the new data are displaced by half of a write cycle. Original data is not erased, but remain intact, while new data is written into the spaces which were previously blank. The composite signal thus destroys both new and old data. Rewriting using RZ is only possible if each new bit coincides perfectly with each old bit.

A variation of RZ called return-to-bias (RB) eliminates blank spaces from the recorded data. With RB the negative voltage of RZ is the reference level. When a logic 1 is recorded, the voltage goes to the same positive level as in RZ, but when no signal or a logic 0 is recorded, the recording is at the negative bias value. Thus RB magnetizes the entire record. New data is recorded over old data regardless of whether or not there is overlap. Because no blank spaces exist, old data is automatically erased when new data is written. Figure 8.12 shows the RB version of 1011.

Maximum and minimum RB write voltages are the same as for RZ, but with RB the zero write level never occurs. RB also requires strobe pulses. However, precise synchronizing is not as critical because all data is magnetized at nonzero levels.

Time and recording space are wasted in returning to a reference level. Nonreturn-to-zero (NRZ) recording eliminates the reference level portion of each bit. With NRZ, a sequence of 1s is recorded as a constant voltage at the level which results in positive saturation. Similarly, a sequence of 0s is recorded at the negative saturation value. Flux change only occurs when data changes from 1 to 0, or vice versa. Since fewer flux changes result, NRZ stores data in a more compact form. Figure 8.13 shows the waveforms and circuit required to write and read 1011 in NRZ form. NRZ circuits are more complicated than RZ or RB circuits. The flip-flop is reset before read begins. When the first positive-going flux change occurs, the flip-flop sets and Q goes high. As soon as a negative-going flux occurs, the inverter resets the flip-flop. Thus, the flip-flop output is the same as the original data. NRZ also requires strobing an AND gate to recover the original data. NRZ is at least twice as fast as RB recording.

Figure 8.13
NRZ recording. (*a*) Waveforms; (*b*) read circuit.

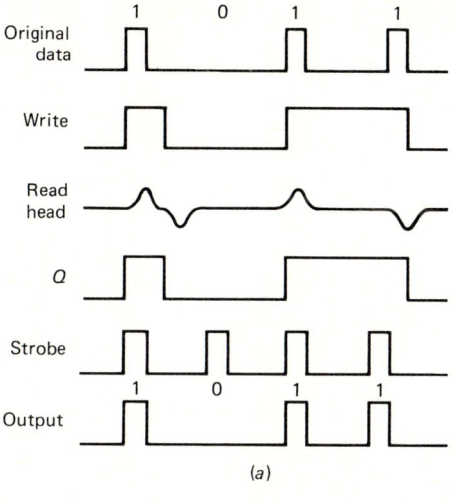

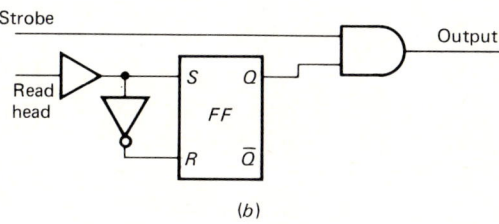

Two variations of the basic NRZ exist, nonreturn-to-zero inverted (NRZI) and split-encoded, the latter also called *split-phase* or *Manchester* recording. With NRZI, recording 1s results in flux changes while recording 0s does not. With phase encoding, 1 is recorded as a positive half-bit pulse width followed by a negative half-bit pulse width. A phase-encoded 0 is a negative half-bit pulse width followed by a positive half-bit pulse width. NRZI and phase encoding result in higher recording speed.

Many small table-top computers use conventional audio cassette recorders for magnetic data. Cassettes are slower and suffer from speed variation. Low cost is the dominant concern, with speed and bit packing density secondary considerations. Reading and writing data on audio cassettes is different from return-to-reference or nonreturn-to-reference techniques. Digital standards for recording data on audio cassettes were adopted at meetings held in the middle 1970s in Kansas City. The resulting format is called *KC Standard*.

With KC Standard an audio signal is converted to a square wave by overdriving an amplifier. Logic 1 is represented by eight cycles of a 2400-Hz signal and logic 0 by four cycles of 1200 Hz.

Example 8.7 In KC Standard, how much time is required to record (*a*) a logic 1 and (*b*) a logic 0?

Solution Since period is the reciprocal of frequency, the period t_1 required to record a logic 1 is

(*a*) $\quad t_1 = 8 \times \dfrac{1}{2400} \simeq 3.33$ ms

(*b*) Similarly, the period t_0 for recording a logic 0 is

$$t_0 = 4 \times \dfrac{1}{1200} \simeq 3.33 \text{ ms}$$

Thus KC Standard is limited to 300 baud. Like teletype, KC Standard uses 11 bits to record alphanumeric characters, the first bit and last 2 bits being transmission line signals. The remaining 8 bits are the ASCII representation of the character. With 11 bits per character, about 37 ms is needed per character.

The standard audio cassette tape transport speed is 1.875 in/s. Thus the packing density of KC Standard is a little less than 15 characters per inch. KC Standard is about 50 percent more efficient than punched paper tape. However, within the requirements of audio cassette recording, KC Standard is acceptable for smaller computers.

8.5 CRT TERMINALS

While permanent records are obtained from printers, punched tape, and cards, paper must be replenished. Besides consuming paper, the machines contain moving parts which limit speed. Magnetic machines also store data permanently but do not consume input or output material. Disks and tapes can be written on and rewritten as often as necessary. However, disk and tape drives also contain moving parts with speed limitations. The cathode-ray tube (CRT) does not consume material and contains no moving parts. In fact, the CRT is the only surviving vacuum tube in a computer which once consisted entirely of vacuum tubes. A CRT terminal displays input data, which are entered from the keyboard or a light pen. Once on the screen, input data can be verified by the operator. A CRT terminal also presents output data.

Vacuum tubes, including the CRT, contain a filament. When the filament is heated, electrons are emitted into the vacuum. Grids shape the electrons into a narrow beam and accelerate the beam towards the face. The inside face of a CRT is a phosphor coating, which emits light when activated by electrons. Typical CRT phosphors emit either white or green light. If the electron beam position remains fixed, only a single spot emits light, but if the

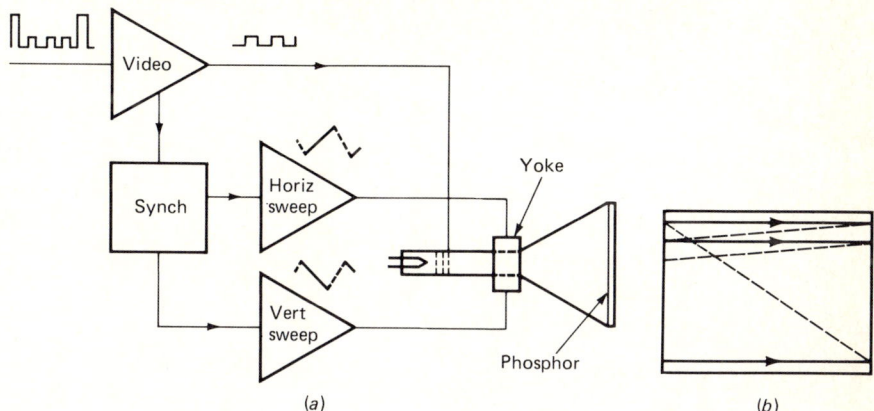

Figure 8.14 CRT terminal. (*a*) Block diagram; (*b*) beam deflection.

beam is deflected, additional spots glow. When the beam is deflected horizontally, a line across the face of the CRT terminal glows. When the beam is deflected either up or down, a vertical line glows. In a CRT, as in a television set, the beam is moved both horizontally and vertically to cover the entire screen. As shown in Fig. 8.14*a*, the beam is deflected by an external yoke around the CRT neck. The yoke contains two mutually perpendicular coils, which are insulated from each other.

An increasing-voltage waveform across the horizontal deflection coil creates a magnetic field, which moves the beam across the screen from left to right. When the beam reaches the right-hand edge, a decreasing horizontal-deflection voltage returns the beam to the left. The retrace beam is cut off, or *blanked*, so retrace is not visible. Retrace lines are shown dashed in Fig. 8.14*b*. During retrace a decreasing voltage across the vertical deflection coil shifts the beam slightly below the previous trace. Horizontal and vertical sweep voltages continue until the entire screen has been covered. The waveforms required to sweep the entire screen are called a *field*. If the field is completely illuminated, a *raster* is visible. The intensity of the electron beam can be varied as the field is swept. This variation in light intensity is the picture and is called the *video*.

Horizontal and vertical synchronizing signals prevent the video from drifting or tearing. For a CRT terminal, horizontal synchronizing pulses for a CRT occur at a 15.6-kHz rate. Television sets have a slightly larger horizontal sweep rate. In either case, the vertical sweep rate is 60 Hz. Thus, regardless of screen size, a field consists of 15,600/60, or 260 horizontal lines.

While the brightness of a television video signal varies, a CRT video signal only has two levels: a 1 causes a bright spot and a 0 causes a dark spot. The appropriate combination of 1s and 0s displays data on a CRT screen.

These data can be stored in the main RAM. However, video data are usually stored in a separate video RAM. A CRT display is organized into rows and columns to form areas each of which can display one alphanumeric

Figure 8.15
Data display. (*a*) CRT locations; (*b*) 5 × 7 pixel data.

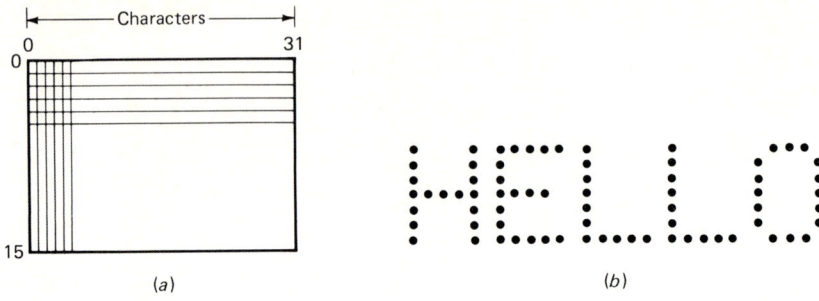

character. A screen with 16 rows and 32 columns can display 512 characters, as shown in Fig. 8.15*a*. Each CRT character location has an address, which corresponds to the location in video RAM. Other popular CRT formats are 25 × 40 and 24 × 80. An 80-column display is especially useful when CRTs work as word processors.

Each video RAM address contains the ASCII code for the character displayed at the same screen address. The full ASCII code contains 128 characters, 96 of which can be displayed. Each of the 96 characters requires a 7-bit code. If parity is used, 8 bits per character are needed. A full ASCII 16 × 32 CRT display with parity needs a 512 × 8 video RAM. By eliminating lowercase letters and some characters, a 64-character ASCII results. This reduced version can use 6-bit characters.

Alphanumeric characters are displayed in matrix form. The dot-matrix format shown in Fig. 8.15*b* is the same as for printers. Each video dot is called a *picture element* or *pixel*. A 5 × 7 character contains 35 pixels and is sufficient for the 64-character ASCII code. When the full ASCII code is used, more pixels per character are needed. Quality display of upper- and lowercase letters requires at least 7 × 12 pixels per character. Since the horizontal sweep is continuous, pixels in the same row of a character "blend" together. However, each vertical sweep lowers the beam slightly. Separation between pixels in the same column is visible.

A video RAM contains the ASCII code for each character on the screen. However, it does not contain the actual character. ASCII characters are stored in a special ROM called a *character generator*. In the previous chapter a trigonometric ROM was described, which contained sines and used the angles as addresses. Character generators use a similar principle. A pixel format is a row-by-row output using the appropriate ASCII code as an address. Figure 8.16 illustrates how 4_{10} is prepared for CRT display.

The ASCII code for 4_{10} is 0110100. This word is in video RAM at some location, and 0110100 in the character generator is addressed. Beginning at the top, each row of the character is clocked out. Character generator output is parallel. Since the electron beam sweeps one line at a time, parallel character generator data must be converted to serial. Conversion is performed by a PISO and the output goes to a video amplifier.

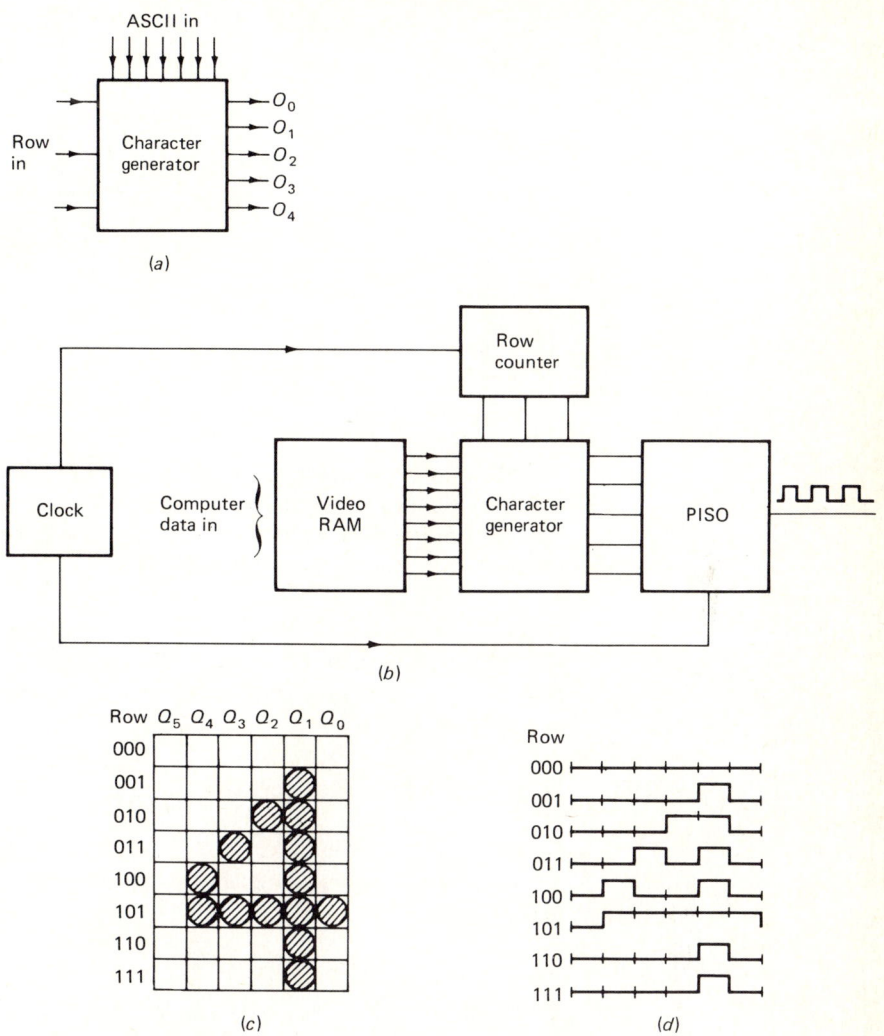

Figure 8.16 Generating video characters. (*a*) Video ROM; (*b*) block diagram; (*c*) 4_{10} pixel format; (*d*) PISO output.

The serial data for 4_{10} are shown in Fig. 8.16*d*. This is only a single character. For 32-column presentation, a complete timing diagram includes the serial data for all 32 characters. After five sweeps have been completed, one row of characters is displayed on the CRT. Each remaining row is sequenced in the same manner to display the entire video RAM contents.

In this discussion a 5×7 character is actually shown as 6×8. A blank row along the top and a blank column along the left provide separation between adjacent characters. In some computers extra blanks are clocked into the PISO for additional separation between characters. Two vertical spaces between characters and three horizontal spaces between rows are common. In this case a 5×7 format becomes 7×10 when displayed.

Figure 8.17 CRT video display system.

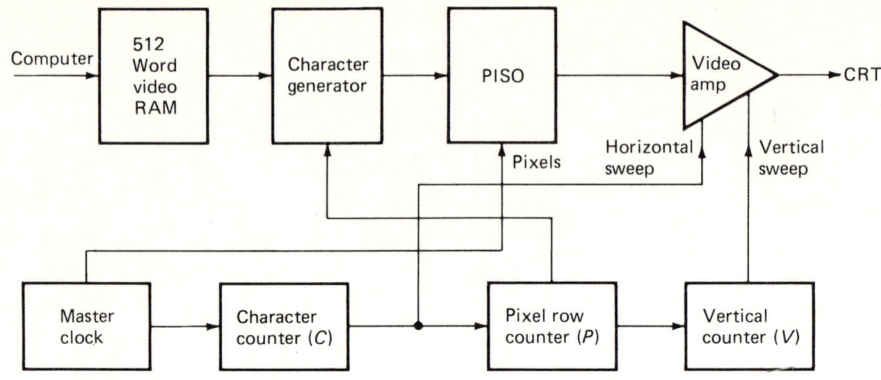

CRT display is controlled by the clock. As shown in Fig. 8.17, the clock operates a chain of frequency dividers. Horizontal sweep rate, vertical sweep rate, and pixel format are related to clock frequency. These three factors are integral multiples of each other.

Example 8.8 A 16 × 32 CRT display uses a 7 × 10 pixel format. Determine: (*a*) the master clock frequency; (*b*) the division ratio of each stage in the frequency counter chain.

Solution (*a*) With 32 characters per row and 6 pixels per character there are

$$7 \times 32 = 224 \text{ pixels per row}$$

Since a horizontal row is swept at a 15.6-kHz rate, the master clock frequency is

$$224 \times 15.6 \text{ kHz} = 3.4944 \text{ MHz}$$

(*b*) At 10 rows of pixels per character, the frequency of the pixel row counter is

$$\frac{15.6 \text{ kHz}}{10} = 1.56 \text{ kHz}$$

The output of the vertical sweep must be 60 Hz. Therefore

$$\frac{1.56 \text{ kHz}}{60} = 26$$

Referring to the symbols used in Fig. 8.17

$$V = \div 26$$
$$P = \div 10$$
$$C = \div 224$$

Frequency dividers automatically reset at the required counts. Since the vertical sweep rate is 60 Hz, displaying an entire video field requires $\frac{1}{60}$ s. This is too fast to be seen. Thus the display continually refreshes the screen with video RAM data. Whenever new data replace old data in the video RAM, the new data are displayed.

The method of generating alphanumeric characters also applies to graphics. The various graphic symbols require a larger character generator. Optimum resolution results from individual control of each pixel and is called *fine* graphics. A 16 × 32 display with 5 × 7 format contains about 18,000 pixels. An 18K character generator is possible but is larger than the RAM of many computers. *Coarse* graphics is a compromise based on controlling groups rather than individual pixels. For example, graphic characters can be generated by controlling columns of pixels. In this case a 2.5K character generator is sufficient.

8.6 D/A CONVERSION

Computer circuits only process binary information, but some peripheral devices require continuous voltages. A continuous voltage is called an *analog* signal. The output of a speech synthesizer is an analog signal to operate a loudspeaker. An *x–y* curve plotter also has an analog output. In this case, the digital input is changed to analog voltage to position a pen motor. Similarly, a programmable power supply accepts digital information to adjust a continuous dc output. A *digital-to-analog* (D/A) converter accepts input voltage pulses and outputs an analog voltage. The analog output is proportional to the numerical equivalent of the input pulses.

Figure 8.18a shows a 4-bit D/A converter. Digital input is obtained from a buffer, and the output is a single voltage representing the input. The analog output voltage V_a tracks binary input and generates a continuous signal.

One circuit for converting digital to analog signals uses *weighted resistors*. A 4-bit weighted resistor converter is shown in Fig. 8.18b. The resistor connected to the MSB has a value of $2^0 \cdot R\,\Omega$. Each lower-order bit is connected to a resistor which is higher by a power of 2: B_2 to $2^1 \cdot R$, B_1 to $2^2 \cdot R$, and B_0 to $2^3 \cdot R$. If R_1 is 1 kΩ then, R_2 is 2 kΩ, R_4 is 4 kΩ, and R_8 is

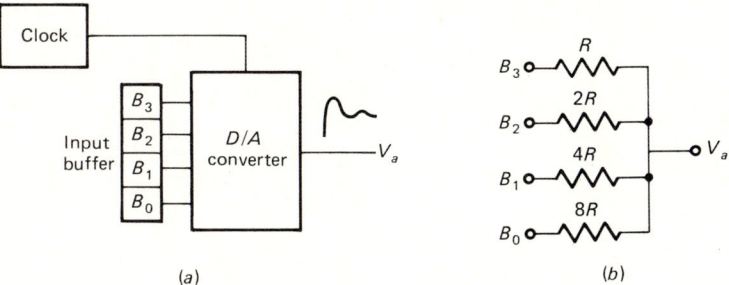

Figure 8.18 Digital-to-analog conversion. (*a*) Block diagram; (*b*) weighted resistor converter.

350 Computer Circuit Concepts

8 kΩ. The analog output is obtained at the junction of the binary weighted resistors.

With a 4-bit D/A converter, the input range is from 0000 to 1111. Maximum analog output results when the input is 1111, and there are proportionate outputs for smaller inputs. If V_a is 15 V when the binary input is 1111, then V_a is 12 V when the input is 1100, etc.

Example 8.9 Determine V_a when the inputs to a D/A converter are (*a*) 1000 and (*b*) 0100. A voltage of V represents a binary 1 and zero voltage a binary 0.

Solution (*a*) An input of 1000 means that the MSB resistor is connected to V and all other resistors are connected to ground. This circuit is shown in Fig. 8.19a. V_a is taken across the parallel combination R_p of 2R, 4R, and 8R. Thus:

$$\frac{1}{R_p} = \frac{1}{2R} + \frac{1}{4R} + \frac{1}{8R}$$

or

$$\frac{8}{R_p} = \frac{4}{R} + \frac{2}{R} + \frac{1}{R} = \frac{7}{R}$$

and

$$R_p = \frac{8}{7} R$$

The equivalent circuit is a two-resistor voltage divider. Thus the analog V_a voltage is

$$V_a = \frac{V \times \frac{8}{7} R}{R + \frac{8}{7} R} = \frac{V \times \frac{8}{7} R}{\frac{15}{7} R} = \frac{8}{15} V$$

Thus $V_a = \frac{8}{15} V$ when the input is 1000_2.

(*b*) Similarly, when the input is 0100, the 2R resistor is connected to V and all other resistors to ground. This condition is shown in Fig. 8.19b.

Thus

$$\frac{1}{R_p} = \frac{1}{R} + \frac{1}{4R} + \frac{1}{8R}$$

$$\frac{8}{R_p} = \frac{8}{R} + \frac{2}{R} + \frac{1}{R} = \frac{11}{R}$$

and

$$R_p = \frac{8}{11} R$$

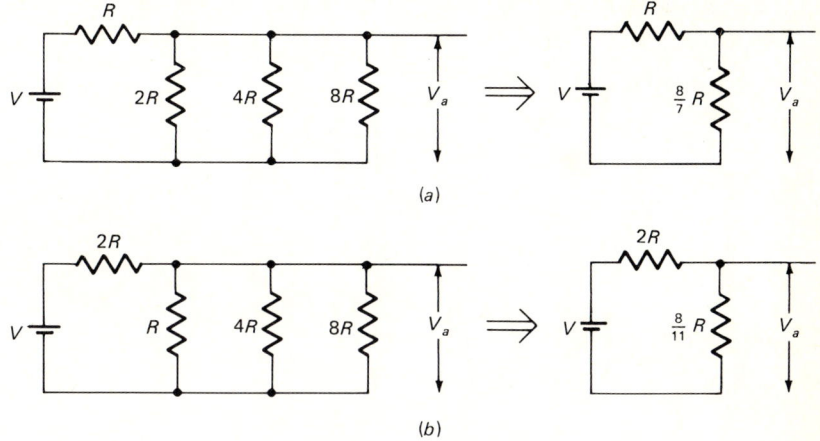

Figure 8.19 Weighted resistor data. (a) 1000_2 input; (b) 0100_2 input.

In this case

$$V_a = \frac{V \times \frac{8}{11}R}{2R + \frac{8}{11}R} = \frac{V \times 8}{30}$$

and $V_a = \dfrac{4}{15}V$ when the input is 0100_2.

Continuing the weighted resistor analysis yields

$$V_a = \frac{2}{15}V \text{ when the input is } 0010_2$$

and $V_a = \dfrac{1}{15}V$ when the input is 0001_2.

If the load resistance is much greater than $8R$, analog outputs can be added. For example, when the input is 1111, the analog output is

$$\frac{8}{15}V + \frac{4}{15}V + \frac{2}{15}V + \frac{1}{15}V = V$$

In particular, if V is selected as 15 V, V_a is the decimal equivalent of the binary input. In this case, V_a is 0 to 10 V if the input is BCD.

For TTL circuits the maximum value of V_a is 5 V. Thus TTL D/A conversion has an output which is one-third of the decimal equivalent. As a practical matter, 5 V is a nominal TTL value; the high output from TTL can be as low as 3.4 V and the low can be as much as 0.4 V. Therefore, if TTL outputs are directly connected to a D/A converter, the accuracy is very poor.

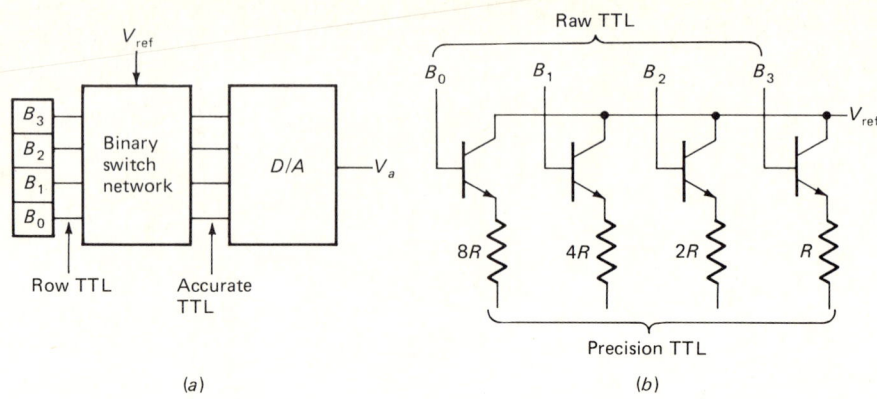

Figure 8.20 Improving D/A accuracy. (*a*) Block diagram; (*b*) binary switch schematic.

As shown in Fig. 8.20, accurate conversion requires connecting TTL inputs to switches operated from a reference voltage. A high TTL output saturates the transistor. Similarly, a low TTL keeps the transistor cut off. The outputs of the binary switches are accurate TTL level or whatever other reference is selected.

The smallest output voltage change, ΔV, of a D/A converter is an important design parameter. For an N-bit D/A converter, ΔV depends on the reference voltage.

$$\Delta V = \frac{V_{\text{ref}}}{2^N - 1} \tag{8.1}$$

Example 8.10 How many bits are required if a D/A converter must detect a 1-V change when the reference voltage is 15 V?

Solution Since
$$\Delta V = \frac{V_{\text{ref}}}{2^N - 1}$$

$$2^N = \frac{V_{\text{ref}}}{\Delta V} + 1$$

In this case
$$2^N = \frac{15}{1} + 1 = 16$$

$$N \log 2 = \log 16$$

$$N = \frac{\log 16}{\log 2} = \frac{1.20412}{0.30103} = 4$$

A 4-bit register is required.

D/A converter accuracy relates measured and theoretical outputs. Precision-voltage switch accuracy is determined by the tolerance of the

Figure 8.21
R-2*R* 4-bit ladder.

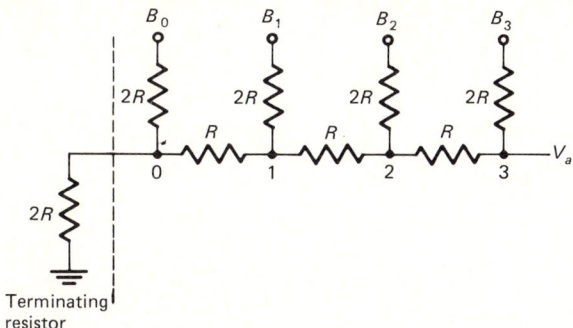

weighted resistors. Resolution is related to but not the same as accuracy. Resolution defines the smallest detectable analog output change in percent or in *parts per million* (ppm)

$$\text{Resolution (\%)} = \frac{100\%}{2^N} \tag{8.2a}$$

$$\text{Resolution (ppm)} = \frac{10^6}{2^N} \tag{8.2b}$$

Accuracy and resolution should be compatible. For example, the resolution of a 4-bit D/A converter is 6.25 percent. In this case 0.1 percent weighted resistors are useless. Conversion would be quite accurate, but the analog output cannot resolve the accuracy. On the other hand, an 8-bit converter has a resolution of less than 0.4 percent, and accurate weighted resistors are required. Single-chip D/A converters exist with 8 or more bits.

An 8-bit weighted resistor converter uses eight different resistors. If B_7 is connected to R, then B_0 is connected to $128R$. As the number of bits increases, weighted resistor conversion becomes increasingly difficult and costly. The *ladder network* is an alternate method of D/A conversion. Since each bit of a ladder requires two resistors, a 4-bit ladder contains eight resistors. But a ladder only contains two different resistor values regardless of the number of bits. One resistor is twice as large as the other, and a ladder is called an *R*-2*R* network. Figure 8.21 shows a 4-bit ladder network. Nodes are numbered to correspond with the bit positions. Ladder resistors connected to register bits are 2*R*, and resistors connected between nodes are *R*. Ladder networks also contain a 2*R* terminating resistor. V_a is obtained from the MSB node. The resistance from any node looking towards the bit register is 2*R*, and the resistance from node 0 looking toward the terminating resistor is also 2*R*.

Conditions at node 1 are shown in Fig. 8.22*a*. The LSB can be high or low. In this discussion B_0 is at ground. However, when B_0 is connected to a binary 1, the output resistance of the bit register is extremely low compared with 2*R*. In either case the B_0 resistor is always effectively in parallel with the terminating resistor. Since both resistors are 2*R*, the parallel equivalent is *R*.

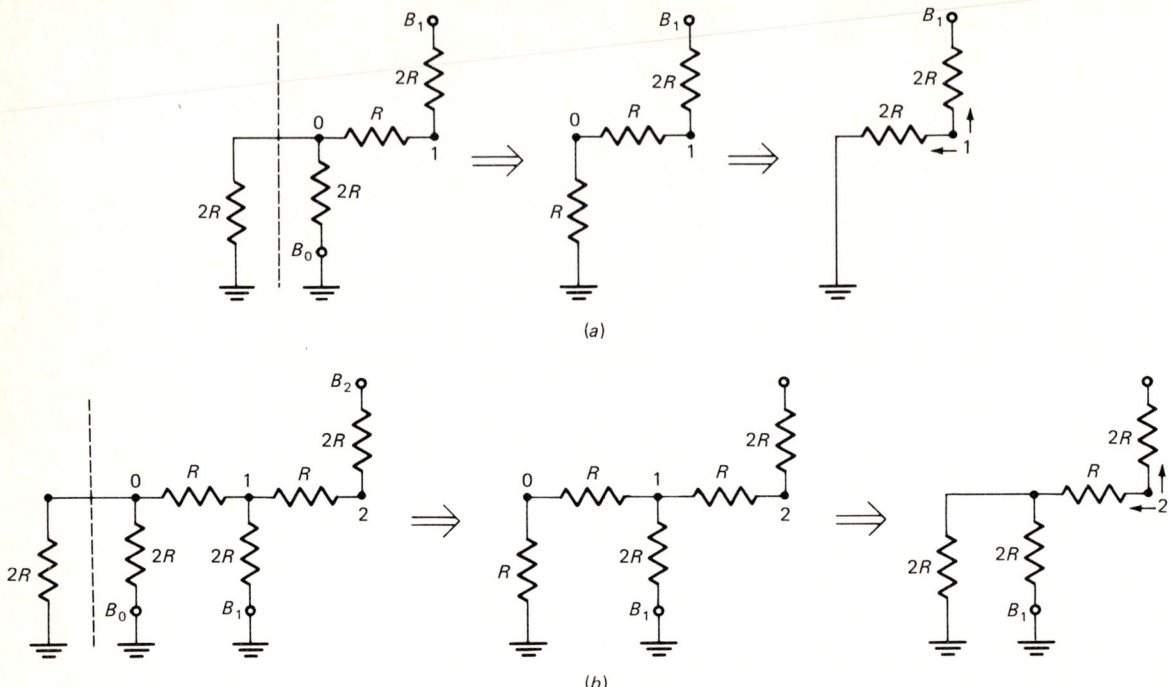

Figure 8.22 Ladder equivalent circuit. (*a*) Node 1; (*b*) node 2.

This equivalent resistor is in series with the R resistor connected between nodes 1 and 0. Therefore, the conditions at node 1 are the same as those at node 0. Resistance from the node towards the bit register and towards the terminating resistor is $2R$.

Conditions at node 2 are shown in Fig. 8.22b. Additional analysis is required because there are more resistors between node 2 and the terminating resistor. However, results at node 2 are identical to those at nodes 1 and 0—there is $2R$ in both directions. The same result is obtained at node 3. Furthermore, the same conditions exist at each node regardless of the number of bits in a ladder.

Example 8.11 Find the analog output when the input to an R-$2R$ ladder is (*a*) 1000 and (*b*) 0100.

Solution (*a*) For a binary input of 1000, the voltage at B_3 is V, and all other bits are at ground. The resistance from any node towards the terminating resistor is $2R$, and Fig. 8.23a shows the equivalent circuit. This circuit is a voltage divider

Figure 8.23
R-2R equivalent circuits. (a) 1000 input; (b) 0100 input; (c) Thevenin V and R for 0100 input; (d) simplified circuit.

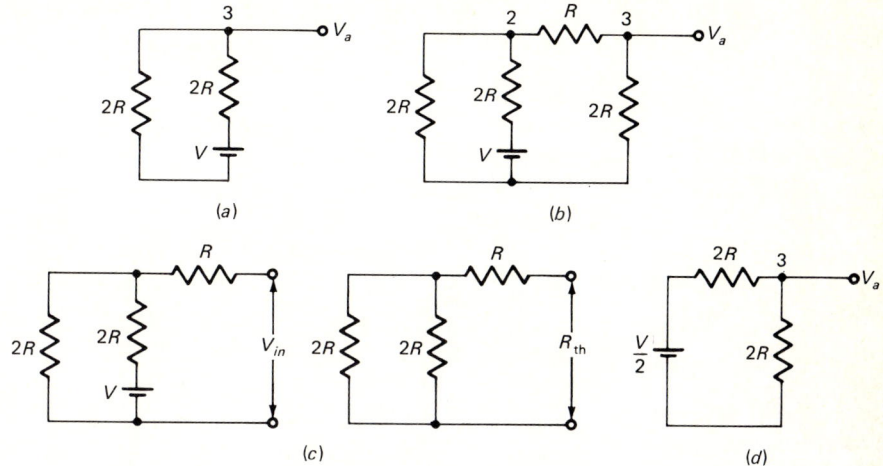

and the output is

$$V_a = \frac{V(2R)}{2R + 2R} = \frac{V}{2}$$

When the binary input is 1000, the analog output is $V/2$.

(b) Figure 8.23b shows the equivalent circuit for 0100, which is more complicated. The Thevenin equivalent voltage and resistance are shown in Fig. 8.23c. Since no current flows through R, the Thevenin voltage is the same as in (a), and the Thevenin resistance is $2R$. Figure 8.23d shows the simplified circuit. This circuit is also a voltage divider

$$V_a = \frac{\frac{V}{2} \times 2R}{2R + 2R} = \frac{V}{4}$$

Thus the binary input is 0100, the analog output is $V/4$.

Continuing this analysis for lower-order bits shows a pattern. The output at each bit is half the output of the bit of next higher order. Table 8.4 summarizes these results for a 4-bit ladder.

TABLE 8.4
4-Bit Ladder Output

Bit	V_a
2^3	$V/2$
2^2	$V/4$
2^1	$V/8$
2^0	$V/16$

As with weighted resistor converters, ladder analog outputs can be added when output load resistance is much greater than R. Similarly, a reference voltage is needed to obtain accurate conversion.

Example 8.12 A 4-bit ladder D/A converter uses a 15-V reference. Find: (*a*) V_a for an input of 1010; (*b*) the reference voltage to obtain corresponding decimal output voltages.

Solution Since voltage outputs decrease by a factor of 2 for each bit

(*a*) $$V_a = V_{ref}\left(\frac{1}{2^1} + \frac{0}{2^2} + \frac{1}{2^3} + \frac{0}{2^4}\right)$$

$$V_a = 15(0.5 + 0.125) = 9.375$$

(*b*) For a decimal output voltage V_a is 10 V when the input is 1010. Therefore

$$10 = V_{ref}\left(\frac{1}{2^1} + \frac{0}{2^2} + \frac{1}{2^3} + \frac{0}{2^4}\right)$$

$$10 = V_{ref} \times 0.625$$

$$V_{ref} = 16 \text{ V}$$

Although output voltages differ, weighted resistor and ladder converters have similar features, as follows:

- Resolution is the same.
- Accuracy is determined by resistor tolerance.
- Connecting V_a to an operational amplifier in the voltage follower mode is required for high output resistance.

8.7 A/D CONVERSION

Transducers convert physical measurements into electrical signals. For example, a thermocouple converts a temperature change into a proportional voltage change. A strain gauge converts a pressure change into a corresponding change in resistance. Similarly, when light intensity changes, a photoresistor such as cadmium sulfide decreases in resistance with increasing light intensity. Other transducers produce changes in inductance and capacitance. If negative feedback is used, the transducer output regulates the measured quantity. Chemical and manufacturing plants use computers to process the transducer data. Since transducer outputs are analog voltages, *analog-to-digital* (A/D) conversion is required.

IC differential input voltage comparators are the basis of A/D conversion. Figure 8.24*a* shows a voltage comparator connected to a dual-polarity power supply. Input voltage V_x is connected to the inverting terminal and V_y is connected to the noninverting terminal.

If V_x is greater than V_y, the output V_0 is negative. Similarly, if V_y is greater than V_x, the output is positive. In particular, small differences between V_x and V_y result in saturated outputs of opposite polarity.

Example 8.13 A comparator with a voltage amplification of 200,000 is connected to ± 15-V supplies. Find the differential input which causes saturation at the output.

Solution Voltage amplification A_v is the ratio of output to input voltage. For a differential input

$$A_v = \frac{V_0}{V_x - V_y}$$

Since power is obtained from 15-V sources, V_0 saturates at either $+15$ or -15 V. Thus the minimum input to cause saturation is

$$V_x - V_y = \frac{V_0}{A_v} = \frac{15\text{ V}}{200{,}000} = 75\ \mu\text{V}$$

Saturation occurs whenever the differential input exceeds 75 μV. Increased gain or lower supply voltage results in saturation at lower inputs.

Very small changes at the input drive the comparator from positive to negative saturation. Since the output of a saturated comparator has two possible values, it is a binary device. If negative saturation is taken as logic 0, then positive saturation is logic 1. A differential amplifier is a 1-bit comparator. The output indicates which input is larger.

Figure 8.24b shows a comparator circuit to indicate polarity of an

Figure 8.24 Basic voltage comparator. (*a*) Circuit; (*b*) polarity detector; (*c*) 1-bit comparators.

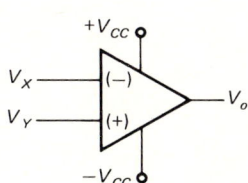

(a)

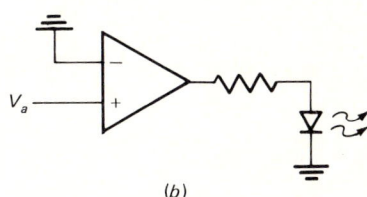

(b)

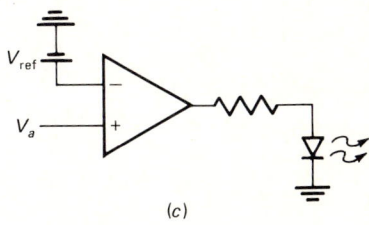

(c)

unknown voltage. The inverting input is connected to ground. If the unknown voltage is positive, the comparator output is positive saturation, and the LED is on. Similarly, if the unknown voltage is negative, the output is negative saturation, and the LED is off. Comparators are not limited to determining the unknown voltage polarity. In Fig. 8.24c the inverting input is connected to a reference voltage. In this case the LED is on when the unknown voltage is greater than V_{ref} and off when the unknown voltage is less than V_{ref}.

Comparators connected to reference voltage are the basis of A/D conversion. In Fig. 8.25a one reference voltage is connected to three comparators. The inverting inputs are connected in series to V_{ref} through an equal-resistor voltage divider. Since all four resistors are equal, the voltage at the lowest comparator is $V_{ref}/4$. Similarly, the voltage is $V_{ref}/2$ at the middle comparator, and $\frac{3}{4}V_{ref}$ at the upper comparator. All three noninverting inputs are connected in parallel to the analog input. Thus V_a compared with the fraction of V_{ref} at each comparator input determines the output. For example, if V_a is less than $V_{ref}/4$, the outputs of all three comparators are negative. In this case the output is 000. Similarly, if V_a is more than $V_{ref}/4$ but less than $V_{ref}/2$, C_0 will be positive while the other comparator outputs remain negative. In this case the output is 001. If V_{ref} is 10 V, a 001 output means V_a is more than 2.5 but less than 5.0 V. The complete A/D conversion table is shown in Fig. 8.25b. Although the output contains only 0s and 1s, it is not in

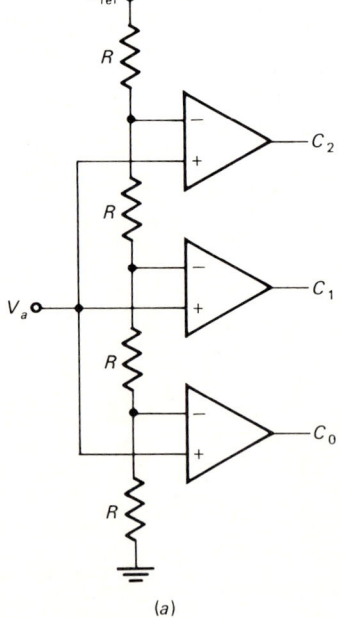

Figure 8.25
3-comparator parallel A/D converter. (a) Basic circuit; (b) voltage conditions; (c) complete circuit.

Input	C_2	C_1	C_0
$V_a < V_{ref}/4$	0	0	0
$V_{ref}/4 < V_a < V_{ref}/2$	0	0	1
$V_{ref}/2 < V_a < \frac{3}{4}V_{ref}$	0	1	1
$\frac{3}{4}V_{ref} < V_a$	1	1	1

(b)

Figure 8.26
Staircase A/D converter.

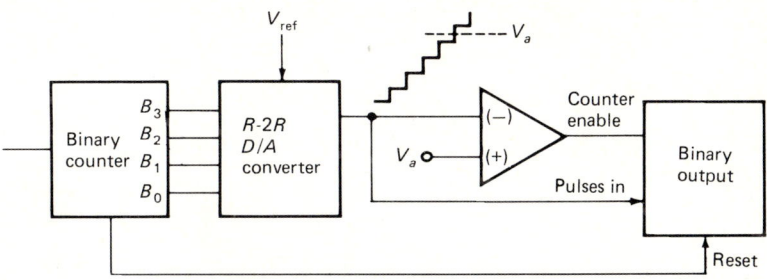

true binary form. However, the output can be converted to true binary with logic gates.

The comparator shown in Fig. 8.25a has four possible output combinations. Thus 2 bits, B and A, are required for true binary. Similarly a four-comparator parallel D/A converter has five different output combinations. Each doubling of comparators results in one additional bit. In principle, the number of comparators can be increased until acceptable resolution results, but as a practical matter, variations in comparator performance restrict parallel input D/A converters to 3 or 4 bits. When such limited resolution is acceptable, parallel conversion is extremely rapid and is sometimes called *flash* conversion.

Greater resolution results when a single comparator is used with a sequentially increasing reference voltage. In this case the analog input is serially compared with incremental increases in the reference voltage. A voltage which increases in equal steps is called a *staircase* waveform, and a staircase A/D converter is shown in Fig. 8.26. As long as the increments are less than V_a, the comparator output is high and the steps continue. Finally V_{ref} exceeds V_a and the comparator goes negative. This disables the counter and no more steps are generated. The highest count is the A/D value. An R-2R D/A generates an accurate staircase voltage. The binary ladder is driven from a clock rather than an analog voltage, and each count increases the R-2R output by one voltage unit. If an 11-bit counter is used with a 1-V reference, the staircase contains 1/2048-V increments. Accurate staircase A/D conversion requires precision resistors in the ladder.

A/D conversion time is determined by the clock frequency and the number of staircase steps. If the analog voltage is low, comparator saturation is fast, but if the analog voltage is near the top staircase value, conversion is slow. Maximum conversion time t_{max} is given by

$$t_{max} = N \times T \tag{8.3}$$

where T is the clock period and N is the number of staircase steps.

Example 8.14 A 4-bit staircase A/D converter is driven by a 1-MHz clock. Determine the maximum conversion time.

Solution The period of a 1-MHz clock is

$$T = \frac{1}{f} = \frac{1}{10^6} = 1 \; \mu s$$

Since $N = 4$ there are $2^N = 16$ steps. With $N = 16$

$$t_{max} = N \times T = 16 \times 1 \; \mu s = 16 \; \mu s$$

While 16 μs is a relatively short time, the resolution of a 4-bit counter is only 6.25 percent. High-quality resolution requires 8- or 10-bit staircase converters, in which case conversion times are in the millisecond range. Staircase resolution is more accurate than flash conversion but is approximately 100 times slower.

Another A/D conversion technique produces true binary output. In this case one comparator input is inverted and connected to a linearly increasing ramp voltage. As shown in Fig. 8.27a, the ramp is generated by RC integration of a constant voltage using an operational amplifier. The operational amplifier has high gain and high input resistance. Thus a constant voltage connected to R results in a constant current. Capacitor C is connected between the inverting input and the output terminal and therefore is charged by the constant current flowing through R. The voltage across a capacitor charged from a constant current is a linearly increasing ramp. Ramp slope is determined by values of R and C as follows:

$$\frac{V_0}{t} = -\frac{V_{in}}{RC} \tag{8.4}$$

Example 8.15 The input to the integrator shown in Fig. 8.27a is -1 V, $R = 5$ kΩ, and $C = 0.2 \; \mu F$. Determine ramp slope.

Solution Equation (8.4) is directly applicable:

$$\frac{V_0}{t} = \frac{-(-1)}{5 \times 10^3 \times 2 \times 10^{-7}} = \frac{1}{1 \times 10^{-3}} = \frac{1 \; V}{ms}$$

As shown in Fig. 8.27b, the ramp is connected to the inverting input of the comparator, and V_a is connected to the noninverting input. The comparator output is connected to an AND gate, an oscillator being the other AND gate input. The AND gate output is connected to a counter. As long as V_a is greater than V_{ramp}, the comparator is high and the AND gate is enabled. While the AND gate is enabled, the oscillator increases the count, but when V_{ramp} equals V_a, the comparator output goes low and disables the AND gate. At this time the counter contains the digital equivalent of the analog input voltage. For example, if the oscillator runs at 1 MHz, the count increases by

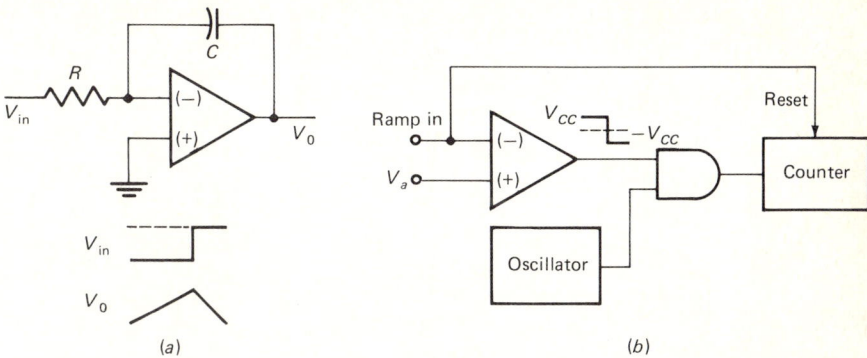

Figure 8.27 Single-ramp A/D converter. (*a*) Integrator; (*b*) block diagram.

1000 counts for each millisecond that V_a is greater than V_{ramp}. If the count stops at 1530 and the ramp slope is 1 V/ms, then the analog voltage is 1.53 V. When the counter output is connected to a 7-segment display, the result is a digital voltmeter.

Single-ramp A/D conversion is useful for resolutions up to 0.01 percent. If greater resolution is required, stability of R and C are critical. A dual-ramp integrator compensates for changes in R and C. Integration using V_a as the integrator input takes place for a fixed time, at the end of which the integrator is discharged through a reference voltage. The time required to discharge C is proportional to V_a. Since the same R and C are used for charge and discharge, component variations are cancelled.

Other A/D conversion techniques exist. The trade-offs are speed and accuracy for cost and complexity. In all cases A/D conversion is based on a voltage comparator.

8.8 DATA LINKS

Depending on circumstances, peripheral devices may be separated from the computer. Input CRTs and output printers may be in an office while the main frame is in a central computer room. In other cases, such as airline reservation and banking systems, I/O devices and the computer are separated by many miles. Cables called *data links* connect remote devices to the main computer. The electrical characteristics of a data link determine information transfer rate.

Data links can be either parallel or serial. Parallel transfer is faster because an entire word is transferred simultaneously rather than bit by bit. However, parallel transmission is only possible for short distances; because of shunt capacitance between wires, it is restricted to a length of 5 ft or less. Most data links are considerably longer, and transmission is limited to the slower serial rates.

Frequently I/O devices and computers are manufactured by different companies, and component compatibility must be considered. A standard,

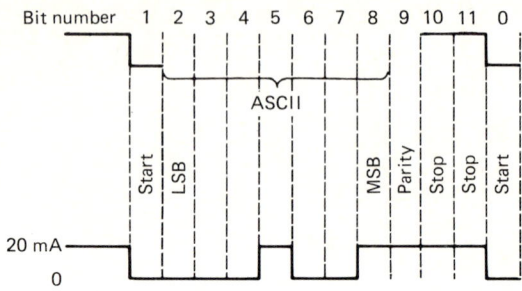

Figure 8.28 Teletype transmission of H.

such as ASCII, is the result of coordination between interested parties. In other instances, a major manufacturer develops a system which becomes a standard because other manufacturers follow along. EBCDIC from IBM is such an example. Regardless of origin, data transmission standards are very detailed. All parameters from format down to pin connections for each wire in the data link must be specified.

Data transmission is actually older than computers. At the beginning, many existing standards were simply adopted by the computer industry. The 20-mA constant-current signal level used for teletype transmission is one example. In this case, a 20-mA pulse is a logic 1 and no current is a logic 0.

When no message is transmitted, a teletype line is high. Every character begins with a low, which is called the *start bit*. Data transmission consists of alphanumeric characters, which are all 11 bits long. The next 7 bits are the ASCII code for the character. A parity, odd in this case, bit is next. The remaining 2 bits are 1s and are called stop bits. Figure 8.28 shows the letter H transmitted in 20-mA teletype format.

The standard bit time of a 20-mA pulse bit is 1/110 s. Since a character contains 11 bits, 0.1 s is required to transmit a single character. In this case, the transmission rate is 10 CPS. The twelfth bit in Fig. 8.28 is 0 and is the start of the next character. After the final character of a message has been transmitted, the transmission line returns to the normal 20-mA condition.

Teletypes are solenoid-operated. Noise spikes generated by the solenoid are transmitted on the data link. These noise spikes are short, and normal teletype transmission is not affected. However, computers interpret spikes as bits, and thus noise must be eliminated. An *optical coupler* eliminates noise and also converts 20-mA pulses to TTL levels. As shown in Fig. 8.29a, there is no electrical connection between the input and output of an optical coupler. The input is an LED and the output is a *phototransistor*, that is, a transistor with a light-activated base. The LED and phototransistor are sealed in a light-tight DIP package.

When a 20-mA pulse occurs, the LED emits light. This light is detected by the base of the phototransistor and saturation occurs. When no current flows, the LED is off and the phototransistor is cut off. The 20-mA pulses are translated to TTL voltage levels. Since the phototransistor is a common

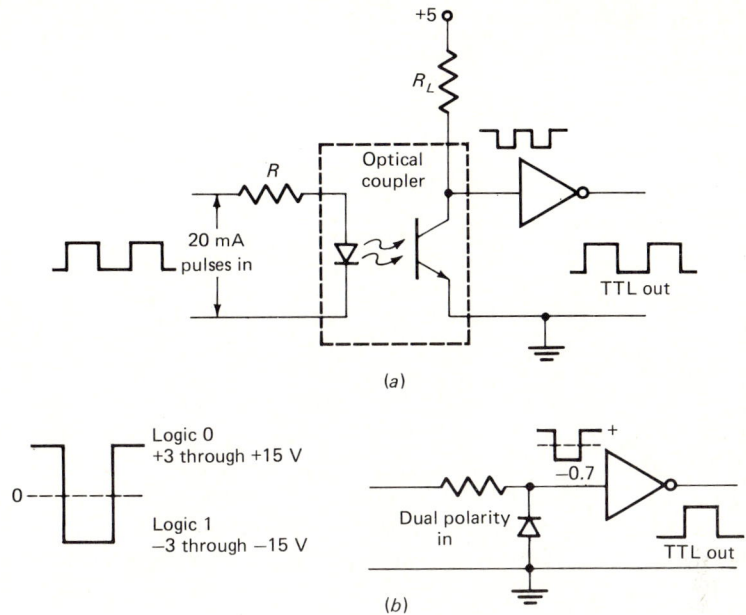

Figure 8.29
Voltage translation circuits. (*a*) 20 mA to TTL; (*b*) RS-232C to TTL.

emitter, input and output are out of phase, highs and lows being reversed. Original logic levels are restored by connecting the phototransistor to an inverter. An optical isolator can also translate TTL levels to 20-mA pulses. This situation arises when computer data are transmitted on teletype data links.

The Electronic Industries Association RS-232C is another data transmission standard, which while newer than 20-mA teletype, was developed before ICs. RS-232C also requires voltage translation to TTL. Figure 8.29*b* shows the RS-232C logic levels. A logic 0 is more than $+3$ and less than $+15$ V and, similarly, a logic 1 is between -3 and -15 V. Since RS-232C signals are electronic rather than electromechanical, noise is low. The noise reduction provided by optical isolation is not needed and voltage translation is easier. The diode clips negative voltages at -0.7 V and is reverse-biased for positive pulses. Thus, the voltage across the diode is low for logic 1 and high for logic 0. Connecting an inverter to the diode converts these levels to TTL; -0.7 V translates to $+5$ V, and positive voltages translate to 0 V. RS-232C data links are about 20 times faster than teletype links.

These older standards are still in use, but TTL-compatible standards are preferred. IEEE 488 is particularly useful for transmission between computers and scientific instruments such as DVMs and signal generators. RS-422 is an improved version of RS-232C, in which line impedance has been lowered from 600 to 50 Ω, and speed of data transmission is in the megabaud

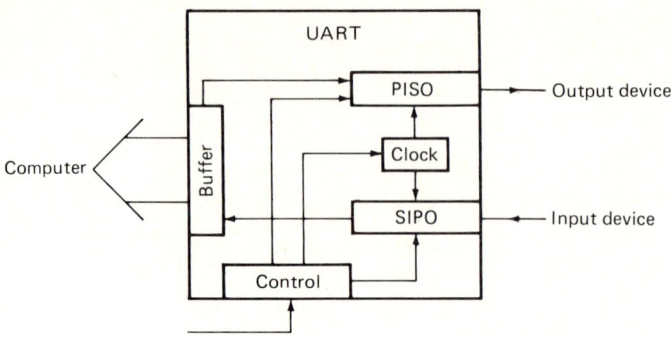

Figure 8.30
Simplified UART.

range. S-100 is useful for byte-sized computers with RAMs up to 64K, and IEEE S-100 is compatible with 16-bit words and larger RAMs.

Some I/O units either receive or transmit data, while others perform both functions. A printer is a undirectional output device; information received from the computer can be printed, but the printer cannot send data to the computer. Undirectional devices are called *simplex*. Bidirectional devices such as CRT terminals are called *duplex*. A full-duplex device can transmit and receive data simultaneously. A *half-duplex* device can transmit and receive data but not at the same time.

Simplex or duplex data transmitted with individual start and stop bits for each character are called *asynchronous*. The functions needed to transmit and receive asynchronous data are combined in a single LSI chip called a universal asynchronous receiver-transmitter (UART). Figure 8.30 shows a UART block diagram. The UART control section adds start and stop bits to each character. Different codes use various numbers of start and stop bits, and the control section is programmed for the correct number. The control section also determines whether the UART is a receiver or a transmitter. Received data enter the SIPO register and are clocked into the parallel buffer. Similarly, when the UART is a transmitter, parallel computer data are stored in the buffer and clocked into the PISO register. UART clock rate is adjustable. Typically the UART clock runs at a multiple of the main computer clock. A higher clock frequency permits sampling I/O data at specific times, which minimizes transmission line noise. UARTs can be programmed for either odd or even parity checks and send data in either direction at upwards of 20 kilobaud.

When faster data transfer is required, data are sent in *synchronous* form. While an asynchronous line is always open, synchronous data transmission requires "handshake" signals to open the data link. Individual start and stop bits for each character are not used. Therefore synchronous transmission is faster by at least a factor of 11/8, or about 40 percent, than asynchronous transmission. The additional functions and clock synchronizing circuits for synchronous transmission are combined in an even more sophisticated IC

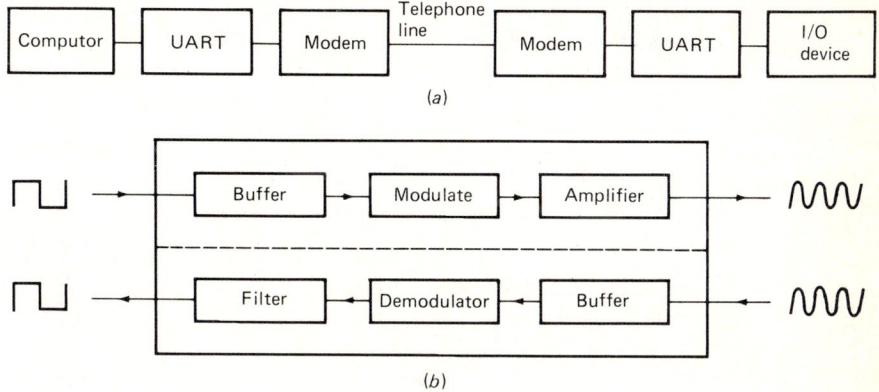

Figure 8.31
Digital data transmission.
(*a*) Complete system; (*b*) modern block diagram.

called a universal synchronous/asynchronous receiver-transmitter (USART), which can be programmed to perform in either synchronous or asynchronous mode.

At increased distances, telephone lines are the data link between computer and peripheral devices. Sending digital data by telephone requires conversion, as the telephone was originally designed for voice communication and has a frequency range of 300 to 3300 Hz. While a 3 kHz bandwidth is adequate for transmitting human speech, high-speed data transmission is not possible. At the transmitting end, computer data must be serialized and then converted to the telephone frequency range. The reverse process is performed at the receiving end—information in the telephone frequency range must be converted back to 1s and 0s. With a duplex device both functions are performed in a modulator/demodulator, or *modem*. The modulator section converts from digital to audio, and the demodulator reverses the procedure.

There are two modem types. The *acoustic* modem uses the telephone transmitter microphone as a pickup to convert each bit into an audio tone. At the receiving end, the telephone speaker detects the audio tones. Acoustic modems operate at either 110 or 300 baud. Logic 0s are converted to 1070 Hz and logic 1s to 1270 Hz. When full duplex is used, the other unit uses 2025 Hz for logic 0 and 2255 Hz for logic 1. Thus simultaneous transmission along a single line is possible. Figure 8.31*a* shows a telephone transmission system.

A *direct wire* modem bypasses the handset, since digital data are converted into audio tones in the telephone itself. Direct wire modems are less susceptible to noise and faster than acoustic modems, sending data at up to 1800 baud. At the faster rates, transmission must be half-duplex. Direct wire modems use 2200 Hz for logic 0 and 1200 Hz for logic 1.

SUMMARY

1. Punching holes in paper is a method used to perform input as well as output operations. Punched cards are Hollerith-coded and punched tape is binary-coded.

2. Because of the greater number of code combinations, ASCII is preferred to Baudot and Hollerith. Hard-copy printers use formed character or dot-matrix.

3. Magnetic tape and disks combine permanent memory with erasability. Packing density is greater and access time is faster for magnetic I/O devices than for paper.

4. Data are stored on disk and tape in either return-to-reference or nonreturn-to-reference form. Audio cassette recording of digital information uses the less sophisticated KC Standard.

5. CRT terminals display input and output information. Alphanumeric character display is standard, and graphics are possible.

6. D/A converters use weighted resistor or *R-2R* ladders. In both cases, bit count determines resolution.

7. A/D converters change analog voltages to pulse form. All A/D conversion methods use a voltage comparator circuit.

8. Data links connect remote peripheral devices to the computer. Characteristics of the data link determine data transfer rate.

PROBLEMS

8.1 What is the storage cpacity of a Hollerith card which is organized into nibbles?

Answer: 240 nibbles.

8.2 Code 6:30 A.M. into Hollerith.

8.3 How many symbols exist in the Baudot code?

Answer: 58.

8.4 Excluding line feed and stop signals, code 6:30 A.M. into Baudot.

8.5 A punched tape program consists of 2000 characters. Find: (*a*) tape length; (*b*) pin sense read time; (*c*) photoelectric read time.

Answer: (*a*) 16.7 ft; (*b*) 40 s; (*c*) 2 s.

8.6 Show the word *List* in ASCII.

8.7 Show the word *List* in ASCII using MSB odd-bit parity.

Answer: 0100 1100, 1110 1001, 0111 0011, 1111 0100.

8.8 The entire contents of a 2K × 8 ROM are printed by using a dot-matrix printer. Find the time required for printing.

8.9 If each page of a book is separated by a gap, how many 300-page books can be stored on a 2400-ft long tape?

Answer: 41.5 books.

8.10 A 1200-ft tape is loaded onto a 125-in/s tape transport unit. The last record is on the last inch of tape. How much time is needed to locate the last record?

8.11 A 200-ft digital cassette contains 250 kilobytes. Find the bit packing density.

Answer: 833 bits per inch.

8.12 Estimate the bit packing density of: (*a*) a floppy disk; (*b*) a minifloppy.

8.13 Determine the (*a*) baud rate and (*b*) character rate of KC Standard.

Answer: (*a*) 300 baud; (*b*) 27.3 CPS.

8.14 Show the writing waveform for magnetically recording 1010 in (*a*) RZ and (*b*) NRZ.

8.15 How much time is needed to read a 6K-character program in KC Standard?

Answer: 3.7 min.

8.16 How many characters can be written onto a 30-min audio cassette tape?

8.17 How many addresses are required for (*a*) 25×40 and (*b*) 24×80 video RAMs?

Answer: (*a*) 1000; (*b*) 1920.

8.18 Determine the size of the video RAMs for full ASCII display of (*a*) 25×40 and (*b*) 24×80 displays.

8.19 Find the number of pixels in: (*a*) the reduced ASCII 16×32 format; and (*b*) the full ASCII 24×80 format.

Answer: (*a*) 17,920; (*b*) 161,280.

8.20 A 24×80 CRT display uses a 7×13 pixel format to display the full ASCII code. Find the: (*a*) master clock frequency; and (*b*) characteristics of the timing chain.

8.21 Express the outputs of a 4-bit weighted resistor D/A converter in terms of V_{ref} when the inputs are: (*a*) 0010; and (*b*) 0001.

Answer: (*a*) $\frac{2}{15} V_{ref}$; (*b*) $\frac{1}{15} V_{ref}$.

8.22 A weighted resistor D/A converter operates from a 25.5 V reference. How many bits are required to detect a voltage change of 0.1 V at the output?

8.23 Find the resolution in (*a*) percent and (*b*) parts per million of an 8-bit *R-2R* D/A converter.

Answer: (*a*) 0.39 percent; (*b*) 3906 ppm.

8.24 An 8-bit *R-2R* D/A converter operates from a 16-V reference. Find the output when the input is: (*a*) 1010 0000; and (*b*) 0000 1010.

8.25 A comparator has a gain of 100,000. Find the differential input which causes saturation when (*a*) a 10-V and (*b*) a 15-V power supply is used.

Answer: (*a*) 100 μV; (*b*) 150 μV.

8.26 A three-comparator flash converter operates from a 4-V reference. Prepare a table showing input range versus binary output.

8.27 Find the maximum conversion time for a 10-bit staircase A/D converter when the clock rate is: (*a*) 500 kHz; and (*b*) 1 MHz.

Answer: (*a*) 2.048 ms; (*b*) 1.024 ms.

8.28 A single-ramp A/D converter uses an integrator with a 10 kΩ resistor and a 0.1 μF capacitor. Find the rate at which the ramp increases when the input voltage is: (*a*) -1 V; and (*b*) -5 V.

8.29 A teletype system transmits 10 CPS. Determine the baud rate.

Answer: 110 baud.

8.30 Show the teletype transmission of the letter n.

8.31 If asynchronous data are transmitted at 20 kilobauds, determine the corresponding synchronous rate.

Answer: 27.5 kilobauds.

8.32 Synchronous data are transmitted by modem at (*a*) 110, (*b*) 300, and (*c*) 1600 baud. Find the times required to transmit a 5-character word.

9
A COMPLETE COMPUTER

9.0 INTRODUCTION

This book presents a sequential development. Boolean algebra is discussed in the first chapter because it is the basis for understanding digital computers. The second chapter extends Boolean algebra to describe binary arithmetic. In the third chapter, gates are introduced to implement Boolean and binary arithmetic operations. This process continues, each chapter building on the previous ones. With all concepts and circuits in place, this last chapter combines the previous material to discuss a complete computer.

A complete computer contains the systems which have been developed. RAM, ROM, ALU (Arithmetic Logic Unit), input, and output are vital, but operation of these systems must be coordinated. Different circuits are active during each step in computer operation. Data move from input to RAM for storage, from RAM to ALU for processing, and finally to an output device for display. At each step some circuits must be turned on while others are turned off. Outside intervention is not required. Each step is performed automatically under the direction of the control circuits, which generate the signals for proper sequencing. If the computer can be said to have a "brain", it is the control circuits.

The control circuits permit high-speed error-free operation. While computers are fast, they are also stupid. Each operation must be broken down into small segments, which have to be performed one step at a time. Every step has the same form—there is an instruction and there are data. First an instruction is accessed and decoded, and then data is processed according to the decoded instruction. A sequence of instructions and data constitutes a computer program. Processing a computer program consists of repeating similar steps in the correct order.

This chapter completes a long chain, in which the first link is Boolean algebra and the final link is automatic program processing.

9.1 COMPUTER ARCHITECTURE

Computers obtain data from a variety of input devices. A relatively simple numeric keypad is the input to a pocket calculator. More sophisticated computers use alphanumeric keyboards, as well as tape drives, modems, and A/D converters. Once data have been entered, processing occurs in the ALU. This is where arithmetic, sorting, alphabetizing, etc., are performed. Two memory sections, RAM and ROM, work in conjunction with the ALU. User instructions, data, and results are stored in the RAM. RAMs are volatile, and information can be altered or removed. ROMs store permanent information such as operating instructions and mathematical tables. After processing, results are presented on output devices. Pocket calculators use LCDs or LED displays. Larger computers have CRTs, tape drives, modems, and D/A converters. These systems have been discussed, and all computers, regardless of size, perform the entire sequence of operations from input to output under the direction of a *control unit*. Data transfer, processing, and sequencing are regulated by the control unit.

The systems which have been described are combined in an orderly manner to arrive at a complete computer. One method of connecting these systems is shown in Fig. 9.1. The blocks represent functions and are not related to chip count. In a pocket calculator, all operations are performed in a single IC. Larger computers have several input and output devices, several RAMs, and several ROMs. The largest computers contain several ALUs, which process independent programs at the same time.

Figure 9.1 shows separate data lines between functional blocks. Separate lines permit simultaneous communication. New data can move from input to memory at the same time as ALU results move to the output. Data stored in

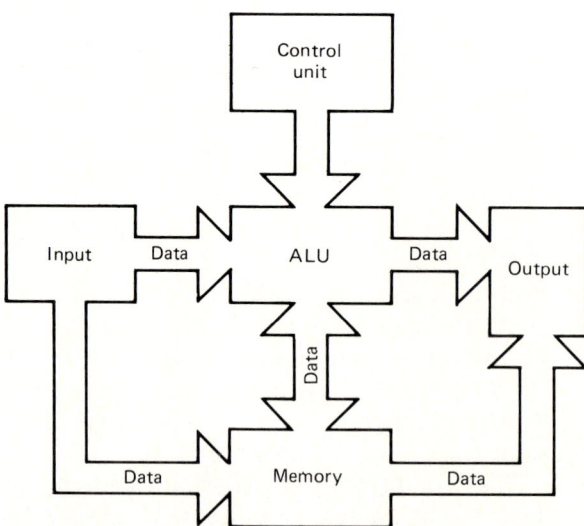

Figure 9.1 Multiple data bus computer.

memory can be output while new data is input. Simultaneous data transfer is efficient and was used in vacuum tube computers. The disadvantages of independent data lines are the physical size and complexity of cables needed to interconnect computer circuits. Modern computers use a simpler connection method, which reduces size and complexity.

The single-data-bus system is shown in Fig. 9.2. As described, a *bus* is a set of parallel lines on a printed circuit board. The number of lines in a bus determines computer word size. For example, a nibble-organized computer has a four-line data bus. While a single data bus simplifies connections, it has a major disadvantage in that only two blocks can communicate at any given time. When an input device transmits information to RAM, no other data can be transferred. Similarly, when the ALU transfers data to an output device, no new information can be input, etc. The penalty for the simplicity of a single data bus is loss of simultaneous data transmission.

Blocks involved in a data transfer must be selected and turned on. At the same time all other blocks must be disconnected from the data bus. Three-state gates enable and disable blocks from the data bus. Enabling and disabling are regulated by the control unit, which does not carry data and is not connected to the data bus, but must be connected to the ALU to regulate data processing. These two blocks are closely related, and IC manufacturers combine control and arithmetic circuits into a single LSI chip. The *central processing unit* (CPU) contains the control unit and ALU. A CPU chip is called a *microprocessor*, and a microprocessor-based computer is called a *microcomputer*.

There are a variety of microprocessors, and all perform similar operations. However, design details differ from one CPU to another. Each manufacturer furnishes compatible support chips; selecting a specific CPU obligates a user to the specific support chips designed for that CPU. For example, the ROM contains internal instructions which the CPU must be able to interpret. Each manufacturer provides detailed information about compatible chip families in catalogs. Byte, 16-bit, and 32-bit microprocessors are available.

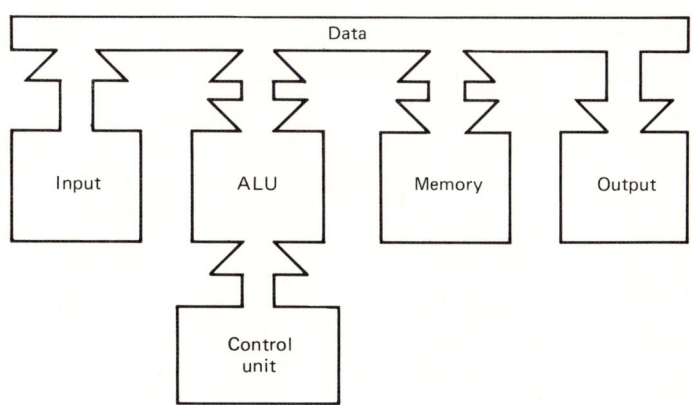

Figure 9.2 Single data bus computer.

Of necessity a CPU chip has many leads. Clearly, power supply leads are required. CPUs for desk-top computers are MOS and originally required several operating voltages. Newer CPUs operate from a single voltage source. Thus 2 leads, +5 V and ground, are required for CPU power. A CPU requires clock pulses, and these pulses are derived from a crystal-controlled oscillator, which is external to the CPU and is controlled by a separate support chip. Two leads are needed between oscillator and CPU.

Address lead count is related to memory capacity. A CPU with 8 address lines can access 2^8, or 256, addresses. The typical CPU contains 16 address lines, and 2^{16}, or 64K, addresses are possible. These 16 address lines are the *address bus*.

Control signals are required within the CPU and also to regulate the other chips. Selecting particular devices while disabling others, selecting write or read, and routing data are control functions requiring external leads. A control bus has at least 12 leads. Thus a byte-size computer with 64K RAM requires the following leads:

16 address bus

12 control bus

8 data bus

2 oscillator

2 power supply
_____

for a total of 40 leads

In principle a 16-bit computer requires 8 additional leads for a total of 48 leads. Similarly, a 32-bit computer should have 64 leads. However, practical manufacturing, testing, and handling problems usually limit DIPs to 40 pins. Thus 16- and 32-bit computers usually transfer data in byte-size pieces. Segmenting 16- or 32-bit data words slows information transfer between CPU and other chips but is generally preferable to using chips with more than 20 pins on each side.

Figure 9.3 shows the structure, or *achitecture*, of a CPU-based computer. Input and output are combined to emphasize the single data bus design. A common data bus allows input and output, but not at the same time. The I/0 devices are connected to an *interface adapter* (IA), which is an LSI chip for controlling input and output data flow. IA control signals originate in the CPU and determine direction of data flow. Previously described UARTs and USARTs are serial IAs. Typical IAs have 20 to 24 leads.

There are two techniques for connecting peripheral devices. Sampling input devices in fixed sequence is called *polling*. For example, a process control computer might monitor pressure, temperature, and flow rate. In this case, three A/D converters are polled by a single IA. Since inputs are sequenced, polling, or *multiplexing*, uses a significant amount of computer

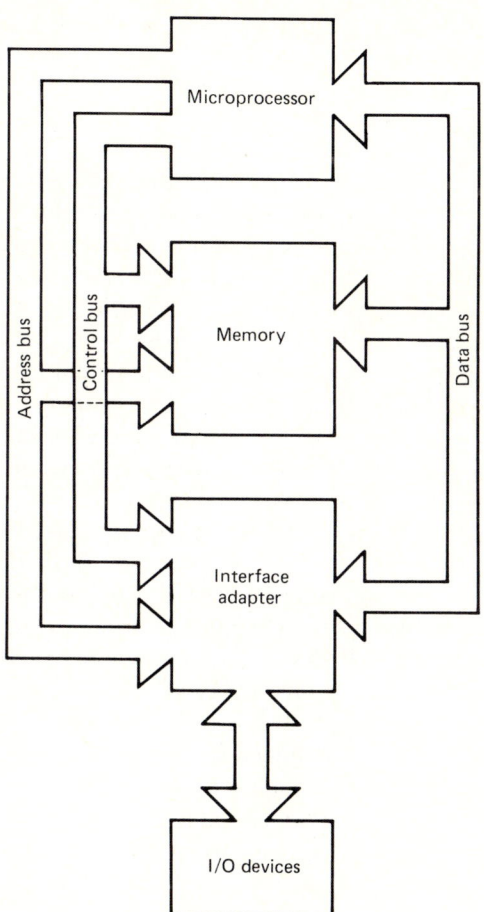

Figure 9.3 Microcomputer architecture.

operating time. However, a process control computer does not "waste" time because there is nothing else to do except output control signals in response to input variations.

In contrast, general-purpose computers handle large quantities of data. Polling is too slow because peripheral data transfer takes hundreds of milliseconds while internal operations only require microseconds. An *interrupt* is the alternative to polling. During an interrupt, internal operation is suspended and an entire block of data is either input or output. When peripheral data transfer is completed, the interrupt is disabled and high-speed internal operation resumes. Interrupts are used with tape, keyboard, printer, and CRT data transfers.

Figure 9.3 shows control and address buses originating in the CPU. As far as the CPU is concerned, RAM and ROM are other peripheral devices. Each device has an address. After the control unit selects a device, the address bus decodes that particular location; then information is transferred along the data bus.

9.2 CPU OPERATIONS

Actually the term "central processing unit" is a carry-over from vacuum-tube days. At that time, all arithmetic and control circuits were housed in a centrally located cabinet. Cables between CPU and other units ran under the floor. The cables were independent, using the private telephone line system described in Chap. 3. Independent cables allow simultaneous data transfer. IC computers use the single-bus party line, or time sharing, to connect data lines. As described, time sharing reduces complexity, but only two units can communicate at any given time. Computers with a common data line have *single-bus architecture*. The word "single" refers only to data transmission; address and control buses are also required for single bus architecture.

Processing data requires communication between CPU and RAM. Information is stored in RAM but processed in the CPU, and therefore instructions and data must be transferred from RAM to the CPU. After being processed, data move in the reverse direction from CPU to RAM along the bidirectional data bus. During processing, control signals generated in the CPU enable the RAM and select the read mode. Similarly, storing results generated in the CPU requires enabling the RAM in the write mode. During read, data enter the CPU through the *data register*, and during write, data leave the CPU through the same data register. The data register is thus a buffer between data bus and CPU.

Figure 9.4 shows the decoding of RAM address $A7_H$. If the RAM is in the read mode, then 02_H is transferred along the data bus into the data register.

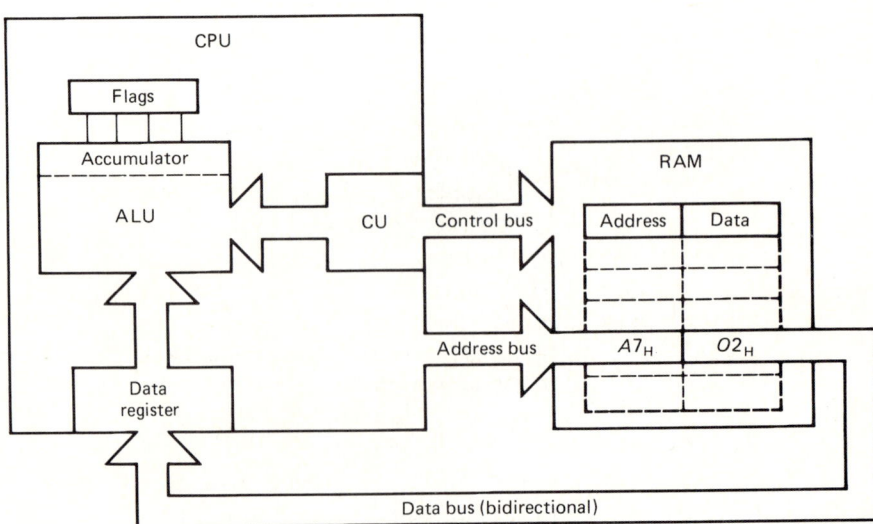

Figure 9.4 CPU/RAM interaction.

Similarly, if the RAM is in the write mode, then 02_H is transferred from the data register along the data bus to RAM address $A7_H$. This information, 02_H, may be data or an instruction.

Figure 9.4 also shows *flags*, or *condition code registers*, connected to the accumulator. Each condition register is a flip-flop which monitors a different accumulator state. One flip-flop is connected to the accumulator MSB. When this flip-flop is set, the signed number is negative; conversely, if it is reset, the signed number is positive. Another flag tests the C_0 bit for overflow. If the C_0 flag is set, operations are discontinued.

The remaining flags test other accumulator conditions. Each flag status has a different effect depending on the operation. For example, the MSB sign flag is important in arithmetic operations. However, when the accumulator word is an alphabetic or special character, the MSB does not indicate sign.

A more detailed picture of CPU architecture is shown in Fig. 9.5. Two

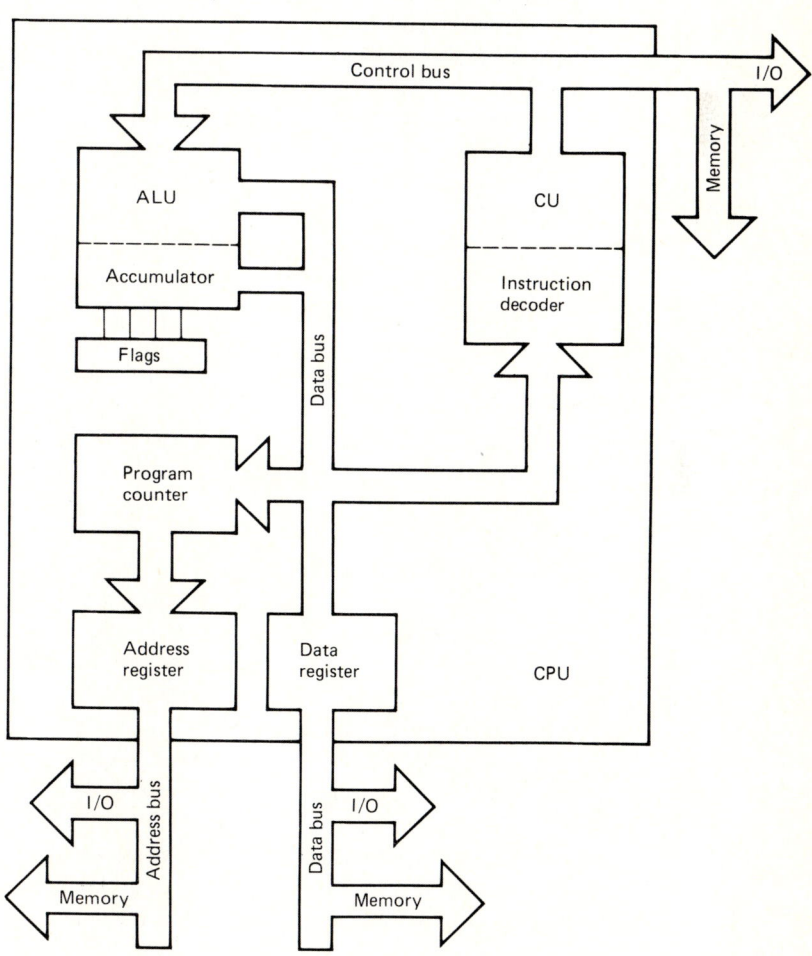

Figure 9.5 Typical CPU architecture.

more registers, a *program counter* and an *address register*, are included. A program counter initiates the entire sequence of operation and is reset prior to use. Thus the program counter is at 00_H when processing begins. This count is transferred into the address register, and the address register output is the address bus. Therefore 00_H is the first decoded address when RAM is accessed. The first user instruction is in location 00_H, and this word is transferred from RAM to data register along the data bus. Next, the data register transfers this word into the *instruction decoder* part of the control unit, which interprets the word. Then the control unit generates control signals for performing the instruction, such as addition, subtraction, or data transfer. The control signals are transmitted on the control bus, and information is transferred on the data bus.

While an instruction is decoded, the program counter is incremented to 01_H. Typically, location 01_H contains data pertaining to the instruction in 00_H. When the address bus is again enabled, the data in location 01_H are transferred from RAM to data register. Next, the appropriate control signals transfer the data register contents into the accumulator. When the program counter is stepped to 02_H, the next instruction is obtained. Processing continues, one location at a time, until the desired results are obtained.

Operation of the program counter, address bus, and data register in conjunction with the ALU and control unit is not a new concept. It was originally used in vacuum tube computers, and the basic procedure is still the same. User instructions and data are processed by repeating similar steps until results are obtained. Each step involves an instruction and data; first the instruction is obtained, and then the accompanying data is processed. This procedure is called *fetch-execute*. Several events occur during each fetch and execute. During fetch, an instruction is transferred from RAM to CPU, the instruction is then interpreted, and necessary control signals are generated. The program counter is also stepped during fetch. The fetch operation is performed in one *machine cycle*. Machine cycle time is constant for any given computer.

When fetch is completed, execute begins. During execute, the data is obtained and processed. While fetch time always requires one machine cycle, execute time depends on instruction type. The time to complete a fetch-execute sequence is called an *instruction cycle*. Instruction cycles vary in duration but always require an integer multiple of the basic machine cycle. A typical instruction cycle is shown in Fig. 9.6. Machine cycle period is determined by chip type and clock frequency. Typical clock frequencies for byte-organized MOS CPUs are in the megahertz range, which means that machine cycles are of the order of microseconds, e.g., 2-MHz clock results in a 0.5-μs machine cycle.

Most chip families also use single-bus architecture within the CPU. However, some CPUs contain multiple data buses to increase operating speed. Figure 9.7 shows three possible ALU data bus configurations. ALU inputs are *A* and *B*, and *S* is the output.

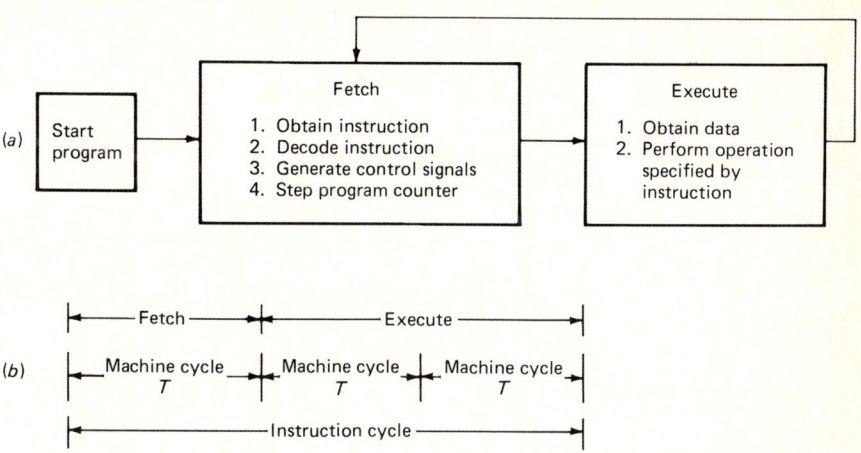

Figure 9.6
Processing an instruction.
(a) Fetch-execute sequence; (b) time relationship.

Conventional single-bus architecture is shown in Fig. 9.7a. At a given time, data only flows from or to a single ALU port. When data enters port A, no data can enter B, and S is also disabled. If each data transfer and each internal ALU operation are assumed to last 0.1 μs, then the time to process data in a single-bus ALU is

0.1 μs to transfer data into port A

0.1 μs to transfer data into port B

0.1 μs to process ALU data

0.1 μs to transfer data out of port S

---

and   0.4 μs is needed for each step

The 2-bus architecture shown in Fig. 9.7b is faster. There is a common input bus and a separate output bus. Times for each operation are the same as in single-bus architecture; however, data is output at the same time as new data enter port A or B. In this case, results are obtained in 0.3 μs rather than 0.4 μs.

Figure 9.7c shows the ultimate in multiple-bus architecture. Each port has a separate bus, and data enter A and B at the same time as the previous result is output from S. ALU processing time is reduced to 0.2 μs, which makes 3-bus architecture twice as fast as single-bus. Multiple-bus architecture is used when maximum speed is the major consideration, but extra buses need more CPU chip area, and multiple bus-control signal circuits are more complicated. The penalty for increased speed is thus increased size and cost. While some large computers have multiple buses, most desk-top units are single-bus.

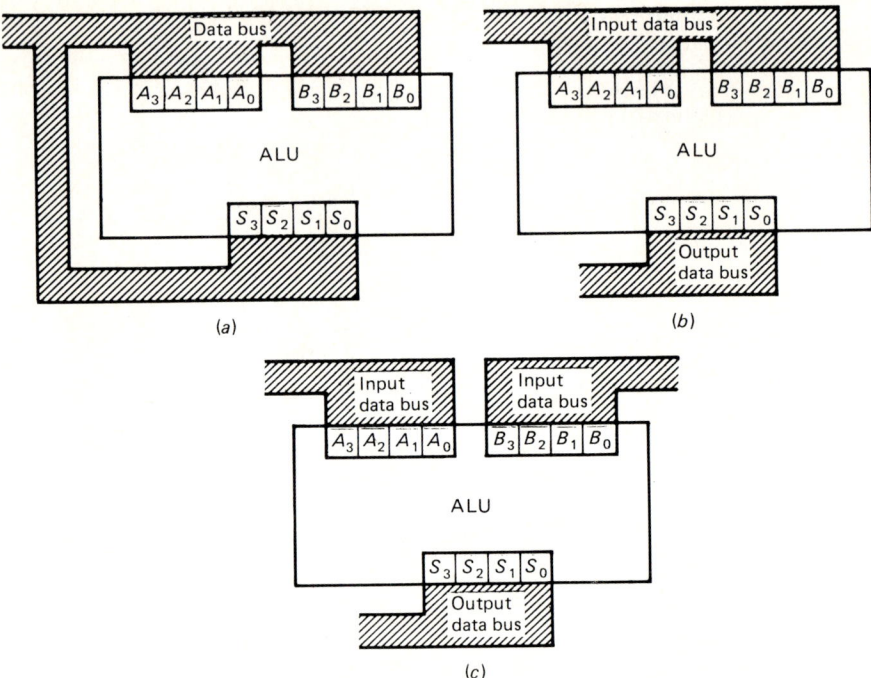

Figure 9.7
ALU bus architectures.
(a) One-bus; (b) two-bus; (c) three-bus.

9.3 CONTROL UNIT

The control unit regulates data processing by generating gating and timing signals. Gating signals activate appropriate circuits, and timing signals synchronize operations. Different registers must be connected and disconnected during each step of a fetch-execute cycle. The control unit generates a sequence of gating and timing signals during each machine cycle. Figure 9.8 shows control bus regulation of a 3-register computer. For simplicity A, B, and C are shown as 4-bit registers. Each register has separate input and output terminals, but there is only one data bus. Thus every input and output must be connected to the common data bus. A data bus is bi-directional; data can enter or leave any register.

All registers are also connected to the common unidirectional control bus. Control signals travel from control unit to registers but not in the reverse direction. Regardless of how many registers exist, the following three control signal leads are connected to each register:

- CLK is the common timing pulse.
- E is the enable signal; enable controls a register output.
- L is the load signal; load controls a register input.

Figure 9.8
Control unit connections.

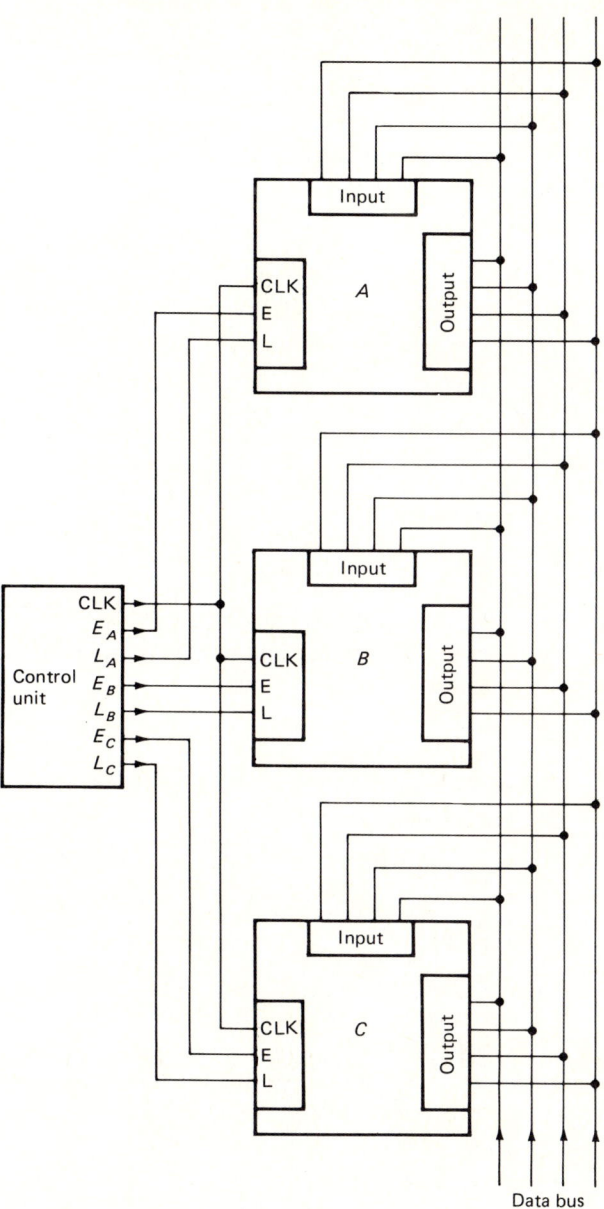

Three-state gates connect register terminals to the data bus; logic 1 connects a 3-state gate and logic 0 disconnects.

The output from the control unit is transmitted along the control bus and is called a *control word* (C_w). A control word ANDs CLK with E and L leads from all registers. The control word for a 3-register computer is

$$C_w = \text{CLK} \cdot E_A \cdot L_A \cdot E_B \cdot L_B \cdot E_C \cdot L_C \qquad (9.1)$$

Combinations of 1s and 0s in a control word activate the appropriate registers. Data transfer between registers requires a control word with three 1s. For example, transfer of data from register A to register C requires that CLK, E_A (the enable line of register A), and L_C (the load line of register C) be 1 and that all other bits be 0. Thus the control word for transferring data from register A to register C is

$$C_w = \text{CLK} \cdot E_A \cdot L_A \cdot E_B \cdot L_B \cdot E_C \cdot L_C$$
$$\underbrace{1\ 1}\ \underbrace{0\ 0\ 0\ 0\ 1}$$

or $\quad C_w = 61_H$

Similarly, the control word for the reverse operation, transferring data from C to A, is

$$C_w = \underbrace{101}\ \underbrace{0010}$$

or $\quad C_w = 52_H$

Example 9.1 Assume that the original data in Fig. 9.8 are $A = 1010$, $B = 0001$, and $C = 0110$. Determine the contents of each register after control words (a) 24_H and (b) 64_H are executed.

Solution (a) $24_H = 010\ 0100_2$.
Since the CLK bit is low, no data transfer is possible. All registers retain the original data.
(b) $64_H = 110\ 0100_2$.
Here CLK, E_A, and L_B are high. Data in register A is transferred along the data bus into B. Original data in B is lost. The original data in register A remains in A. Register C is not involved and retains original data. The results are tabulated below:

Register	Original	After 24_H	After 64_H
A	1010	1010	1010
B	0001	0001	1010
C	0110	0110	0110

The number of registers determines the number of bits in a control word. Each register requires one E and one L line. Thus the number of bits B_w in an n-register control word is

$$B_w = 2n + 1 \qquad (9.2)$$

where 1 accounts for the common CLK line. A 3-register computer requires a

7-bit control word, 4 registers require a 9-bit control word, etc. Practical computers contain at least 6 registers and thus require a 13-bit control word. In some computers CLK activates half the registers, the other half being activated by $\overline{\text{CLK}}$. This makes more efficient use of clock cycles.

Figure 9.9 emphasizes control bus operation and shows 7 registers, including ALU, RAM, ROM, and I/O. While program counter, address register, and data register are part of the control unit, each is activated by different control bits. Consider processing information which is contained in RAM. The first step is resetting, which clears the program counter and begins processing with 00_H in the program counter. Next 00_H is transferred into the address register with the control word

$$C_w = \text{CLK}, \ldots, L_{\text{Address register}}, \ldots, E_{\text{PC}} \ldots$$

and all other bits are low. The address register output is the address bus. When this address is decoded, data are transferred from RAM to data register with the control word

$$C_w = \text{CLK}, \ldots, L_{\text{Data register}}, \ldots, E_{\text{RAM}} \ldots$$

The first RAM location should always contain an instruction. The control word to transfer data register contents to the instruction decoder is

$$C_w = \text{CLK}, \ldots, E_{\text{Data register}}, \ldots, L_{\text{Instruction decoder}} \ldots$$

The next control word transfers the instruction decoder contents into the sequencer. Control words from the sequencer are transmitted along the control bus. For a byte-organized computer, the instruction decoder has 8-bit outputs. However, the sequencer output is at leat 13 bits long. Circuits with more output bits than input bits are not new. The 1-of-n decoder described in Chap. 3 fulfills this requirement: 2 inputs control 4 outputs, 3 inputs control 8 outputs, and n inputs control 2^n outputs.

During the fetch phase the program counter is also increased by 1 to 01_H. When fetch is completed, execute begins. Each part of the execute cycle also requires a different control word. During execute, data are transferred from the incremented address location to the data register and then into the accumulator. If the control word is

$$C_w = \text{CLK}, \ldots, L_{\text{Accumulator}}, \ldots, E_{\text{Data register}}, \ldots$$

addition is performed. On the other hand when the control word is

$$C_w = \text{CLK}, \ldots, \overline{L}_{\text{Accumulator}}, \ldots, E_{\text{Data register}}, \ldots$$

the 2s complement of the data register contents is used. A control word with $\overline{L}_{\text{Accumulator}}$ performs subtraction instead of addition. Obtaining a result in the ALU is the last part of the execute phase.

At the beginning of the second fetch-execute cycle, the counter is at 03_H

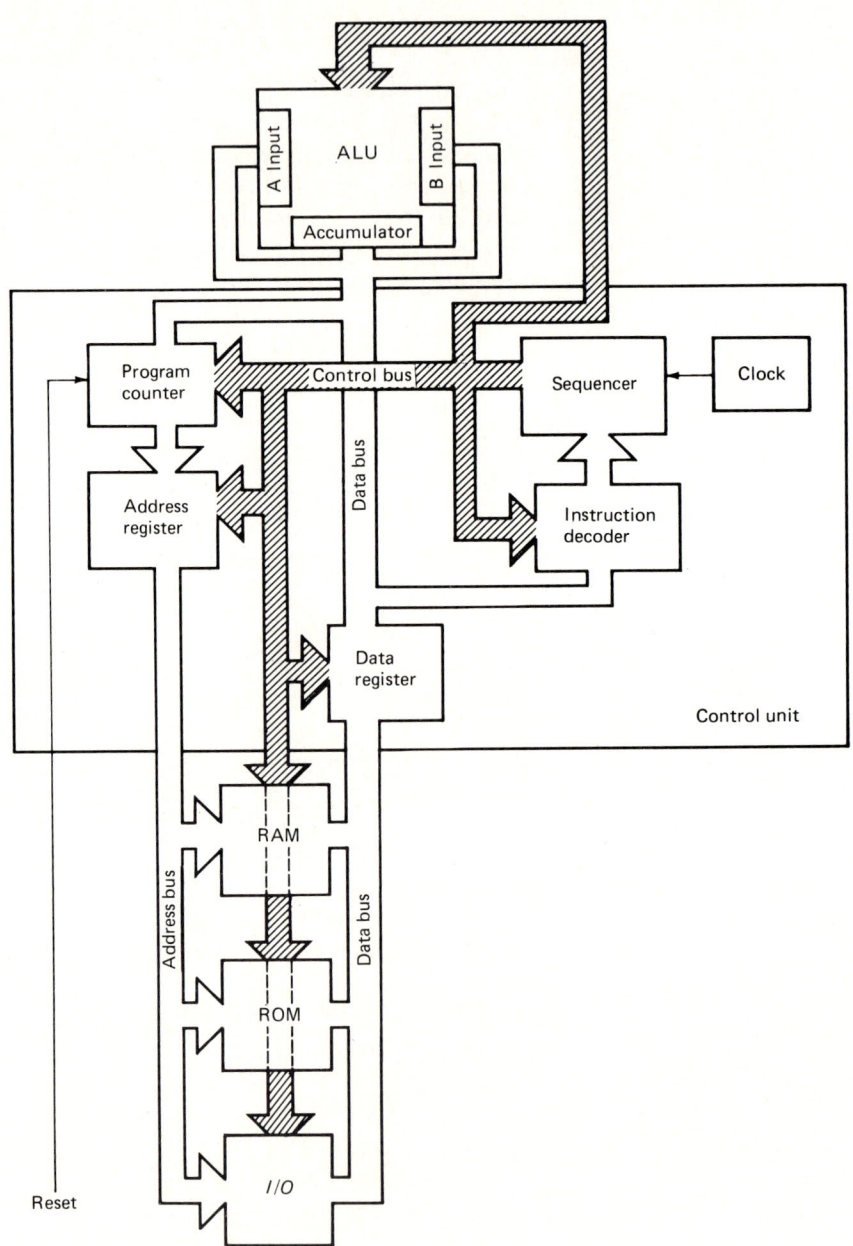

Figure 9.9 Control bus architecture.

and this address contains the next instruction. All instructions have the same fetch sequence and therefore use tne same control words; however, execute phases are different for each instruction and control words differ accordingly.

Fetch-execute cycles continue until final results are obtained. The final control word for terminating computer operation disconnects the sequencer from the control bus, and no further processing is possible. Timing pulses generated in the control unit determine computer speed. TTL circuits operate in nanoseconds and MOS circuits in microseconds. Thus, realizing the high speed benefits of TTL requires a TTL control unit working with an all-TTL computer. Similarly, the operating speed of an MOS control unit is compatible with other MOS chips.

9.4 MICROPROGRAMMING

Control words regulate automatic data processing. In this sense, computer design has not changed since the 1940s, when then cables carried control words from vacuum-tube CPUs to other vacuum-tube circuits. Now parallel wires on printed circuit boards carry control words from an LSI CPU to other ICs. Reductions in computing time, cost, size, and power consumption are a result of new components. However, control words still regulate data processing in the same way.

There are two methods of generating control words, wired logic gates and control ROMs. In either case, design is based on the same truth table. The choice between gates and ROMs was discussed in Chap. 7. If gates are used, data constitute the input and a control word is the output. If a ROM is used, then an address is the input and a stored control word is the output.

Figure 9.10 shows the gate version of a 4-bit instruction decoder. A decoder contains one inverter for each bit in the data register. Thus input bits as well as complements are available. With a 4-bit data register, the decoder requires 4-input AND gates. The output of each AND gate is a different instruction. The inputs to all 16 AND gates are connected in parallel, but bit combinations for every instruction are different and activate a different AND gate. If the data register contains 0001, only this gate is activated, if 0010, only that gate is activated, etc.

An instruction decoder is like the address decoder discussed in Chap. 3. An address decoder enables addresses and an instruction decoder enables instructions, but the circuits are identical. AND gate outputs from the instruction decoder connect to the *sequencer*, the output of which is the control bus. While control words are different for each instruction, some control words are used in several instructions. In particular, all instructions have the same fetch sequences. A fetch sequence contains three operations, called *timing states*, namely, address, memory, and increment.

During the address timing state T_{addr} the program counter contents transfer into the address register. Loading the address register by enabling the

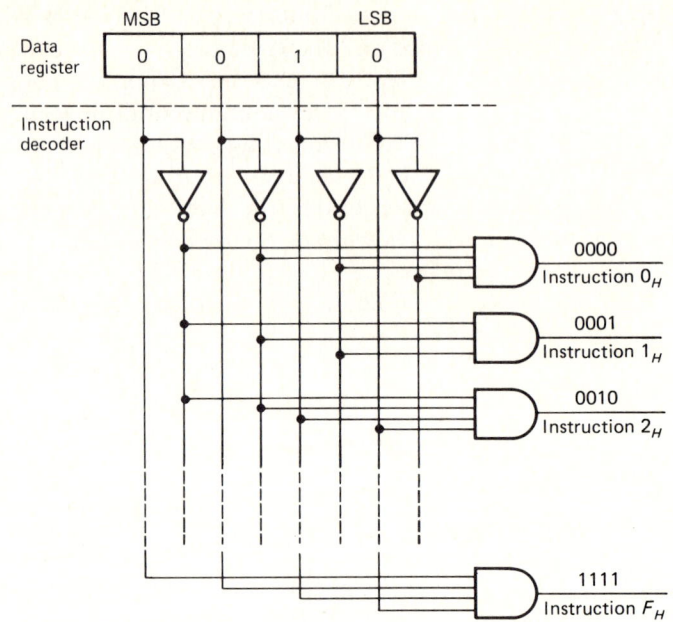

Figure 9.10 Gated instruction decoder.

program counter sends a number onto the address bus. The control word for T_{addr} is

$$C_w = \text{CLK}, \ldots, L_{\text{Address register}}, \ldots, E_{PC}, \ldots$$

and all other control bits are 0.

During the memory state, data from the decoded address is transferred from RAM to data register along the data bus. The control word for T_{mem} is

$$C_w = \text{CLK}, \ldots, L_{\text{Data register}}, \ldots, E_{\text{RAM}}, \ldots$$

Incrementing T_{inc} is the third and final timing state of a fetch phase. During T_{inc} the program counter is increased by one count. The control word is

$$C_w = \text{CLK}, \ldots, L_{PC}$$

While all fetch phases are identical, the timing states of each execute phase are different. If the instruction requires data transfer, appropriate registers are enabled and loaded; if the instruction is arithmetic, the accumulator is enabled, etc. While control words differ, each execute phase also has three timing states.

For example, the timing states for addition are T_{data}, T_{arith}, and T_{sum}. During T_{data}, the number in the incremented RAM address location is transferred into the data register. Then during T_{arith} the data register contents are added in the ALU. Finally, during T_{sum} the ALU result is transferred into

TABLE 9.1 Addition Timing States

	State	Name	Operation
Fetch	T_0	T_{addr}	Counter into address register
	T_1	T_{mem}	Instruction into data resister
	T_2	T_{inc}	Counter is increased by 1
Execute	T_3	T_{data}	Data into data register
	T_4	T_{arith}	Data register into ALU
	T_5	T_{sum}	Sum into accumulator

the accumulator. Table 9.1 summarizes the six timing states for performing addition.

Differences between instructions occur during T_3, T_4, and T_5. When subtraction is performed, only T_4 is different. As described, the 2s complement of the operand rather than the operand is transferred into the data register. Other instructions differ by either two or all three timing states of the execute phase.

Figure 9.11 shows sequencer operation. A sequencer has two inputs:

- Output gates from the instruction decoder.
- Timing pulses for each timing state.

Each address decoder output connects to an entire row of sequencer AND gates. Every pair of AND gates in a row is connected to a different timing state signal. During each timing state, one AND gate output connects to an enable bit and the other to a load bit.

For example, E_{PC} and $L_{Address register}$ are active during T_0. Appropriate combinations are activated during each timing state. The last timing state is T_5 when the active control bits are E_{ALU} and L_{Acc}. After T_5 is completed, a new fetch-execute cycle begins with T_0. Since many op codes require the same control bits, OR gates combine corresponding AND gate outputs, L_{ALU}, E_{RAM}, L_{PC}, etc. These OR gate outputs are the control bus.

While any 6-bit counter can generate timing state pulses, the ring counter is especially useful. As described in Chap. 5, a ring counter output contains a single 1, which shifts 1 bit for each clock pulse. At reset a 6-bit ring counter is at 000001, and this is T_0. On the first clock pulse the counter advances to 000010, which is T_1, etc. After T_5, reset to T_0 is automatic, and the next fetch-execute cycle is processed. Figure 9.12 shows counts and waveforms for a 6-bit ring counter. A reset is required to initiate operation. Observe that the CLK goes high during the middle of each timing state. This delay provides time for gates to stabilize and also minimizes noise along the buses.

Most computers use about 100 different instructions. With 12 AND gates per fetch-execute cycle, a sequencer requires 1200 two-input AND gates. This count can be cut by half because fetch is the same for all instructions.

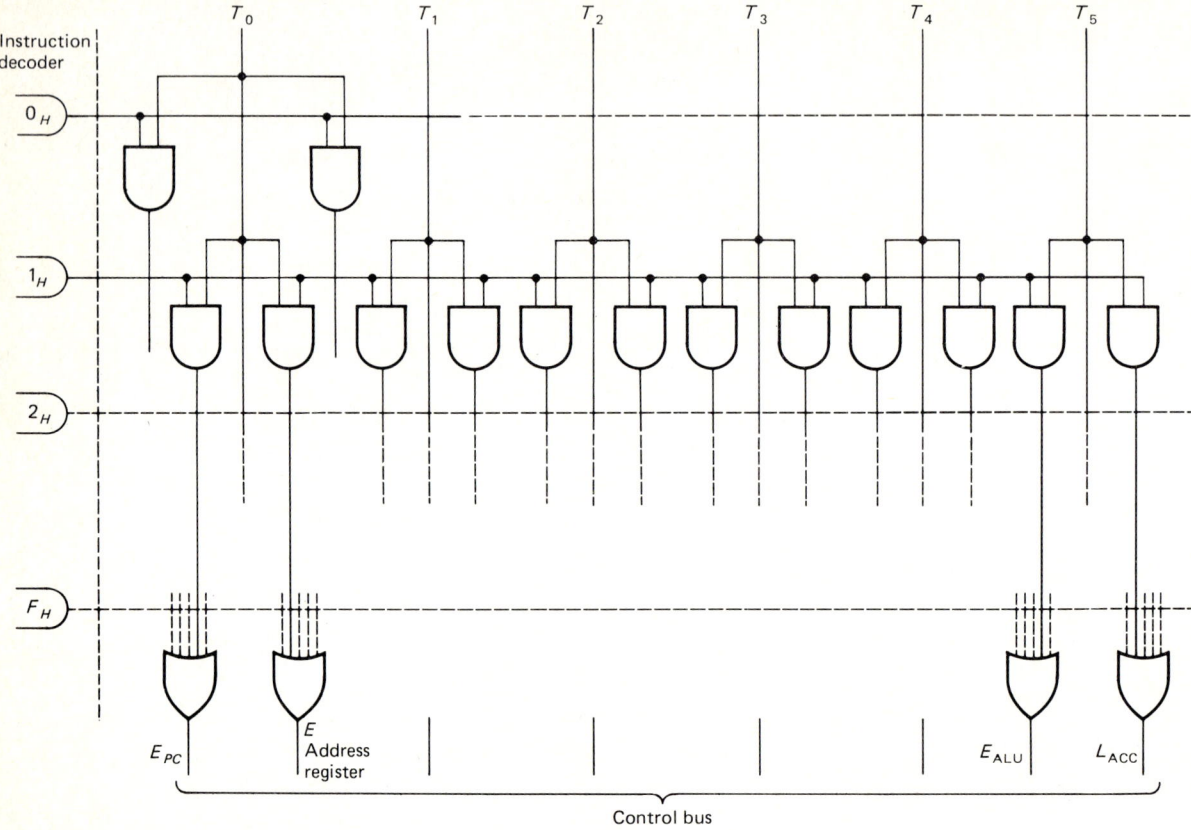

Figure 9.11 Control word sequencer.

Nevertheless 600 AND gates plus the necessary multi-input OR gates and wiring still constitute a complicated sequencer.

A *control ROM*, which is a matrix method of generating control words, is the alternative to the gate type of sequencer. Every ROM address stores an entire control word, each of which is called a *microinstruction*. The first three control ROM locations contain the following fetch cycle:

- 00_H stores the T_{addr} microinstruction.
- 01_H stores the T_{mem} microinstruction.
- 02_H stores the T_{inc} microinstruction.

These three microinstructions constitute the fetch *microprogram*.

All other control ROM addresses store execute microprograms for the various op codes. The addition execute microprogram could be stored in 03_H,

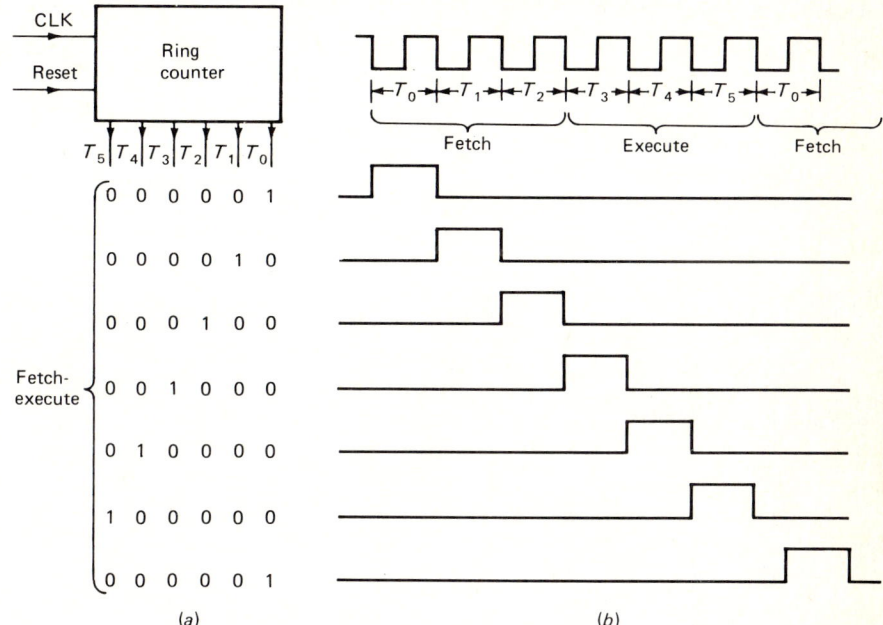

Figure 9.12 Ring counter circuit. (*a*) Block diagram; (*b*) waveforms.

04_H, and 05_H. Similarly, the subtraction microprogram could be in 06_H, 07_H, 08_H, etc. Every execute microprogram stores T_3, T_4 and T_5 in three sequential locations.

Table 9.2 shows the outline of a truth table for designing a control ROM. The first three addresses contain control words for the fetch microprogram, and each subsequent group of three addresses contain an execute microprogram. Active control bits for each address are 1s, and all other control bits are 0s.

Figure 9.13 shows how a control ROM generates control words. The presettable counter is cleared at the beginning of each microprogram. At this time the ring counter is at 000001, and preset is not yet enabled. Thus the presettable counter steps through 00_H to 02_H on the first three CLK pulses. Since these three addresses store the fetch microprogram, a fetch is the first operation which is performed. On T_3 the presettable counter is enabled, and data register contents are transferred into the counter. In this case the contents are 06_H, the first address of the subtraction microprogram. The first microinstruction of subtraction is executed, and the next two CLK pulses step the counter through 07_H and 08_H. During these timing states microinstructions for T_4 and T_5 of subtraction are the control ROM outputs. After T_5 the ring counter resets to 000001; this begins the next fetch microprogram. At T_3 the first address of the next execute microprogram is enabled in the presettable counter, etc.

Generally it is inconvenient to separate instructions by three numbers.

Computer Circuit Concepts

TABLE 9.2 Control ROM Truth Table

	Address	E_{Acc}	L_{Acc}	$E_{Addr\,reg}$	$L_{Addr\,reg}$	⋯	E_{PC}	L_{PC}	E_{RAM}	L_{RAM}
Fetch	00_H		1				1			
	01_H				1			1		
	02_H								1	
Addition execute	03_H				1	1				
	04_H				1	1				
	05_H	1			1					
Subtraction execute	06_H				1	1				
	07_H				1	1				
	08_H	1			1					
⋮										

Thus an address ROM is required between the data register and presettable counter. The address ROM shown in Fig. 9.13b converts the instructions into the first address of each execute microprogram. For example, the subtraction instruction might be $0D_H$. When $0D_H$ is the input to the address ROM, the output is 06_H, and this is the first location of the subtraction microprogram. Similarly, if the addition instruction 02_H is the input to the address ROM, the output is 03_H, and this is the first location of the addition microprogram, etc.

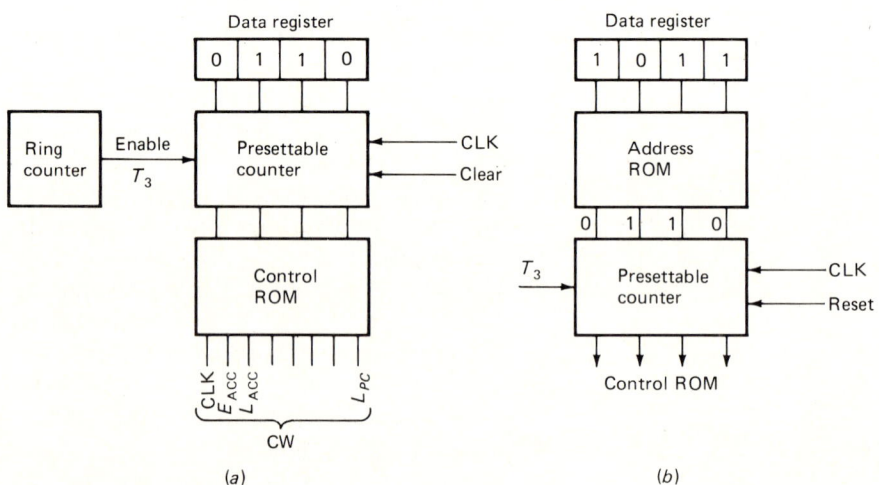

Figure 9.13 Microprogramming circuit. (*a*) Block diagram; (*b*) address ROM.

The address ROM is a matrix decoder. Input from the data register results in an output which is the first address of the appropriate microprogram.

Storing microprograms in MROMs is easier than constructing decoder/sequencers for large instruction sets. Flexibility is another advantage of using a control ROM. If one control ROM is replaced by another and a few A/D converters are added, a general-purpose computer becomes an industrial process controller or else a controller for heating, cooling, and lighting a large office building. Applications for the same computer with different ROMs are limitless.

9.5 STORED PROGRAMS

It is the control word which makes automatic data processing possible. Since the first electronic computers of the 1940s, results have been obtained with sets of instructions called *programs*. A set of self-contained instructions is called a *stored program*. When one stored program is completed, another one can be entered. Although each stored program is different, the same circuits are used to solve all problems. The earliest computers were general-purpose mathematical machines.

Today computers are even more flexible. A single machine can perform mathematical calculations, compose music, play chess, and measure voltage. This flexibility is possible because different programs can be stored. Computers have changed since the 1940s. New machines are smaller, faster, more versatile, and less expensive, but improvement is a result of newer components, especially transistors and ICs. The original concept has remained the same. Computers were and still are general-purpose because different problems are stored and solved by use of the same parts and control words.

Processing a stored program is more systematic than the human approach. When we perform a calculation using pencil and paper, experience permits us to "take shortcuts". Working without positive signs, shifting decimal points, etc., are routine. Solving a problem by using a computer is somewhat different—each step must be performed and nothing can be taken for granted. Using a number in a stored program can destroy the number. If the number is required for later use, it must be saved. Similarly, the existence of an answer in the ALU is not the same as displaying the answer on an output device. Stored programs must be very detailed.

A program contains instructions needed to perform the required task. Programs stored in RAM are in the only language computers understand. Instructions and data are groups of 1s and 0s. While it is possible to enter a program into RAM manually by operating binary switches, less painful methods exist. Figure 9.14 illustrates a stored program. Program steps are loaded into sequential RAM addresses.

Computer word length determines how many instructions are possible. Since each instruction is represented by a different combination of bits, a

390 Computer Circuit Concepts

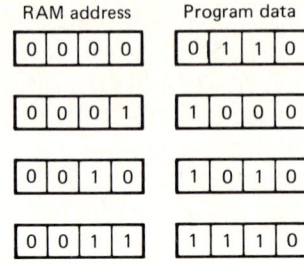

Figure 9.14
Stored program format.

nibble-size computer can contain 16 different instructions. However, programs can be much longer than 16 instructions. Program length is limited by RAM size. Thus a nibble-organized computer might have a $1 \text{ K} \times 4$ RAM, in which case program and data can be up to 1024 nibbles long. However, the stored program is restricted to 16 different instructions.

While a program is stored in RAM, mathematical operations are performed in the ALU. Thus a program must contain data transfer as well as mathematical instructions. Instructions and data are located in sequential addresses. If we assume a computer which uses the code 0001 as the instruction to transfer a number from RAM into the accumulator, a program to add 6 to 4 begins as follows:

RAM Address	Binary Code	Program Description
0000	0001	Move into the accumulator
0001	0100	the number 4_{10}

The first instruction, in RAM location 0000, moves data into the accumulator. The data must be in the address immediately following the instruction.

When a desired number is in the accumulator, addition can be performed. If the code for performing addition is 0010, the program continues

RAM Address	Binary Code	Program Description
0000	0001	Move into the accumulator
0001	0100	the number 4_{10}
0010	0010	Add to the accumulator
0011	0110	the number 6_{10}

Again, data must be in the RAM address immediately following the instruction. When this fourth step is completed, the correct result is in the accumulator, not in RAM. The sum in the accumulator can be displayed, and the device on which the display is presented must be specified. If the instruction for display is 0100 and the identification code for the CRT terminal is 0001, the program continues as follows:

RAM Address	Binary Code	Program Description
0000	0001	Move into the accumulator
0001	01000	the number 4_{10}
0010	0010	Add to the accumulator
0011	0110	the number 6_{10}
0100	0100	Display the accumulator on
0101	0001	the CRT

The sum should now appear on the CRT, but one more instruction is required. When a computer is first turned on, each RAM location contains random combinations of 1s and 0s. Each 4-bit combination represents either a valid instruction or valid data. The next location, 0110, might contain the code for clearing the accumulator, blanking the CRT, adding or subtracting another number, etc. To eliminate undesirable possibilities, stored programs end with an instruction to terminate further operations. If the instruction to stop processing instructions is 1111, the complete program to add 6 to 4 is

RAM Address	Binary Code	Program Description
0000	0001	Move into the accumulator
0001	0100	the number 4_{10}
0010	0010	Add to the accumulator
0011	0110	the number 6_{10}
0100	0100	Display the accumulator on
0101	0001	the CRT
0110	1111	Stop the program

What we do "automatically" in adding two numbers requires seven RAM locations as a stored program. However, the stored program runs at electronic speed. If each step takes 50 ns, the program is completed in 350 ns. Adding more numbers requires a few more locations. The addition of a third number requires one more add instruction as well as the number to be added. Similarly, a complete program to add and display the sum of four numbers occupies 11 RAM locations.

Thus addition programs can be extended to add many numbers. Two limitations exist:

- Accumulator size determines the largest possible sum.
- The complete program must fit into the available RAM addresses.

All stored programs are written in the same manner by using appropriate instructions. For example, subtracting requires a subtraction instruction. Since ALU subtraction is performed by adding the binary complement, the subtract instruction might be the complement of the add instruction. In this case the subtract code is 1101. It is also possible that the code for a printer is 0010. In this case a program to subtract 4 from 11 and print the result is:

RAM Address	Computer Code	Program Description
0000	0001	Move into the accumulator
0001	1011	the number 11_{10}
0010	1101	Subtract from the accumulator
0011	0100	the number 4_{10}
0100	0100	Display the accumulator on
0101	0010	the printer
0110	1111	Stop the program

Just as with addition, a subtraction program occupies seven RAM locations. Similarly, subtraction programs also can be extended to perform several subtractions.

If the RAM is large enough, many programs can be read into RAM; a subtraction program can follow an addition program, etc. Instructions are listed in numerical order in Table 9.3. This table also contains two peripheral device codes. The addition and subtraction programs are relatively simple,

**TABLE 9.3
Introductory Computer Codes**

Code	Description
0001	Move into the accumulator
0010	Add to the accumulator
0100	Display the accumulator
1101	Subtract from the accumulator
1111	Stop the program
0001	The CRT
0010	The printer

and the basis of all stored programs is the principle: instruction, data, instruction, data, instruction, ..., stop.

Instructions, data, and device codes can have the same binary sequence. In these sample programs the add instruction, 1_{10}, and the code for CRT are all 0001. Similarly, 4_{10} and the code for display the accumulator are the same. Given the nature of binary numbers, multiple meanings for computer words are inevitable. Multiple meanings are not a limitation to writing stored programs, but the need for writing stored programs in proper sequence is apparent. Even a single incorrect word in a single RAM location invalidates an entire stored program.

More complicated operations such as multiplication and division cannot be accomplished with a single instruction. Rather, existing instructions are combined. For example, multiplication can be performed by repeated addition and division by repeated subtraction. This of course necessitates instructions for repetition. Table 9.3 contains five instructions. If instructions are restricted to 4-bit sequences, only 11 other instructions are possible. A 4-bit computer is extremely limited.

As a practical matter, a true 4-bit computer has never been used. Nevertheless, the first LSI computer was nibble-organized. The 4004 was introduced by Intel in 1971. Data were organized into 4-bit words, but this was a limitation of then existing IC manufacturing technology. Actually, two nibbles were combined and treated as a single byte. The relatively simple 4004 contained 46 different instructions.

Intel also introduced the first byte-size LSI computer chip, the 8080, in 1975. Less than 1 year later Motorola came out with the 6800, which was also byte-organized. Since then Intel and Motorola have introduced improved LSI chips, and other companies have developed 8-bit computer chips. At the present time 16-bit chips are readily available and 32- and 64-bit chips exist. Generally, larger word chips are faster.

Regardless of word size, the stored program concept is the same: each computer has a specific list of instructions; a programmer writes a program using existing instructions and necessary data; instructions and data are input and stored in RAM; the program stored in RAM is processed in the ALU; and the results are presented at an output device.

9.6 CODES

Each step in a program contains two parts, an *operation* and an *operand*. An operation is an instruction to perform a task. Move into the accumulator, display the accumulator, and add to the accumulator are operations. The binary word for each operation is called an *operation code*, or *op code*. Operands are the information on which op codes act. Data to be added, subtracted, etc., are operands, as are also the identification numbers for output devices such as the CRT and printer.

A stored program is a sequence of op codes and operands, and the binary version is called a *machine language* program. Typical op codes are byte size, and most operands are at least byte size. The nibble-size op codes used in the previous section can be converted to byte size by adding 4 leading 0s. For example, the op code for move into the accumulator changes from 0001 to 0000 0001. Similarly, the code for add to the accumulator changes from 0010 to 0000 0010, etc. As can be appreciated, writing a machine language program is quite tedious. If a programmer transposes only a single 1 or 0, a word is incorrect. If the incorrect word is an op code, the wrong operation is performed; if it is an operand, the data are wrong. Thus a single incorrect word invalidates an entire machine language program.

Writing and loading machine language programs can be simplified by converting the op codes and operands to a higher number base, for example, by replacing binary with octal. As discussed in Chap. 2, each binary sequence has an octal equivalent. Although a byte is 8 bits, imagining 0 in front of the MSB results in 9 bits. This permits writing a byte as three groups of 3 bits each. Consider the op code for add to the accumulator. The byte-size op code is

0000 0010

Adding a leading 0 and grouping by 3s results in

000 000 010

and the octal equivalent is

002

Writing a program in octal instead of binary reduces programming effort. Op codes and operands are shorter, and the chances of writing incorrect words are reduced. Moreover electronic octal-to-binary conversion is simple. Programs can be entered on an octal keypad, and switch debouncer and a few gates are all that is needed for conversion. Thus an octal program automatically converts to binary as the keys are activated, and octal keypad computers are readily available. Table 9.4 shows byte and octal equivalents for op codes which have been discussed.

TABLE 9.4 Some Basic Op Codes

Byte	Octal	Description
0000 0001	001	Move into the accumulator
0000 0010	002	Add to the accumulator
0000 0100	004	Display the accumulator on
0000 1101	015	Subtract from the accumulator
0000 1111	017	Stop the program

Usually RAM address are written in octal when the program is in octal. Thus an octal program to add 6 to 4 becomes:

Address	Op Code	Description
000	001	Move into the accumulator
001	004	the number 4_{10}
002	002	Add to the accumulator
003	006	the number 6_{10}
004	004	Display the accumulator on
005	001	the CRT
006	017	Stop the program

If the computer contains an octal-to-decimal conversion circuit, the CRT displays 10_{10}. Otherwise 12_8 appears on the screen. Octal programs are more compact than binary and simplify writing a program. Hexadecimal numbers are even more compact, and binary to hex conversion is also simple. A byte is considered as 2 nibbles, and each has a hex equivalent. Computers with hex keypads are also available. Table 9.5 shows the basic op codes in binary, octal, and hex form. Op code descriptions are the same.

TABLE 9.5
Op Code Equivalents

Binary	Octal	Hex
0000 0010	001	01
0000 0010	002	02
0000 0100	004	04
0000 1101	015	0D
0000 1111	017	0F

Table 9.5 can be used to write programs in hex. For example, the hex program for 11 − 4 is as listed below:

RAM Address	Hex Op Code
00	01
01	0B
02	0D
03	04
04	04
05	01
06	0F

TABLE 9.6 Extended-Address Op Codes

Byte	Hex	Description
0001 0001	11	Move into the accumulator
0001 0010	12	Add to the accumulator
0001 1101	1F	Subtract from the accumulator

While byte-size words result in sufficient op codes, byte-size operands are somewhat limited. The 256 combinations provided by 8 bits are sufficient for op codes, but 255 for the largest operand restricts computer arithmetic. Also, 256 RAM locations constitute a rather small memory. The limitations of byte size are overcome by using two successive RAM locations for an operand. With 16 bits, the largest RAM address and number is $2^{16} - 1$, or 65,535. Use of 2-byte operands is called *extended addressing*. An extended address instruction occupies three successive RAM locations, one for op code and two for operand. Extended addressing requires a different op code from the corresponding two-location sequences, as otherwise a word in the third address would be interpreted as an op code rather than an extended operand. For our purpose, extended-address op codes can be formed by using 0001 as the 4 highest-order bits. Thus the extended address op code for add to the accumulator becomes 0001 0010, or 12_H. Table 9.6 shows extended op codes for the basic arithmetic instructions.

Figure 9.15*a* shows a typical 3-byte instruction. Addresses are consecutive, and operands follow op codes. In this case the op code in address 08_H is 11_H, the operand in 09_H is $1B_H$, and the operand in $0A_H$ is 09_H. The usual operand sequence is higher-order operand followed by lower. Thus the operand is

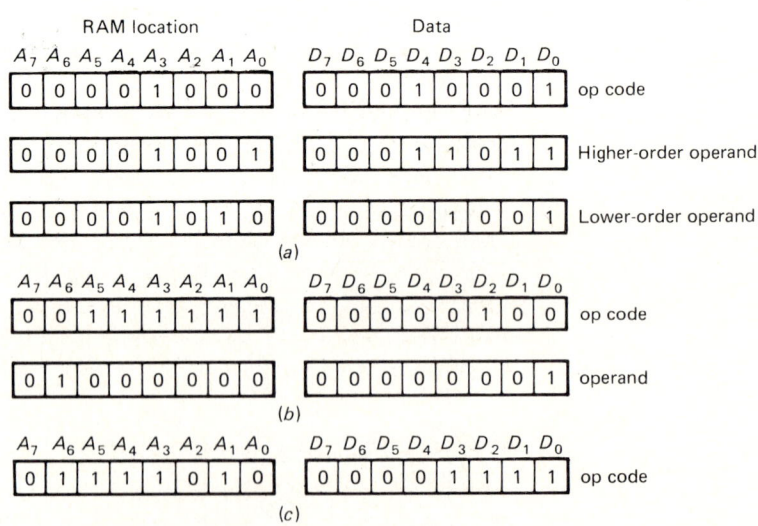

Figure 9.15 Machine language sequences. (*a*) 3-byte instruction; (*b*) 2-byte instruction; (*c*) 1-byte instruction.

1B09. However, some chips use reverse order. Either method is acceptable, and the programmer must know which convention applies.

All instructions do not require extended addressing. For example, displaying the accumulator is a 2-byte instruction and is illustrated in Fig. 9.15b. In this case the op code is located at address 3F and the operand is at 40_H, the next address. Also, 2-byte instructions can be used for arithmetic operations when numbers are less than 256. Instructions with only 1 byte also exist. Stop the program is an example of a 1-byte instruction and is illustrated in Fig. 9.15c.

While any system is preferable to writing and entering programs in binary, octal or hex is a modest improvement. Practical programs require thousands of RAM locations. A long program consisting entirely of number groups is difficult to write and tedious to load. It is also difficult to interpret or alter a program consisting entirely of number groups. Something closer to our everyday experience is helpful.

The first step in program simplification is replacing numeric op codes with *mnemonics*. A mnemonic is a "shorthand" alphabetic group related to the code. Computer mnemonics contain two, three, or four letters. The code MOV might be the mnemonic for move into the accumulator. Adding E could be the mnemonic for the extended-address version of an op code, in this case, MOVE for move into the accumulator extended. Table 9.7 contains a reasonable set of mnemonics. It is easier to remember mnemonic codes than number groups. In fact the description column of a mnemonic program can be very brief. A program such as that listed below is easier to follow than a string of octal or hex numbers.

Address	Op Code	Description
00	MOVE	
01	1A	High-order data
02	2B	Low-order data
03	ADDE	
04	03	High-order data
05	4C	Low-order data
06	OUT	
07	01	On the CRT
08	HLT	Stop the program

Mnemonics simplify writing a program but must be converted to binary before processing is possible. Fortunately, the computer itself can perform alphabetic to numeric conversion. In other words, a mnemonic program can be entered on an alphanumeric keyboard and the computer will perform an internal translation into machine language. Translation is performed by a

TABLE 9.7 Mnemonic Codes

Hex	Mnemonic	Description
01	MOV	Move into the accumulator
02	ADD	Add to the accumulator
04	OUT	Display the accumulator
0D	SUB	Subtract from the accumulator
0F	HLT	Stop the program
11	MOVE	Move into the accumulator (extended)
12	ADDE	Add to the accumulator (extended)
1D	SUBE	Subtract from the accumulator (extended)

special program called an *assembler*. The mnemonic program is read in and stored in RAM. Then each location is searched by the assembler. As each mnemonic code occurs, a table look-up is performed by using a word comparator, and the corresponding binary equivalent is then loaded into RAM. Step-by-step translation continues until the entire program is converted into machine language. A program written in mnemonics is called a *source program*, and the translated binary version is called the *object program*.

Selection of mnemonics for each op code is a matter of choice. The chip designer selects convenient symbols. Once a mnemonic has been selected, it cannot be changed. Assemblers are stored in LSI chips and each manufacturer provides a complete set of mnemonic op codes. Unfortunately, mnemonics differ between manufacturers. Changing from one chip family to another requires a different set of mnemonics to write the same program.

The mnemonics used in this chapter are illustrative and do not represent those of any specific manufacturer. After assembly language programming is understood, codes for a specific assembler can be investigated in detail. This is preferable to learning an actual set of mnemonics only to discover that another family must be used.

9.7 INSTRUCTION SETS

Regardless of complexity, all programs follow the same procedure. Programs consist of op codes and operands. Thus, a computer does not think. Problems are "solved" by following a sequence of instructions, which a programmer selects in advance.

A complete list of op codes is called an *instruction set*, and a typical instruction set contains about 100 instructions. Understanding computer operation requires some familiarity with instruction sets. There are six instruction types in an instruction set:

1. Arithmetic

2. Data transfer

3. Input/output

4. Interrupt

5. Logic

6. Skip

Arithmetic instructions are classified as: addition/subtraction, increment/decrement, left/right shift, and clear. ADD, ADDE, SUB, and SUBE are arithmetic instructions. With these op codes, data in RAM are located immediately after the op code. There are other arithmetic instructions for which an address in memory rather than actual data are the operands. We will use ADM, or 20_H, as the code for add the data from a specific memory location. SBM, or 21_H, will represent the code for subtract the data from a specific memory location. Usually memory address operands are located after the instruction to stop the program.

Example 9.2 Using memory address operands, write an assembly language program for adding 6 to 4.

Solution The format is the same. Any program must show RAM address and op code. The description column is helpful but not necessary.

RAM Address	Op Code	Program Description
00	MOV	Move into the accumulator
01	00	the number 00
02	ADM	Add to the accumulator the data in
03	09	RAM address 09
04	ADM	Add to the accumulator the data in
05	0A	RAM address 0A
06	OUT	Display the accumulator on
07	01	the CRT
08	HLT	Stop the program
09	04	Data, the number 4
0A	06	Data, the number 6

A program using address locations as operands is longer than a program in which operands are the data. However, the longer program has advantages. All data are grouped in a block. Thus, a general-purpose program can be written to which data are entered as they become available.

Generally, ADM-type instructions are more practical when data will be used more than once and also when working with variables.

Increment/decrement instructions are a special type of addition/subtraction. An increment instruction adds 1 to the accumulator. A decrement instruction subtracts 1 from the accumulator. These operations can be performed with ordinary addition and subtraction instructions, but specific op codes are convenient. Increment/decrement instructions are used to count how many times a particular operation is performed.

Left/right shift instructions exist for both circulate shift and shift out operations. A circulate shift-left command moves each accumulator bit one position to the left. B_7 shifts into C_0, B_6 shifts into B_7, etc., and C_0 is returned to B_0. This process occurs each time that a circulate shift-left instruction is performed. Figure 9.16a shows the result of three circulate shift-left operations. A circulate shift right instruction shifts bits in the opposite direction. After nine consecutive shifts the accumulator word returns to its original form. The number of times C_0 changes from 0 to 1 and back can be used to determine parity.

The shift left out instruction is shown in Fig. 9.16b. One bit enters the C_0 position each time the instruction is executed. These emerging bits can be stored in a 4-bit buffer register, and this is one method of separating a byte into two nibbles. In larger computers the shift-left out instruction is used for performing accumulator/MQ division, as described in Chap. 6. Similarly, shift-right out instructions perform accumulator/MQ multiplication.

Clear instructions are used to set all 0s into the accumulator. Clearing can also be accomplished with MOV 00 into the accumulator, but this requires an op code and an operand. On the other hand CLR or $2A_H$ is a single-byte instruction.

Data transfer instructions move information inside the computer. MOV and MOVE are op codes for transferring data from memory immediately after the op code to the accumulator. We will use STO or $2C_H$ and STOE or

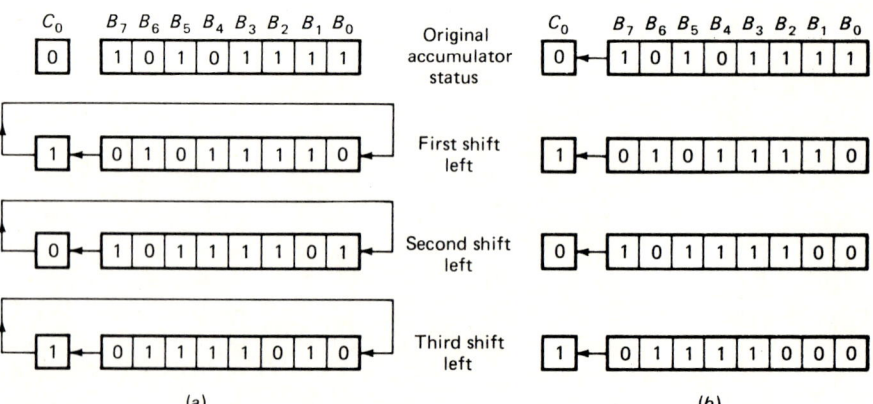

Figure 9.16 Left-shift instructions. (a) Circulating; (b) shift left out.

$2D_H$, as store and extended store op codes, respectively. The operand for these op codes are RAM addresses where accumulator data are placed.

Load instructions are another data transfer for which the operand is a RAM address. A load instruction transfers data from a specific RAM address into the accumulator. Loads are similar to moves, but MOV/MOVE operands are data, not addresses. We will use LOD or $2E_H$, and LODE or $2F_H$ as load and extended load op codes, respectively.

Example 9.3 Write a general purpose program for

$$s = x + y$$

Solution A general-purpose program uses operands as memory addresses rather than data. Thus

RAM Address	Op Code	Program Description
00	LOD	Load the accumulator with data in
01	07	RAM address for x
02	ADM	Add to the accumulator data in
03	08	RAM address for y
04	STO	Store the sum in
05	09	RAM address for s
06	HLT	Stop the program
07	—	Data for x
08	—	Data for y
09	—	Location of s

Locations 00 through 06 are the general program. Sums can be obtained whenever x and y are available and stored in RAM location 09.

Input/output instructions transfer data between the computer and peripheral devices. OUT, a 2-byte instruction, has been used. The operand identifies a specific peripheral device such as a CRT or printer. We will use IN, or 05_H, as the input op code, 41_H as the disk drive code, and 42_H as the modem code. Disk drive and modem can be either input or output devices; the choice is made when either IN or OUT is selected as the op code.

While computers require peripheral devices, some instruction sets do not have separate input and output codes. In these cases, memory locations are assigned as the address of each peripheral device. Data are entered by using the load instructions as the op code and the device address as the operand.

Similarly, data are output by using the store instruction together with the device address.

Interrupt instructions stop program processing. HLT is an example of an internal interrupt. The normal state of affairs in a computer is continuous internal operation. Thus an instruction to enter or display data is a temporary interrupt. Input and output instructions can be considered as external interrupts. Some interrupts require separate instructions to restore internal operation.

Logic instructions implement the basic Boolean operations AND, OR, and NOT. These op codes can be used to evaluate Boolean equations. An operand in memory is logically combined with the accumulator contents on a bit-by-bit basis. The result is in the accumulator. For example, the result of ANDing 9_H and 8_H is

```
  1001
  1000
  ————
  1000
```

or 8_H. The result of AND is the logical product, not the arithmetic sum.

Boolean op codes have other applications besides evaluating logic equations. When a word is ANDed with 0111 1111, all bits remain the same except for the MSB. This method of forcing a 0 onto the MSB changes a signed number from negative to positive. Similarly, an OR op code results in a 1 whenever either or both corresponding bits are 1. Thus ORing with 1000 0000 changes a signed number from positive to negative. The EOR op code generates 1s when corresponding bits are different and 0s when bits are the same. EORing can be used for comparing two words. Whenever the result of EORing is an all-0 byte, the words are equal. The NOR op code complements each bit in the accumulator. This result is the 1s complement. Some computers use NOT and then add 1 to obtain the 2s complement. Other computers generate the 2s complement directly with circuits rather than using op codes.

Skip instructions are also called *branch* or *jump*. Instructions in each of the preceding categories result in *linear* programs. Each instruction in a linear program is performed in numerical order according to RAM address. A skip instruction alters the linear sequence. Whenever an unconditional skip occurs, the program moves out of sequence to a specific address. The location of the skip can be either ahead of or behind the current address. We will use GO TO, or $3A_H$, as the unconditional skip op code.

Among other applications, GO TO can reduce the number of steps in programs which require counting. Consider a program which prints the numbers 0 through 225. One method of accomplishing this is to clear and then print the accumulator contents. Then 1 is added and another print command given, which is in turn followed by another command to add 1, etc.

This method requires writing and entering a program of more than 500 RAM locations. If GO TO is used, the same result occupies only seven RAM locations

RAM Address	Op Code	Program Description
00	CLR	Clear the accumulator
▶ 01	OUT	Display the accumulator contents on
02	02	the printer
03	ADD	Add to the accumulator
04	01	the number 1
05	GO TO	Unconditional skip to
└ 06	01	RAM address 01

This program also prints the numbers 0 through 255. The use of GO TO eliminates the need for writing a long string of ADD and OUT instructions. As each number is printed, 1 is added and the next number is automatically printed.

When 255 is reached, overflow occurs and stops the program. On the other hand if

ADDE

00

01

is used instead of

ADD

01

the program prints up to 65,535 before stopping.

A return to a previous address is called a *loop*, and skip instructions generate loops. While unconditional skips are executed whenever GO TO occurs, conditional skips require a test procedure. Tests are based on data conditions. Typical tests ascertain whether the accumulator is positive, negative, zero, or more than, less than, or equal to some other number. If the test result is correct, a skip is performed, but if the test fails, the program continues in linear fashion. Instruction sets contain a variety of conditional skip instructions. Our modest needs are satisfied with a single conditional skip. SKZ, or $3B_H$, is the op code for skip if zero. This instruction can stop a counting program at some desired number. For example, each time that a number is printed, it can also be subtracted from 100 and tested with SKZ. After 100 prints, SKZ is 0 and the program stops.

Conditional skips are used to make decisions. This does not mean that a computer "thinks". All thinking was done by the programmer when the conditional skip was written into the program.

9.8 PROGRAMMING TECHNIQUES

Although ALU, RAM, ROM, and other chips differ between manufacturers, basic designs are the same. As a result any computer regardless of chip type performs given tasks in a similar manner and at approximately the same speed. Nevertheless, actual design details impose severe restrictions. The byte-size 8080 and 6800 series chips are incompatible. Machine language programs can only be transferred between computers which use the same chips.

The op codes described in the previous section are complete as to type but incomplete as to variety, containing 19 instructions versus 100 or so for a real instruction set. Thus it might be appropriate to call our computer *d*igital *u*nrefined *m*achine-model 1, or simply DUM-1. Table 9.8 shows the complete DUM-1 instruction set. As with real machines, instructions are listed in alphabetical order.

Besides grouping instructions in terms of functional categories, it is helpful to classify instructions according to *address modes*. An address mode describes how data are obtained. ADD and ADM are both addition instructions but different methods of obtaining data are required. In order of increasing complexity, the four address modes are

1. Implied

2. Immediate

3. Direct

4. Relative

Implied address instructions are complete in a single byte. There is an op code but no operand. RAM is not accessed to obtain data; the operand is understood. CLR is an op code for which the implied address is the accumulator, and HLT is an op code for which the implied address is the control unit. Implied instructions require one machine cycle to execute. Thus, fetching and executing an implied instruction requires two machine cycles.

Immediate address instructions are at least 2 bytes long. The first byte is the op code, and the byte in the next location is the actual data. ADD and SUB are immediate address instructions, and ADDE and SUBE are 3-byte immediate address instructions. Extended immediate address instructions increase the data range from FF_H to $FFFF_H$. ADD and SUB also require two machine cycles. Fetch always requires one machine cycle, and in this case execute also requires one machine cycle.

TABLE 9.8
DUM-1 Instruction Set

Mnemonic	Op Code	Description
ADD	02	Add to the accumulator
ADDE	12	ADD (extended)
ADME	20	ADD from memory (extended)
CLR	2A	Clear the accumulator
GO TO	3A	Unconditional skip to —
HLT	0F	Stop the program
IN	05	Input data from — to accumulator
LOD	2E	Load accumulator with data in —
LODE	2F	LOD (extended)
MOV	01	Move into the accumulator
MOVE	11	MOV (extended)
OUT	04	Display the accumulator on —
SBM	21	Subtract data in memory location —
SBME	22	SBM (extended)
SKZ	3B	Skip if zero (conditional)
STO	2C	Store accumulator in —
STOE	2D	STO (extended)
SUB	0D	Subtract from the accumulator
SUBE	1D	SUB (extended)
Peripheral Devices		
	01	CRT
	41	Disk drive
	42	Modem
	02	Printer

Direct address instructions are 2 or 3 bytes long. The first byte is the op code and the next byte(s) are the RAM location(s) where data is stored. In other words, the operands direct us to the data. ADM is the 2-byte direct address instruction: Add the data which is located in memory address —— to the data already in the accumulator. ADME would be the 3-byte extended version and is required for addresses higher than FF_H.

Direct address instructions take longer to execute than immediate address instructions. This is because data for direct addressing must be located before they can be used. With immediate addressing, the data is always in the next address. In DUM-1 instructions such as LODE and ADME use five machine cycles.

Relative address instructions can also be 2 or 3 bytes long. The first byte is an op code, and the operand(s) are an address which must be calculated. Relative addressing is used for program loops. An offset address is added for looping forward and subtracted for looping back. Relative addressing can move an entire program to a different section of RAM. In this case each address has the same offset. For example, adding 1000_H to each address of a program located below FF_H shifts that program from the single-byte to the extended address region.

The operand of an immediate address op code is obtained by increasing the program counter by 1. But direct address and skip op codes require increasing or decreasing the program counter to obtain a different address, which depends on the operand: GO TO 106_H, SKZ $4B_H$, ADM $10A9_H$, etc. In these cases the accumulator contains the next addresses. During the execute phase the control word transfers the accumulator contents into the program counter. On the next fetch-execute cycle, data is fetched from the address corresponding to the new program count. Processing continues from the new address and the program counter resumes the increment-by-1 routine. Whenever another skip or direct address instruction occurs, the program counter is again adjusted from the accumulator.

While all instruction sets contain addition and subtraction instructions, more advanced arithmetic instructions do not exist as individual op codes. Multiplication, division, square root extraction, etc., are generated by combining existing instructions. Large computers perform multiplication by combining addition and shift instructions. Desk-top computers use repeated addition to perform multiplication. Multiplying by repeated addition requires adding the multiplicand to itself the number of times specified by the multiplier. Thus the result of 5×3 is obtained by adding 5 to itself 3 times: $5 + 5 + 5 = 15$, or in assembly language as tabulated below.

Address	Code	Description
00	MOV	Move into the accumulator
01	05	the number 05
02	ADD	Add to the accumulator
03	05	the number 05
04	ADD	Add to the accumulator
05	05	the number 05
06	OUT	Display the accumulator (product) on
07	02	the printer
08	HLT	Stop the program

This program requires a known multiplicand and multiplier.

A general-purpose program is preferable. Skip instructions make a general-purpose multiplication-by-repeated-addition program possible. Consider

$$P = x \cdot y$$

where P is the product, x is the multiplicand, and y is the multiplier. Each time that the multiplicand is added to the accumulator, 1 is subtracted from the multiplier. The multiplier is then tested for 0 condition. If the multiplier is not 0, the multiplicand is added again, and another 1 is subtracted from the multiplier. When the multiplier reduces to 0, the product is correct. The program then skips to the display instruction. In this version the program "decides" when to stop adding by using a conditional skip.

A general-purpose multiplication program sets aside locations for product, multiplicand, and multiplier. In particular, the product location must be cleared before other operations begin. Clearing a specific location is called *initializing*. After the product is initialized, y is loaded from memory into the accumulator and tested for 0. When the test is satisfied, repeated addition is complete. However if y is not yet 0, x is again added to P. Location P contains a partial product which increases until the program loops the required number of times. See p. 408.

This general-purpose multiplication program contains two loops. An unconditional loop, 16 to 01, stores the increasing partial product. The other loop, 08 to 17, is conditional and skips to the display sequence when the final product has been obtained.

Such a program could be used to multiply 5×3. Since 35 RAM locations are required, it is almost nine times longer than the original version. However, without changing any instructions the general-purpose multiplication program can multiply 756×618 or any other numbers. By contrast, the original version, which requires a known multiplier and multiplicand before writing, occupies almost 2K bytes of RAM to perform 756×618.

As a practical matter this general-purpose program must be refined to include floating-point and signed numbers. As this point it is worthy of being called an *algorithm*. In an algorithm instructions are combined to create a general-purpose form. A division-by-repeated-subtraction algorithm begins by initializing location for the quotient. Then, each time that the divisor is subtracted from the dividend, 1 is added to the quotient. When the remainder is 0, the quotient is correct.

Algorithms can be used as *subroutines*. The same operation may be required many times throughout a program. A particular program may require six different multiplication sequences. Instead of writing the multiplication algorithm six times, multiplication is entered as a subroutine and stored outside the main program. Each time multiplication is needed, the subroutine is accessed with an unconditional skip.

When the multiplication subroutine is completed, return to the main program is necessary. Return is accomplished with relative addressing. When GO TO is executed, the appropriate return address is stored. *Stacks* can store return addresses. A stack is a set of temporary storage registers which cannot be addressed individually. Some chips contain hardware stacks, in which case special stack instructions are included. Other chips reserve a section in RAM for stacks. In this case the first address in a stack is known. Data leaves a stack

Address	Code	Description
00	CLR	Clear the accumulator
01	STOE	Store P in
02	1D	High-order product byte P_h
03	1E	and lower-order product byte P_l
04	LODE	Load the accumulator with
05	21	higher-order multiplier byte y_h
06	22	and lower-order multiplier byte y_l
07	SKZ	If multiplier is 0, skip to
08	17	address to begin display routine
09	SUBE	Otherwise subtract
0A	00	the extended number 0001
0B	01	from the accumulator (multiplier)
0C	STOE	And store the reduced accumulator
0D	21	back in y_h
0E	22	and y_l
0F	LODE	Load the accumulator with
10	1D	P_h and
11	1E	P_l
12	ADME	Add the multiplicand
13	1F	x_h and
14	20	x_l
15	GO TO	Unconditional skip to
16	01	the stored product instruction
17	LODE	Load the final product
18	1D	P_h
19	1E	P_l
1A	OUT	Display the product on
1B	02	the printer
1C	HLT	Stop the program
1D	—	P_h
1E	—	P_l
1F	—	x_h
20	—	x_l
21	—	y_h
22	—	y_l

in either last in–first out or first in–first out order. In either case, stack data is only available from the stack output.

Programs written in assembly language, particularly when subroutines are included, are a significant improvement over writing machine language programs. However, assembly language is still a far cry from everyday experience. *High-level languages* were developed to permit programming without the need to understand internal computer operation. A high-level language program is written in complete lines rather than individual words. A single line can contain an entire equation, mathematical symbols such as +, −, *, /, and = are accepted, and variables are represented symbolically. However, a high-level language program cannot be executed as such. A *compiler*, usually located in ROM, translates each program line into the proper sequence of assembly language instructions. Then the resultant assembly language program is converted into machine language.

FORTRAN, COBOL, and BASIC are some of the more famous high-level languages. FORTRAN, from formula translation, was introduced by IBM in the 1950s, when computers contained vacuum tubes. Since then FORTRAN has been updated and it is used in modern IC computers. APL, PL/1, and PASCAL are sophisticated extensions of FORTRAN. COBOL, from common business-oriented language, uses sentences rather than equations. This language is used mainly for business and financial accounts. BASIC, from beginners all-purpose symbolic instruction code, was developed by Dartmouth University. It was introduced in 1977. While BASIC is not as flexible as FORTRAN, it is easier to learn. BASIC is built into most small computers.

Programming in any high-level language is easier than in assembly language, but high-level language programs require more RAM space and are significantly slower. Programs which cannot be changed are called *firmware*, the assembler being an example. Since an object program can be changed, it is called *software*. Programs written by the user and stored in RAM are software. By way of contrast, the actual chips and related components are called *hardware*.

Fifty years ago there were no computers and no computer texts. Forty years ago computer texts described vacuum-tube circuits for performing arithmetic calculations and programming in binary. At that time transistors and ICs did not exist. What computer texts will describe 40 or 50 years from now is anyone's guess.

SUMMARY

1. All computers contain an ALU, a control unit, I/O memory circuits, address, data, and control buses. Most computers use single-bus architecture.

2. Information is processed in discrete segments called fetch-execute cycles. During fetch an instruction is interpreted, and during execute the data are processed.

3. The control unit regulates fetch-execute cycles by generating control signals and timing pulses. Control signals select the correct registers, and timing signals synchronize operations.

4. Control signals can be assembled from gates or ROMs. When instruction sets are large, ROMs are simpler and more flexible.

5. A stored-program computer is versatile because different problems can be solved with the same circuits. Instructions and data are binary and the program sequence must be in specific order.

6. Instructions in a computer program contain op codes and operands. Mnemonic codes simplify writing and entering programs.

7. Instruction sets are assembly language lists of the available op codes. Skip instructions are used to alter program sequence, and conditional skips are the basis for making decisions.

8. Subroutines are based on the skip instructions. Higher-level language programming is easier to learn but is less efficient than machine language.

PROBLEMS

9.1 An address bus contains: (*a*) 8; (*b*) 10; and (*c*) 12 lines. How many locations can be addressed?

Answer: (*a*) 256; (*b*) 1024; (*c*) 4096.

9.2 Determine the size of a computer word for each condition in the previous problem.

9.3 A 40-pin microprocessor can transfer a byte-size word from RAM to CPU in 10 μs. Assuming the same clock frequency, how much time is needed to transfer a 16-bit word?

Answer: 20 μs.

9.4 Find the address and data bus size for a microprocessor which is connected to (*a*) a 16 × 8 and (*b*) a 65,536 × 8 memory.

9.5 In terms of enable, read, and write, give the control signals for: (*a*) entering data into the ALU; (*b*) observing data in RAM; and (*c*) disconnecting the printer.

Answer: (*a*) enable-write; (*b*) enable-read; (*c*) $\overline{\text{enable}}$.

9.6 A machine cycle lasts 0.5 μs. Find the computer clock frequency.

9.7 A computer is run from a 1-MHz clock. How much time is needed to complete instruction cycles which require (*a*) two; (*b*) three; and (*c*) five machine cycles?

Answer: (*a*) 2 μs; (*b*) 3 μs; (*c*) 5 μs.

9.8 Assume that the time required for entering data, processing data, and obtaining output from a single-bus ALU is 500 ns. Estimate the time for (*a*) 2-bus and (*b*) 3-bus architectures.

9.9 The contents of a 3-register computer are: $A = 1100$; $B = 1111$; and $C = 0000$. $C_w = 52_H$ is executed and then $C_w = 64H_H$. Determine the final contents of each register.

Answer: All registers are reset.

9.10 Show another pair of control words to accomplish the same result.

9.11 The control word for a 4-register computer is:

$$C_w = \text{CLK} \cdot E_A \cdot L_A \cdot E_B \cdot L_B \cdot E_C \cdot L_C \cdot E_D \cdot L_D$$

Determine the result of: (*a*) $C_w = 88_H$; and (*b*) $C_w = A0_H$.

Answer: CLK = 0 for both (*a*) and (*b*). No change.

9.12 Find the hex value of the smallest and largest control words which transfer data in a 4-register computer.

9.13 A 5-bit data register is connected to an instruction decoder. How many (*a*) inverters; (*b*) inputs to each AND gate; and (*c*) instructions can be decoded?

Answer: (*a*) 5;   (*b*) 5;   (*c*) 32.

9.14 Determine control words for T_data, T_arith, and T_sum of the addition execute cycle.

9.15 A ring counter operates from a 750-kHz clock. Determine: (*a*) CLK period; (*b*) timing state period; and (*c*) time needed to complete a fetch-execute cycle.

Answer: (*a*) 0.67 µs;   (*b*) 1.33 µs;   (*c*) 8.0 µs.

9.16 A 100 op code instruction set uses 13-bit control words. Determine control ROM size.

9.17 Using Table 9.3 write a program to perform $1 + 2 + 7$ and display the sum on a CRT terminal. How many RAM locations are needed?

Answer: Nine.

9.18 Write a program to list the numbers 0, 1, 2 on a printer. Assume that only the instructions listed in Table 9.3 are available.

9.19 Write a program to add 3, 4, 5, and 7 in a 4-bit accumulator and display the result on a CRT terminal. Determine (*a*) how many locations are required and (*b*) what result is displayed if no overflow detection exists.

Answer: (*a*) 11;   (*b*) 3.

9.20 Using Table 9.3, write a program for $6 + 9 - 2$ and present the result on a printer.

9.21 Write an object program in hex to add 26_{10} to 108_{10}. This program should start at RAM address 00_H, and the sum should appear on the printer.

Answer:

ADDR	Op
00	01
01	6C
02	02
03	1A
04	04
05	01
06	0F

9.22 Start at RAM address FD_H and write a hex program to print the result of adding 271_{10} to 2738_{10}.

9.23 For the program in the previous problem, find the contents of addresses (a) FF_H and (b) 105_H.

Answer: (a) BC_H; (b) $0F_H$.

9.24 Write the program of Prob. 9.23 using mnemonic codes instead of hex.

9.25 If the program in Prob. 9.17 is written in octal, what are the contents of RAM locations (a) 02; (b) 03; and (c) 04?

Answer: (a) 041_8; (b) 011_8; (c) 041_8.

9.26 Write an assembly language program to print the numbers from 0 through 255 which are divisible by 5.

9.27 Determine the results of: (a) $F3_H$ OR $3F_H$; (b) $F3_H$ AND $3F_H$.

Answer: (a) FF_H; (b) 33_H.

9.28 Write a general-purpose program to add three extended numbers x, y, and z, which are stored after the HLT instruction.

9.29 The following program is shown in address-hex code form: CD-01, CE-06, CF-02, D0-05, D1-41, D3-0F. Determine (a) what the program does and (b) where the result appears.

Answer: (a) Adds $6 + 5$; (b) disk drive.

9.30 Begin at the same address as in Prob. 9.29. Write an equivalent program in assembly language, using direct addressing.

9.31 Write an extended address assembly language program for

$z = x + y$

This program begins at address $9A_H$, and the result is stored in RAM. Determine (*a*) how many bytes are needed and (*b*) where the program ends.

Answer: (*a*) 15_{10}; (*b*) 108_H.

9.32 Write a general-purpose assembly language program for

$s = x + y + z$

The sum is stored in RAM, and extended addressing is not necessary.

INDEX

INDEX

Abacus, 91–92
Access time, 282–283, 301, 318
Accumulator, 242–245, 247–249, 374, 375, 392, 400
A/D (analog-to-digital) conversion, 356–361, 370, 389
Addend, 64, 77
Addition, 62–67
  binary, 64–65
  hexadecimal, 66–67
  logical, 3
  octal, 65–66
Address, 282, 296–299
Address decoder, 105–106, 293, 311–313
Address mode, 404–406
Algorithm, 407
Alphanumeric character, 326, 345–346
ALU (arithmetic logic unit), 268, 369–378, 381, 390
American Standard Code for Information Interchange (see ASCII)
Analog-to-digital conversion (see A/D conversion)
AND, 2, 3
AND gate, 32–35, 95–98
Architecture, 370–374
Arithmetic logic unit (see ALU)
ASCII (American Standard Code for Information Interchange), 331–333, 346, 362
Assembler, 398, 409
Astable multivibrator, 177–178, 186

Asynchronous counter, 183–187
Augend, 64, 77

Base, 46, 51
BASIC, 409
Baud, 212, 328
Baudot, Jean, 328
Baudot code, 328–330
BCD (binary-coded decimal), 268–274
Binary addition, 64–65
Binary-coded decimal (see BCD)
Binary division, 73–75, 261–263
Binary multiplication, 72–73, 257–260
Binary subtraction, 69–70
Bipolar gates, 111–120
Bistable device, 140–142
Bistable multivibrator, 172–174
Bit, 47
Blanked, 345
Boole, George, 1
Boolean algebra, 1, 232, 369, 402
Boolean equations, 11–15, 37, 145, 154
  commutative, 4
  distributive, 5
Boolean theorems, 6, 37
Branch, 402
Bubble notation, 32
Bus, 371
Byte, 160, 265, 281, 283, 288, 371, 394

Calculus, 92
Canonical form, 16
Carry anticipation, 246–249
Cathode-ray tube (see CRT)
Central processing unit (see CPU)
Character generator, 346–349
Chip enable (CE), 299
Clock, 154
CMOS (complementary metal-oxide semiconductor), 124–125
COBOL, 409
Coincidence circuit, 98
Coincident current, 286, 289–291
Common collector, 104, 105
Common emitter, 104–105
Commutative arithmetic, 64, 71
Comparator, 234–235, 282
Compiler, 409
Complementary metal-oxide semiconductor (see CMOS)
Condition code register, 375
Contact conditioner, 153
Control unit, 370–373, 378–383
Control word, 379–381, 383
Core memory, 284–293
Core RAM, 288–293
Counter:
  asynchronous, 183–187
  down, 187–191
  hybrid, 203–210
  Johnson, 209
  modulo, 197–203
  program, 376, 381, 383
  ring, 206–208, 385–387

417

Counter (*Cont*.):
   ripple, 183–185, 198–200, 203, 245
   shift, 209, 210
   synchronous, 183, 191–197, 199, 203
   up, 187
   up/down, 206–208
CPU (central processing unit), 371–373, 375–378
CRT (cathode-ray tube), 344–348
Current sink, 127
Current source, 127

D-type circuit, 158–163, 206, 215–217
D/A (digital-to-analog) converter, 349–356
Daisy wheel, 333–335
Data link, 361–365
Data register, 374, 376, 384
Decoder, 102, 185, 204–205, 381
De Morgan's theorem, 14–15, 37, 148–150, 157, 231, 238–239
Digit, 44
Digit field, 326–327
Digital-to-analog converter (*see* D/A converter)
Diode gate, 100–101
Diode transistor logic (*see* DTL)
DIP (dual in-line package), 125–126
Direct address, 404, 405
Dividend, 73
Division:
   binary, 73–75, 261–263
   frequency, 164–166, 186–189
Divisor, 73
"Don't care" condition, 31, 32, 194, 195
Dot-matrix printers, 334–335
Double-dabble method, 48
Double precision, 266
Down counter, 187–191
Drum memory, 279–283
DTL (diode transistor logic), 112–114
Dual in-line package (*see* DIP)
Duplex, 364
Dynamic logic, 123–125

EBCDIC (extended binary-coded decimal interchange code), 333, 362
Edge trigger, 161, 184, 188
Enable lead, 109, 158–160
Encoder, 102
End-around carry, 78, 81, 82, 251–253, 273, 274
EOR (exclusive-OR), 232–235, 252, 253, 259, 312
EPROM (erasable programmable read-only memory), 318–320
Erasable programmable read-only memory (*see* EPROM)
Excitation table, 193, 200
Exclusive-OR (*see* EOR)
Extended addressing, 396–398
Extended binary-coded decimal interchange code (*see* EBCDIC)

Fan-in, 127–128, 167
Fan-out, 127, 128, 167
Fast adder, 246–249
Fetch-execute, 376–378, 381–383, 385
Firmware, 409
Fixed-point arithmetic, 263, 264
Flag, 375
Flash conversion, 359
Flip-flop, 153–163, 240–242, 285, 290, 293–295
   J-K (*see* J-K flip-flop)
Floating gate, 318–319
Floating-point arithmetic, 263–268, 407
Floppy disk, 338–340
Formed-character printers, 333, 334
FORTRAN, 409
Frequency division, 164–166, 186–189
Full-adder, 235–240, 251, 256–259, 312–313

Gate, 32–37
Gate cost, 35–37
Glitch, 190–191

Half-adder, 229–235, 240, 311
Hard-copy printers, 333
Hardware, 409
Hexadecimal addition, 66–67
Hexadecimal subtraction, 71
High-level languages, 409
Hold time, 167
Hollerith, Herman, 326
Hollerith code, 326–328, 340
Hybrid counter, 203–210
Hysteresis, 174

IA (interface adapter), 372–373
Immediate address, 404
Implied address, 404
Inhibit line, 288–292
Initializing, 407
Input line, 284
Input/output (*see* I/O)
Instruction cycle, 376, 377
Instruction decoder, 376, 381–384
Instruction set, 398–402, 404–405
Interface adapter (*see* IA)
Interrupt, 373, 399, 402
Inverse, 2
Inverter, 32–36
Inverter gate, 103–107
I/O (input/output), 335–339, 361, 364, 381

J-K flip-flop, 168–172, 183, 192–196, 206, 215–217
Johnson counter, 209
Jump, 402

Karnaugh, Maurice, 19
Karnaugh map, 19–31, 37, 193–196, 200–204, 221–222, 230, 236, 237, 269–270
KC Standard, 343–344

Ladder network, 353–356
Large scale integration (see LSI)
Latch, 140–143, 171
LCD (liquid crystal display), 219–223
Least significant bit (see LSB)
Least significant digit (see LSD)
LED (light-emitting diode), 94–96, 104, 184, 219–223, 244, 358
Light-emitting diode (see LED)
Light-foot, 95
Linear program, 402
Liquid crystal display (see LCD)
Loading, 127
Logical addition, 3
Logical multiplication, 3, 402
Look ahead circuit, 246
Loop, 403, 406–408
LSB (least significant bit), 47, 83–84, 207, 209, 211, 215, 240–242, 331
LSD (least significant digit), 46, 78
LSI (large scale integration), 126, 224, 364, 371, 372, 383, 393

Machine cycle, 376, 377, 404, 405
Machine language, 394, 404
Magnetic disk, 337–339
Magnetic drum, 280–283
Magnetic hysteresis loop, 280
Magnetic I/O (see I/O)
Magnetic recording, 339–344
Magnetic tape, 335, 337
Manchester recording, 343
Mask, 316
Maskable read-only memory (see MROM)
Masking, 268
Master/slave, 168–172
Matrix, 285, 311, 389
Maxterm, 17–18, 21
Medium scale integration (see MSI)
Memory:
    core, 284–293
    drum, 279–283
    dynamic, 304–310
    nonvolatile, 287, 311

Memory (Cont.):
    sequential, 287
    static, 304
    volatile, 287
Memory address register, 292
Memory buffer register, 291–293
Memory cell, 293–296, 300–302
Metal-oxide semiconductor (see MOS)
Metal-oxide semiconductor field-effect transistor (see MOSFET)
Microcomputer, 371
Microinstruction, 386–387
Microprocessor, 371
Microprogram, 383–389
Minifloppy, 339
Minterm, 17–18, 21
Minuend, 68
Mnemonic, 397–399
Modem, 365, 370
Modulo counter, 197–203
Monostable multivibrator, 173
MOS (metal-oxide semiconductor), 120–125, 172, 174, 177, 191, 210, 214, 223, 232, 233, 283, 287, 299–310, 372, 383
MOSFET (metal-oxide semiconductor field-effect transistor), 120–125, 279, 299–310
Most significant bit (see MSB)
Most significant digit (see MSD)
MROM (maskable read-only memory), 316–318, 389
MSB (most significant bit), 47, 75, 82, 207, 209, 211, 215, 240–243, 331, 349
MSD (most significant digit), 46
MSI (medium scale integration), 126, 224
Multi-emitter, 114, 295, 296
Multiplexing, 372
Multiplicand, 71, 257, 407
Multiplication:
    binary, 72–73, 257–260
    logical, 3, 402
Multiplier, 71, 257, 407
Multivibrator:
    astable, 177–178, 186

Multivibrator (Cont.):
    bistable, 172–174
    monostable, 173
    one-shot, 174, 175

NAND gate, 107
Natural code, 269
Natural modulus, 197
NDRO (nondestructive readout), 290–292
Nibble, 160, 370, 390, 393, 394
9s complement, 77–78, 272–274
Noise immunity, 132, 133, 167
Noise margin (NM), 132–133
Nondestructive readout (see NDRO)
Nonreturn-to-zero (see NRZ)
Nonreturn-to-zero inverted (see NRZI)
NOR gate, 107
NOT, 2
NOT gate, 32–33
NRZ (nonreturn-to-zero), 342, 343
NRZI (nonreturn-to-zero inverted), 343
Number systems:
    binary, 46–51
    decimal, 44–47
    hexadecimal, 57–62, 394–406
    octal, 51–57, 394
    Roman, 43–45

Object program, 398
Octal addition, 65–66
Octal subtraction, 70
Off resistance, 93–95
Offset, 266–267
On resistance, 93–95
One-shot multivibrator, 174, 175
1s complement, 78–80, 249–252, 272
Op code (operation code), 393–405
Open-collector gates, 110–111, 118
Operand, 393, 394, 406
Operation, 393
Operation code (op code), 393–405
Optical coupler, 362, 363

OR, 2
OR gate, 32–35, 98–102
Overflow, 76, 77, 256, 403

Parallel adder, 244–249
Parity:
   even, 234, 333
   odd, 234, 333
Partial product, 72–73
Phototransistor, 362
Pixel, 346–349
PLA (programmable logic array), 320
Polling, 372
POS (product of sums), 16–17
Product of sums (see POS)
Program, 389
Program counter, 376, 381, 383
Programmable logic array (see PLA)
Programmable read-only memory (see PROM)
PROM (programmable read-only memory), 318–320
Propagation delay, 130, 166, 190, 196
Pull-up resistor, 111
Pulse stretcher, 177

Quotient, 73

Racing, 168
Radix, 46
RAM (random-access memory), 279, 287–310, 330, 345–347, 376, 381, 390–393
Ramp converter, 360, 361
Random-access memory (see RAM)
Raster, 345
RB (return-to-bias), 342
Read-only memory (see ROM)
Reading, 281–283
Refresh, 306–310
Regeneration, 140, 141
Register:
   data, 374, 376, 384
   MQ, 259–263

Register (Cont.):
   parallel input, 214–219
   PIPO (parallel input–parallel output), 218
   PISO (parallel input–serial output), 215, 346, 347, 364
   recirculating, 217
   serial input, 210–214
   SIPO (serial input–parallel output), 212–214, 364
   SISO (serial input–serial output), 213, 214
   shift, 210
   shift-left, 213
   shift-right, 213
   single-rail, 210
Relative address, 404–406
Resistor-transistor logic (see RTL)
Resolution, 353
Return-to-bias (see RB)
Return-to-zero (see RZ)
Right-hand rule, 280, 284
Ring counter (see Counter, ring)
Ripple counter (see Counter, ripple)
ROM (read-only memory), 279–280, 310–320, 331, 346, 381
RS latch, 144–153, 279, 295
RTL (resistor-transistor logic), 112–114
RZ (return-to-zero), 341–343

Schmitt trigger, 173–174
Schottky diode clamp, 119–120
Sector, 282
Select line, 295–298, 300–303, 306–309
Sense line, 284
Sequencer, 382, 383
Sequential circuits, 153
Serial adder, 240–244
Settling time, 167–169
Setup time, 167
Seven-segment display, 221–224
Shannon, Claude, 1
Shift counter, 209, 210
Signed arithmetic, 249–256
Signed numbers, 75–80, 272, 407
Simplex, 364

Single-bus architecture, 374, 376–378
Single precision, 266
Skip, 402–404, 406–408
Small scale integration (see SSI)
Software, 409
SOP (sum of products), 16–17, 230, 231, 236, 247, 320
Source program, 398
Speed-power product (spp), 131
Split-phase recording, 343
SSI (small scale integration), 126, 224, 232
Stack, 407, 409
Staircase, 359–360
Static logic, 123
Stored program, 389–393
Subroutine, 407
Subtraction, 67–71
   binary, 69–70
   hexadecimal, 71
   octal, 70
Subtrahend, 68
Sum of products (see SOP)
Switch, 91–95
Switch debouncer, 152, 153
Synchronous counter (see Counter, synchronous)
Synthesizer, 187

T-type circuits, 163–168, 183
10s complement, 272–274
Three-state gate, 107–111
Timing diagram, 99–103, 166
Timing state, 383–388
Toggle, 163
Totem pole, 116–119, 122
Tracks, 281–283
Transistor-transistor logic (see TTL)
True/complement generator, 234
Truth table, 2–5, 96–97, 383
TTL (transistor-transistor logic), 114–120, 172, 174, 177, 191, 214, 219, 223, 232, 244, 283, 287, 310, 351, 362–363, 383
2s complement, 83–85, 249–250, 252–256, 272, 381, 385

Index **421**

UART (universal asynchronous receiver-transmitter), 364, 372
UL (unit load), 128
Up counter, 187
Up/down counter, 206–208

USART (universal synchronous/asynchronous receiver-transmitter), 365, 372

Video, 345

Weighted resistors, 349–352, 356
Writing, 281–283

Zone field, 326–327